D1201928

Molecular Biology of Mutagens and Carcinogens

Molecular Biology of Mutagens and Carcinogens

B. Singer
University of California at Berkeley
Berkeley, California

and

D. Grunberger
College of Physicians and Surgeons of Columbia University
New York, New York

PLENUM PRESS • NEW YORK AND LONDON

Library of Congress Cataloging in Publication Data

Singer, B. (Beatrice), date–
Molecular biology of mutagens and carcinogens.

Bibliography: p.
Includes index.
1. Carcinogens. 2. Chemical mutagenesis. 3. Molecular biology. I. Grunberger, Dezider, 1922– . II. Title.
RC268.6.S56 1983 616.99′4071 83-17683
ISBN 0-306-41430-9

A Division of Plenum Publishing Corporation
233 Spring Street, New York, N.Y. 10013

Printed in the United States of America

To our children,
Ellen, Mark, George, and Ivan,
and our spouses,
Heinz and Marta

Preface

This book originated in numerous Gordon Research Conferences and many other meetings of scientists working in chemistry, biophysics, biochemistry, and biology related to mutagenesis and carcinogenesis. It seemed the appropriate time to sit back and summarize the results of several decades of research in laboratories in different countries.

We are very grateful to the Rockefeller Foundation for inviting us to formulate and begin writing the book at the Center for International Studies in Bellagio, Italy, where we were Resident Scholars.

We are fortunate to have had the assistance of so many colleagues around the world who cheerfully sent original data, figures, and preprints and listened patiently to us as we worked out the various conflicting ideas in this fast-moving field. The names of these scientists are found within the tables, figures, and references.

There is one person whose contributions we especially wish to acknowledge. Professor Heinz Fraenkel-Conrat was present at the inception of this book and throughout the writing encouraged and criticized in approximately equal proportion. Finally, his editing and amalgamation of our two styles gave us great comfort.

B.S.
D.G.

Contents

I

Introduction

Cancer, as an ultimate result of exposure to chemicals, has been described for at least 200 years and it has probably existed as long as man. When the span of human life was short due to epidemic disease, war, and famine, cancer was not a major concern. However, as most causes of early death have been conquered, cancer emerges as one of the most common causes of death in late life. Evaluation of the extent to which exposure to, or ingestion of, chemicals causes cancer is not the purpose of this book. Rather, we will examine what is known of the chemical and biochemical reactions of different classes of carcinogenic chemicals, or their metabolites, with cellular components. This study will focus primarily on nucleic acids, and most specifically on the mechanism of the initiation stage of carcinogenesis. Initiation in this context is the term for the covalent binding of a reactive group to a nucleotide in an informational macromolecule. For this reason, reaction with other cellular components such as proteins, lipids, and polysaccharides is not included.

However, following this initial binding, there are many factors that influence whether the nucleic acid modification will lead to cancer. Among these factors are repair, promotion, inhibition, and genetic susceptibility. Thus, cancer is termed a multifactorial disease. In spite of the multifactorial and multistage character of cancer, this book will be restricted to the initial events in the process of carcinogenesis and mutagenesis. Enzymatic repair of adducts in eukaryotic cells soon after initiation is one other area where the biochemistry is now advancing, and it is the subject of a chapter.

While the aim of understanding cancer is to prevent it in man, much of the data to be discussed use simpler models, such as nucleosides, synthetic polynucleotides, or natural nucleic acids. Information gained in this way is then used to design experiments in the more complex systems represented by bacteria and mammalian cells in culture, and by whole animals.

It is generally accepted that ultimate carcinogens, whether metabolites or directly acting, are mutagens. That is, using one or more of a variety of test systems, the chemical can be shown to cause a heritable change in the genetic material. However, many effective mutagens, such as sodium nitrite, have not at this time been found to be carcinogenic. This may be due to the fact that mutation can be achieved by direct modification of a biologically active nucleic acid *in vitro*, while carcinogenesis is the end result of a long series of biological processes.

Mutagenesis, whether directly caused by chemicals or as an intrinsic property of nucleic acids, is an inescapable event. Certainly in the context of Darwinian evolution, survival of the fittest implies that mutation can confer a genetic advantage, as well as disadvantage. However, the mutations we observe are usually detected as errors. Genetic diseases that can be shown to result from mutation are by no means uncommon and it has been estimated that as many as 2% of live births include dominant, recessive, X-linked, or chromosomal abnormalities. It has been established biochemically that a single amino acid exchange in hemoglobin is responsible for sickle cell anemia. There are also a considerable number of diseases resulting from the absence or near absence of an enzyme or a transport or receptor protein that are classified as inborn errors of metabolism. These include defects in DNA repair such as xeroderma pigmentosum, ataxia telangiectasia, and Falconi's anemia.

As was previously discussed for carcinogens, only certain well-studied mutagenic chemicals will be included. We will not consider intercalating agents, which are generally frameshift mutagens, ultraviolet or other radiation damage, nor the effects of base analogs.

Research in chemical modification of macromolecules and carcinogenesis has expanded enormously in the last decade. It therefore would be virtually impossible to cite all the relevant papers. Fortunately there are many excellent review articles and book chapters that summarize specific areas and contain references to the original literature. We will generally cite, at the end of each chapter, reviews and occasionally a particularly recent or fundamental paper. References to specific data are given in tables and figures. We hope that our colleagues and workers in this field will understand that this type of book cannot give due credit to all of these scientists who have contributed so much. Without their results, this book could not have been written.

II

Chemicals as Environmental Mutagens and Carcinogens

There has been a constant and dramatic increase in the number of chemicals synthesized and encountered in the environment. However, in spite of new chemicals accruing at the rate of 6000 per week, the number of chemicals in "common use" is estimated to be about 65,000 (of the greater than 4,000,000 known) and the number of chemicals for which there is evidence of carcinogenicity is remarkably small.

There is no estimate of the number of naturally occurring compounds in foods, drinks, seeds, nuts, etc. Some of the most potent carcinogens known, such as aflatoxin, safrole, estragole, and cycasin, are natural products. Aflatoxin has been classified as a human carcinogen in epidemiological studies in Africa where large amounts of peanuts contaminated with this mycotoxin are consumed. In other countries the level of exposure to aflatoxin and other natural carcinogens appears to be quite low, but not zero.

While concern has been expressed regarding the potential hazards of synthetic and natural chemicals in the environment, it is not generally realized that some of the best studied and most effective mutagens are synthesized in humans as the result of normal metabolic processes. These include nitrosamines and other *N*-nitroso compounds, bisulfite, and hydroxylamines, some in very considerable amounts. While the presence of these endogenous mutagens may well contribute to the incidence of cancer, they are unavoidable. Their normal occurrence *in vivo* makes the presence in the environment of low levels of such compounds of lesser importance than those that may be avoided if harmful.

The International Agency for Research on Cancer (IARC), which has as its function the continuous review of cancer-causing agents, has several categories of certainty in classifying chemicals as carcinogens. It must be remembered that this classification is based largely on epidemiology. Thus, the chemicals listed in Tables II-1 and II-2 are primarily industrial, or are widely used and many people may become exposed to them. Epidemiology does not deal with individuals but rather arrives at statistical data based on large groups

Table II-1. Chemicals or Industrial Processes Carcinogenic for Humans

Chemicals	Industries or industrial processes
Industrial chemicals	Manufacturing industries
Aromatic amines	15. Auramine
1. 4-Aminobiphenyl	16. Isopropyl alcohol
2. Benzidine	17. Boots, shoes, and other leather goods
3. 2-Naphthylamine	18. Furniture and cabinetry
Others	Other industrial processes
4. Benzene	19. Hematite mining (? radon)
5. Bis(chloromethyl)ether	20. Nickel refining
6. Vinyl chloride	21. Exposure to soots, tars, oils
Inorganic substances	
7. Arsenic and As compounds	
8. Asbestos	
9. Chromium and Cr compounds	
Drugs	
Alkylating agents	
10. Chlornaphazine	
11. Melphelan	
12. Mustard gas	
Hormones	
13. Diethylstilbestrol	
14. Conjugated estrogens	

Table II-2. Chemicals or Groups of Chemicals Probably Carcinogenic for Humans

High degree of human evidence	Lower degree of human evidence
Drugs	Drugs
1. Chorambucil	1. Iron dextran
2. Cyclophosphamide	2. Oxymetholone
3. Thio-TEPA	3. Phenacetin
Inorganic compounds	Industrial chemicals
4. Cadmium and Cd compounds	4. Acrylonitrile
5. Nickel and Ni compounds	5. Auramine
Food contaminant	6. Carbon tetrachloride
6. Aflatoxins	7. Dimethyl carbamoyl chloride
	Alkylating agents
	8. Dimethylsulfate
	9. Ethylene oxide
	Others
	10. Aminotriazole
	11. Polychlorinated biphenyls

of potentially carcinogen-exposed populations as compared to appropriate control groups.

The value of epidemiology has been shown by the striking correlation between the amount and duration of cigarette smoking and incidence of cancer of the lung and other squamous cell tissues. Predictably, the increase in such cancers in women, who became heavy smokers later than men, started decades after a similar increase in men. At this time, the slopes of the cancer rates of the two sexes are similar.

Figures II-1 and II-2 show some epidemiological data for a variety of cancers, and it is comforting that, notwithstanding our living in this "sea of chemicals," all but the cancers attributed to cigarettes are decreasing, or not increasing at a significant rate.

There are many interesting illustrations of the application of epidemiology to the identification of carcinogens and two are cited here. In an old study, the high incidence of stomach cancer in Japan was compared with that of Japanese in Hawaii (before statehood) and in the United States. It became evident that this incidence, which is the highest in the world, was lower for the Japanese who had moved to Hawaii and even lower for those who had moved to the United States. American-born Japanese have an even lower stomach cancer rate. Since the same racial group was being compared, the

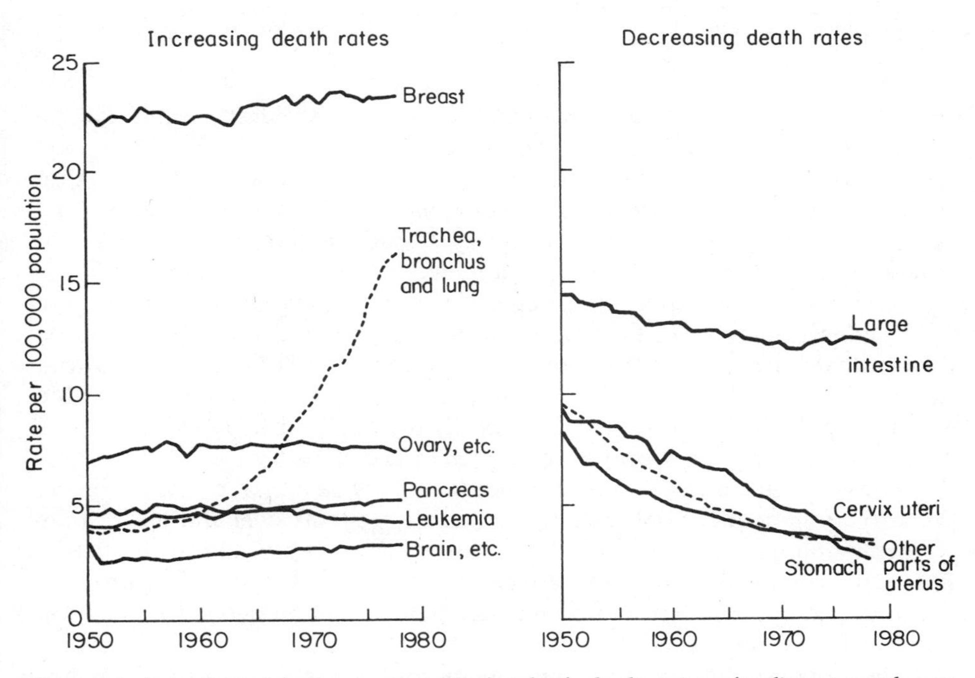

Figure II-1. Age-adjusted death rates for white females for leading sites of malignant neoplasms: United States, 1950–1980.

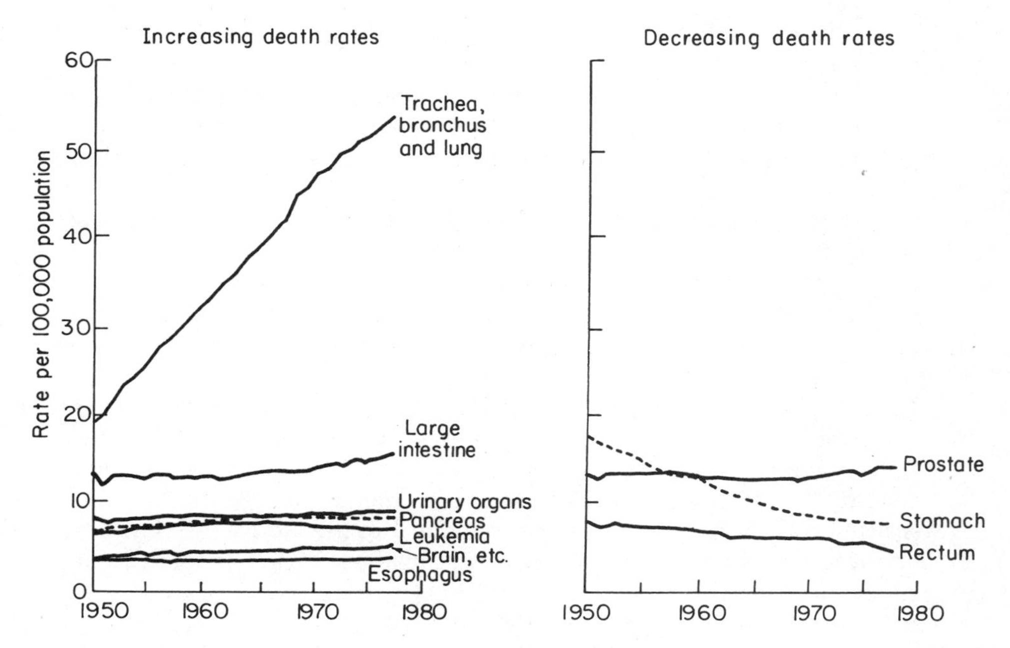

Figure II-2. Age-adjusted death rates for white males for leading sites of malignant neoplasms: United States, 1950–1980. Note that the scale for death rates for males is 0–60 while that for females is 0–25.

most likely conclusion was that these cancers had a dietary origin and diminished when the Japanese lived among other populations and their food habits changed. The intake of nitrate in Japan is four to seven times higher than the intake by Europeans or Americans. Since nitrate is enzymatically converted to nitrite, which is a precursor to nitrosamines, it has been considered that high nitrate consumption is involved in gastric cancer.

In China, where cancer statistics are extremely good and populations are stable, a high incidence of esophageal cancer in a particular area was correlated with fungal infections of the esophageal epithelium. The fungus produced an acid that lowered the pH in the esophagus. It was hypothesized that the endogenous nitrosamines were then locally metabolized to a carcinogen at a much higher rate than found at the normal pH.

These specific success stories of epidemiology as a tool for cancer detection are not generally applicable to explain the general level and wide range of cancers in mixed populations. There are, however, a few specific industrial situations where a correlation between high and prolonged exposure to a chemical and increased incidence of a particular cancer has led to the chemical being termed a human carcinogen.

The impossibility of using humans as test subjects for carcinogens has led to the development of animal testing of suspected cancer-causing chemicals. Since the number of animals in any one group (level of chemical, sex, age, controls) is necessarily limited for economic and other reasons, high doses, which may be four to six orders of magnitude greater than the potential human exposure levels, are needed to be able to detect tumors. Such tests have led to a considerable number of chemicals becoming identified as possible carcinogens. It is, however, not certain that extrapolation from very high doses used in animal studies to very low amounts ingested by humans, possibly over long time periods, is justified and valid as a basis for concluding that a chemical is a human carcinogen.

Examples for this are the concerns raised about the use of cyclamates and saccharin, which in an experiment with 80 rats fed together at very high doses for 78 weeks led to bladder tumors. Two government commissions examined all the data in this and every other related study. At the end it was concluded that the carcinogenicity of cyclamate had not been established. Nor does saccharin convincingly act as a carcinogen.

The important question that such issues raise involves the so-called "threshold theory." Hypothetical results that might be obtained with an animal carcinogenesis test are illustrated in Figure II-3. The dotted line indicates

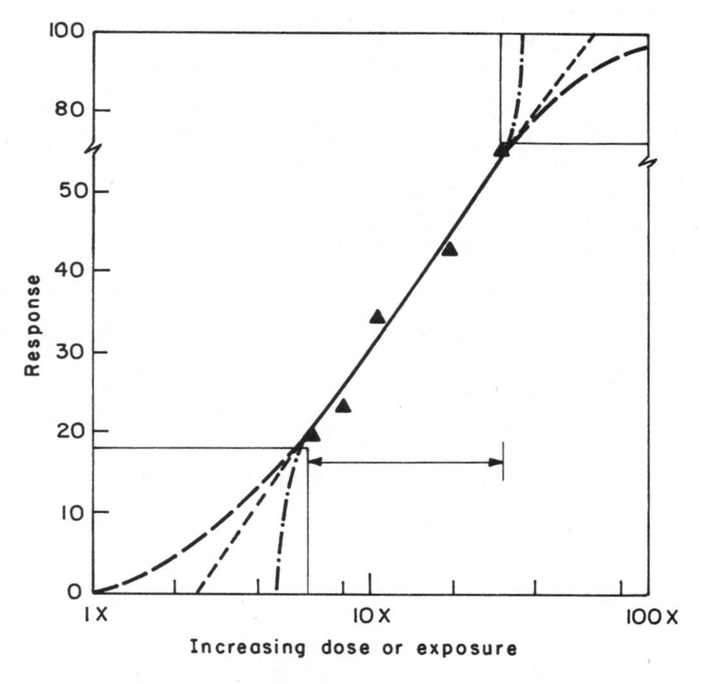

Figure II-3. Data from a typical animal experiment designed to test the carcinogenicity of a chemical. The boxes indicate extrapolation to very low or high doses using different mathematical models. Linear extrapolation (- - - -), sigmoid extrapolation (– –), and no extrapolation (– · –).

a simple mathematical extrapolation which would lead to the conclusion that since the intercept is below zero, a chemical such as saccharin is not a carcinogen at normal levels of exposure. The threshold would be a dose level that was proven to be harmless. On the other hand, one can hypothesize that no safe or threshold level exists and therefore arbitrarily extrapolate to zero by drawing a sigmoid curve or by choosing a slightly different slope toward zero, which is always possible since the points are rarely on a straight line.

There are several additional difficulties in classifying chemicals as carcinogens on the basis of animal experiments. Different results are obtained when different sexes, age groups, and modes of introducing the chemical are compared. Particularly dramatic differences are noted when comparing different species. An example is shown in Table II-3. A single dose of 20 mg/kg ethylnitrosourea administered to 10-day-old BD IX rats will ultimately lead to $> 95\%$ of the animals developing specifically brain tumors. In contrast, the adult rat does not develop any malignancy with a single dose but only upon long-term repeated administration, and in this case tumors are noted at many sites. Other animals also developed tumors on sites that are related to species or sex. The toad *Xenopus laevus,* after several years of observation, is resistant to the analogous carcinogen, methylnitrosourea, even though it can be shown that the expected chemical reactions have occurred. It is likely that the variations in occurrence or sites of tumors are a function of the ability of the specific animal tissue or cell to metabolize or hydrolyze the chemical and repair the damage.

The high cost of performing carcinogenesis tests on animals, and the problems in interpreting animal data, have led to another type of test, namely short-term bacterial or mammalian cell assays. The basic premise is not to test for transformation or other morphological changes related to malignancy, but to use mutagenesis as a criterion. While not all mutagens need be carcinogens, it appears that, as mentioned in Chapter I, most ultimate carcinogens exert their biological effect by first modifying nucleic acids, and this

Table II-3. Effect of Single Exposure to Ethylnitrosourea

Animal	Primary site(s) of tumors
Perinatal	
Rat	Brain, CNS
Mouse	Liver, kidney, lung
Syrian hamster	PNS
Rabbit	NS, kidney
Gerbil	Melanocytes
Adult	
Rat	Brain, hemopoietic system
Mouse F	Lymph nodes, lung
M	Liver, kidney, lung

initial event or damage, when expressed, would be termed mutation. Before discussing specific tests and their use, it is necessary to understand how we classify a chemical as a mutagen.

Historically, a mutation was observed as a result of producing a phenotypic change. Testing was primarily done by measuring forward mutations of *in vitro*-mutagenized phage, transforming DNA, seeds, TMV RNA, or *Drosophila*. No metabolic or enzymatic activation of chemicals occurred and there was no toxicity to consider, but a danger did exist that the nucleic acid might be degraded by the conditions of chemical treatment and lose its biological activity, if any. With these tests, mutagenesis could theoretically be related to specific chemical modification of nucleic acid, even though the detection of the mutational events required replication *in vivo*. Using *in vitro*-treated nucleic acids, the observed mutagenicity and resultant amino acid exchanges generally correlated with either a base change (i.e., nitrous acid deamination) or a tautomeric shift (i.e., hydroxylamine or methoxyamine replacement of the amino group of cytidine). This simple explanation of mutation as a function of a single base change, a point mutation, cannot be extended to *in vivo* systems where the reagent must penetrate the cell wall and may need to be activated, and where the modification of the nucleic acid may be repaired by several pathways that themselves may be mutagenic (i.e., induced error-prone repair). Under these conditions it is not surprising that, just as in animal testing, the detection and quantitation of mutation is greatly affected by the test system. The use of back mutation, i.e., restoration of a specific gene function lost through a mutational event, has been found to be particularly useful. In this way very large numbers of bacteria can be "mutagenized," but only those with the mutation at a locus that is nonlethal under specific conditions will be replicated. Reversants that regain viability under the nonpermissive conditions can then be easily detected, isolated, or counted.

Regardless of whether forward or back mutations are being studied, bacteria certainly do not metabolize carcinogens in exactly the same way as animal cells and they may therefore fail to detect the mutagenicity of a proximate carcinogen. It was therefore a great advantage in mutagen testing when a rat liver extract was included in the *Salmonella typhimurium* system developed by Ames and co-workers. In this way it became possible to test for the necessity for metabolic activation of a potential carcinogen by a comparison of mutation with and without the rat liver enzymes. When the mutation frequency was greatly increased, or if mutation occurred only in the presence of the liver extract, it was apparent that one or more metabolites of the potential carcinogen were responsible for mutation. Other animal species and organs have also been used as the source of activating enzymes and the mutagenicity of some carcinogens is quite different under such varying conditions. Similarly, the various tester strains have given different results as a consequence of their particular genetic characteristics.

In the best studied short-term tests employing *Salmonella* strains or the Chinese hamster cell lines CHO and V79, mutation assays are performed with

Table II-4. Mutagenic Efficiency of Alkylating Agents[a]

Reagent	Test system: TA 100, McCann *et al.*[b]		TA 100, Bartsch *et al.*[c]		HGPRT-TG^R V79, Hodgkiss *et al.*[d]		TG^R V79, Suter *et al.*[e]		8 aza G V79, Peterson *et al.*[f]		8 aza G V79, Chu and Malling[g]		HGPRT CHO, Couch *et al.*[h]	
MeMS	4	*0.63*	32	*350*	9	*1.25*	1.1	*15*	10	*5.7*	4.8	*46*	3.8	*140*
MeNU	28	*4.4*	43	*470*	210	*38*	40	*400*					7.5	*243*
MNNG	8600	*1375*	8700	*$\simeq 10^5$*	3.8×10^6	*30*	1560	*70*	330	*79*	1160	*125*	1480	*200*
EtMS	1	*0.16*	1	*$\simeq 11$*	1	*33*	1	*140*	1	*17.5*	1	*107*	1	*1550*
EtNU	7	*1.1*	8	*$\simeq 85$*	165	*44*	11	*300*					1.25	*550*
ENNG	2200	*350*			1.4×10^6	*23*			720	*19.8*			216	*522*

Abbreviations: MeMS, methylmethanesulfonate; MeNU, methylnitrosourea; MNNG, *N*-methyl-*N'*-nitro-*N*-nitrosoguanidine; EtMS, ethylmethane sulfonate; EtNU, ethylnitrosourea; ENNG, *N*-ethyl-*N'*-nitro-*N*-nitrosoguanidine.

[a] Data calculated on the basis that for each series the concentration of each alkylating agent is identical. The first number in each column represent ratios of mutants obtained compared to EtMS. In many instances these figures differ markedly from those, shown in italics, reported by the authors who measured mutation at equal toxicity which in some cases requires the use of very low reagent concentration. One additional factor, not included in the calculations, is the difference in reactivity between methylating and ethylating agents which gives the impression that the ethylating agents are less mutagenic than the corresponding methylating agents. In general, at equal concentration of alkylating agent, methylation is 10–20 times more efficient than ethylation.

[b] McCann, Choi, Yamasaki, and Ames (1975) *Proc. Natl. Acad. Sci. USA* **72,** 5135. In italics are revertants per nanomole.

[c] Bartsch, Malaveille, Camus, Martel-Planche, Brun, Hautefeuille, Sabadie, Barbin, Kuroki, Drevon, Piccoli, and Montesano (1980) IARC Scientific Publication No. 27, 179. In italics are revertants per micromole.

[d] Hodgkiss, Brennard, and Fox (1980) *Carcinogenesis* **1,** 175. In italics are revertants per 10^6 survivors.

[e] Suter, Brennard, McMillan, and Fox (1980) *Mutat. Res.* **73,** 171. In italics are mutants per 10^5 cells at D_{37} dose.

[f] Peterson, Peterson, and Heidelberger (1979) *Cancer Res.* **39,** 131. In italics are mutants per 10^5 survivors.

[g] Chu and Malling (1968) *Proc. Natl. Acad. Sci. USA* **61,** 1306. In italics are mutants per 10^5 survivors.

[h] Couch and Hsie (1978) *Mutat. Res.* **57,** 209; Couch, Forbes, and Hsie (1978) *Mutat. Res.* **57,** 217. In italics are mutants per 10^6 survivors.

several levels of the chemical. These levels are primarily limited by the toxicity of the compound, and when evaluating mutation data it should be noted that the range of usable concentrations of a series of chemicals can vary by 10^8. Thus, the assays, while giving qualitative data, cannot be used for quantitative comparisons. An example is given in Table II-4. It compares the mutagenicity of several alkylating agents when normalized for concentration with the actual reported data, which use concentrations adjusted to produce equal toxicity or survival.

Given all these caveats regarding short-term tests, they have been exceedingly useful as a preliminary screen of potentially carcinogenic chemicals due to the ease and speed of obtaining results at a much lower cost than animal tests. Their main drawback is that further development is necessary to reduce the number of false-negatives, i.e., animal carcinogens giving few or no mutants in the test system. There are also chemicals that test as positive, usually when it is possible to use very high concentrations because of their low toxicity. These false-positives appear as unlikely to be relevant to environmental situations as are animal tests at very high dosages, both causing agents to be falsely classified as mutagenic.

All statistical and biological data discussed in this chapter are strongly suggestive of the involvement of chemicals in mutation and carcinogenesis. It is most probable that direct modification of nucleic acids (somatic mutation) is the first step, but it must be kept in mind that human carcinogenesis is a multistep process. As molecular genetics has developed, new insights into mammalian genes have resulted. For example, a single amino acid exchange in a small protein, resulting from a point mutation, distinguishes a cell derived from human bladder carcinoma from the naturally occurring cell. We do not rule out that there can be indirect, or epigenetic mechanisms, particularly in regulation of gene expression. Our aim is to understand the mechanism of somatic mutation with the hope of preventing, at the initiation stage, some cancers caused by environmental conditions.

SELECTED REFERENCES

Ames, B. N., and McCann, J. (1981) Correspondence re: S. J. Rinkus and M. S. Legator. Chemical characterization of 465 known or suspected carcinogens and their correlation with mutagenic activity in the *Salmonella typhimurium* system. *Cancer Res.* **39,** 3289–3318, 1979. Validation of the *Salmonella* test: A reply to Rinkus and Legator. *Cancer Res.* **41,** 4192–4196.

Bartsch, H., Malaveille, C., Camus, A.-M., Martel-Planche, G., Brun, G., Hautefeuille, A., Sabadie, N., Barbin, A., Kuroki, T., Drevon, C., Piccoli, C., and Montesano, R. (1980) Bacterial and mammalian mutagenicity tests: Validation and comparative studies on 180 chemicals. In *Molecular and Cellular Aspects of Carcinogen Screening Tests* (R. Montesano, H. Bartsch, and L. Tomatis, eds.), IARC Scientific Publication No. 27, Lyon, pp. 179–241.

Brown, A. L., Bates, R. R., and Thompson, G. R. (1977) FDA-Cyclamates. In *Origins of Human Cancer,* Book C, *Human Risk Assessment* (H. H. Hiatt, J. D. Watson, and J. A. Winsten, eds.), Cold Spring Harbor Laboratory, Cold Spring Harbor, N.Y., pp. 1683–1708.

Druckrey, H., Preussmann, R., Ivankovic, S., Schmähl, D., Afkham, J., Blum, G., Mennel, H. D., Muller, M., Petropoulos, P., and Schneider, H. (1967) Organotrope carcinogene Wirkungen bei 65 verschiedenen N-Nitroso-Verbindungen an BD-Ratten. *Z. Krebsforsch.* **69,** 103–201.

Fisher, P. B., and Weinstein, I. B. (1980) In vitro screening tests for potential carcinogens. In *Carcinogens in Industry and the Environment* (J. M. Sontag, ed.), Dekker, New York, pp. 113–166.

Freese, E., and Freese, E. B. (1966) Mutagenic and inactivating DNA alterations. *Radiat. Res.* **Suppl. 6,** 97–140.

Higginson, J. (1979) Perspectives and future developments in research on environmental carcinogenesis. In *Carcinogens: Identification and Mechanisms of Action* (A. C. Griffin and C. R. Shaw, eds.), Raven Press, New York, pp. 187–208.

Hirayama, T. (1977) Changing patterns of cancer in Japan with special reference to the decrease in stomach cancer mortality. In *Origins of Human Cancer* (H. H. Hiatt, J. D. Watson, and J. A. Winsten, eds.), Cold Spring Harbor Conferences on Cell Proliferation, Vol. 4, Cold Spring Harbor Laboratory, Cold Spring Harbor, N.Y., pp. 55–75.

Hoel, D. (1979) Low-dose and species-to-species extrapolation for chemically induced carcinogenesis. In *Banbury Report 1. Assessing Chemical Mutagens: The Risk to Humans* (V. K. McElheny and S. Abrahamson, eds.), Cold Spring Harbor Laboratory, Cold Spring Harbor, N.Y., pp. 135–145.

Hsia, C.-C., Sun, T.T., Wang, Y.-Y., Anderson, L. M., Armstrong, D., and Good, R. A. (1981) Enhancement of formation of the esophageal carcinogen benzylmethylnitrosamine from its precursors by *Candida albicans. Proc. Natl. Acad. Sci. USA* **78,** 1878–1881.

IARC Working Group (1980) An evaluation of chemicals and industrial processes associated with cancer in humans based on human and animal data: IARC Monographs Volumes 1 to 20. *Cancer Res.* **40,** 1–12.

Linsell, C. A., and Peers, F. G. (1977) Field studies on liver cell cancer. In *Origins of Human Cancer* (H. H. Hiatt, J. D. Watson, and J. A. Winsten, eds.), Cold Spring Harbor Conferences on Cell Proliferation, Vol. 4, Cold Spring Harbor Laboratory, Cold Spring Harbor, N.Y., pp. 549–556.

Loprieno, N. (1982) Mutagenic hazard and genetic risk evaluation on environmental chemical substances. In *Environmental Mutagens and Carcinogens* (T. Sugimura, S. Kondo, and H. Takebe, eds.), University of Tokyo Press, Tokyo, and Liss, New York, pp. 259–281.

Mirvish, S. S. (1975) Formation of *N*-nitroso compounds: Chemistry, kinetics, and *in vivo* occurrence. *Toxicol. Appl. Pharmacol.* **31,** 325–351.

Perera, F. P., Poirier, M. C., Yuspa, S. H., Nakayama, J., Jaretzki, A., Curnen, M. M., Knowles, D. M., and Weinstein, I. B. (1982) A pilot project in molecular cancer epidemiology: determination of benzo(*a*)pyrene-DNA adducts in animal and human tissues by immunossays. *Carcinogenesis* **3,** 1405–1410.

Peto, R. (1977) Epidemiology, multistage models, and short-term mutagenicity tests. In *Origins of Human Cancer,* Book C, *Human Risk Assessment* (H. H. Hiatt, J. D. Watson, and J. A. Winsten, eds.), Cold Spring Harbor Laboratory, Cold Spring Harbor, N.Y., pp. 1403–1428.

Reddy, E. P., Reynolds, R. K., Santos, E., and Barbacid, M. (1982) A point mutation is responsible for the acquisition of transforming properties by the T24 human bladder carcinoma. *Nature (London)* **300,** 149–152.

Rinkus, S. J., and Legator, M. S. (1980) The need for both *in vitro* and *in vivo* systems in mutagenicity screening. In *Chemical Mutagens: Principles and Methods for Their Detection* (F. J. de Serres and A. Hollaender, eds.), Vol. 6, Plenum Press, New York, pp. 365–473.

Rinkus, S. J., and Legator, M. S. (1981) *Salmonella* revisited: A reply to Ames and McCann. *Cancer Res.* **41,** 4196–4203.

Singer, B., and Fraenkel-Conrat, H. (1969) Mutagenicity of alkyl and nitroso-alkyl compounds acting on tobacco mosaic virus and its RNA. *Virology* **39,** 395–399.

Squire, R. A. (1981) Ranking animal carcinogens: A proposed regulatory approach. *Science* **214,** 877–880.

Sugimura, T., and Nagao, M. (1980) Modification of mutagenic activity. In *Chemical Mutagens: Principles and Methods for Their Detection* (F. J. de Serres and A. Hollaender, eds.), Vol. 6, Plenum Press, New York, pp. 41–60.

Sugimura, T., and Sato, S. (1983) Mutagens–carcinogens in foods. *Cancer Res. (Suppl.)* **43,** 2415s–2421s.

Sugimura, T., Kawachi, T., Nagao, M., and Yahagi, T. (1981) Mutagens in food as causes of cancer. In *Nutrition and Cancer: Etiology and Treatment* (G. R. Newell and N. M. Ellison, eds.), Raven Press, New York, pp. 59–71.

Tabin, C. J., Bradley, S. M., Bargmann, C. I., Weinberg, R. A., Papageorge, A. G., Scolnick, E. M., Dhar, R., Lowy, D. R., and Chang, E. H. (1982) Mechanism of activation of a human oncogene. *Nature (London)* **300,** 143–149.

Tannenbaum, S. R., Green, L. C., de Luzuriaga, K. R., Gordillo, G., Ullman, L., and Young, V. R. (1980) Endogenous carcinogenesis: Nitrate, nitrite and *N*-nitroso compounds. In *Carcinogenesis: Fundamental Mechanisms and Environmental Effects* (B. Pullman, P. O. P. Ts'o, and H. Gelboin, eds.), Reidel, Dordrecht, pp. 287–296.

Vesselinovitch, S. D., Rao, K. V. N., Mihailovich, N., Rice, J. M., and Lombard, L. S. (1974) Development of broad spectrum of tumors by ethylnitrosourea in mice and the modifying role of age, sex, and strain. *Cancer Res.* **34,** 2530–2538.

Weinstein, I. B. (1981) The scientific basis for carcinogen detection and primary cancer prevention. *Cancer* **47,** 1133–1141.

III

Intrinsic Properties of Nucleic Acids

A. Introduction

Nucleic acids are polymeric chains of nucleotides consisting of purines and pyrimidines attached to ribose or deoxyribose (nucleosides) and phosphates crosslinking the sugars from 3′ to 5′ position. Double-stranded deoxyribonucleic acids typically occur in a double helix with A opposite T and G opposite C. Ribonucleic acids generally do not occur in such a rigidly defined structure, but have the capacity to form double-stranded regions when the bases can be hydrogen-bonded in a sufficiently long section so that there is thermodynamic stability. In double-stranded nucleic acids, reactions at hydrogen-bonded sites are prevented by the unavailability of protons. However, it should be pointed out that even at temperatures below melting there is sufficient reversible thermal denaturation, so-called breathing, to allow some reactivity in a double-stranded polynucleotide. In addition, near the site of replication all nucleic acids are single-stranded. Thus, the terms single-stranded and double-stranded are relative, not absolute.

Carcinogenic consequences are generally attributed to exogenous chemical modifications of nucleosides followed by replication of the mutational event. However, the existence of spontaneous or, more accurately, background mutations has long been recognized, and it is now clear that these events are not principally or necessarily different from those resulting from mutagen treatments. They are the consequences of the inherent weaknesses of the polynucleotide structure, or of conformational changes as well as the energetic characteristics of these molecules.

Abbreviations used: The letters A, G, U, C, T, HX, I refer to the base moiety in a polynucleotide. Terminology for bases, nucleosides, and nucleotides follows the recommended nomenclature of the IUPAC-IUB. Thus, Ado, Guo, Urd, Cyd, Thd stand for the nucleosides. A small d preceding indicates the deoxy form. N is any nucleoside, while the number of phosphates follows as NMP, NDP, and NTP (mono-, di-, and tri-).

The calculated number of thermodynamically determined modifications would be expected to lead to an unacceptably high level of mutation (≅ 1 per 10^{-2} bases). DNA polymerase selection would decrease this error to 10^{-3}–10^{-5}. However, DNA polymerases and associated repair enzymes from bacteria can correct errors during synthesis and thus maintain genetic identity. In contrast, as discussed in Section E, the level and type of mutagen/carcinogen-induced modifications of nucleic acids often cause much higher levels of misincorporation that, if unrepaired, result in increased mutation rates.

B. Reactions with Water

Many of the endogenous or background mutations are hydrolytic. These include the loss of bases through depurination and depyrimidination, deamination of the exocyclic amino groups, and phosphodiester bond cleavage.

1. *Depurination and Depyrimidination*

The basic facts concerning the stability of the glycosyl bond are now well understood and, while catalyzed by acid, the mechanism shown in Figure III-1 also applies to nucleic acids at neutrality. Glycosyl bond cleavage of deoxyribonucleotides is highly pH-dependent, but the rate constants are very different for deoxypurine and deoxypyrimidine nucleosides (Figure III-2). In the case of deoxycytidine, which exhibits a similar rate constant over the range of pH 0–3, it can be demonstrated that the rate of hydrolysis is tem-

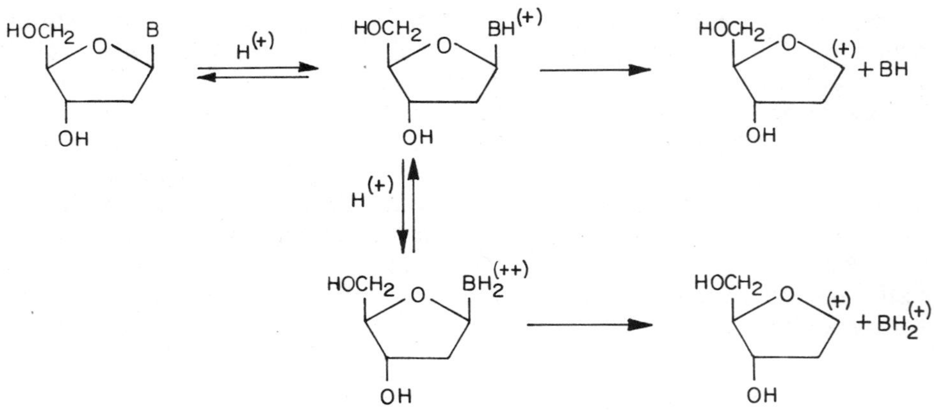

Figure III-1. Postulated mechanism for acid-catalyzed hydrolysis of the glycosyl bond of nucleosides. This mechanism also applies to depurination and depyrimidination of nucleic acids at neutrality. B indicates any base. Figure adapted from Shapiro and Danzig (1973) *Biochim. Biophys. Acta* **319**, 5.

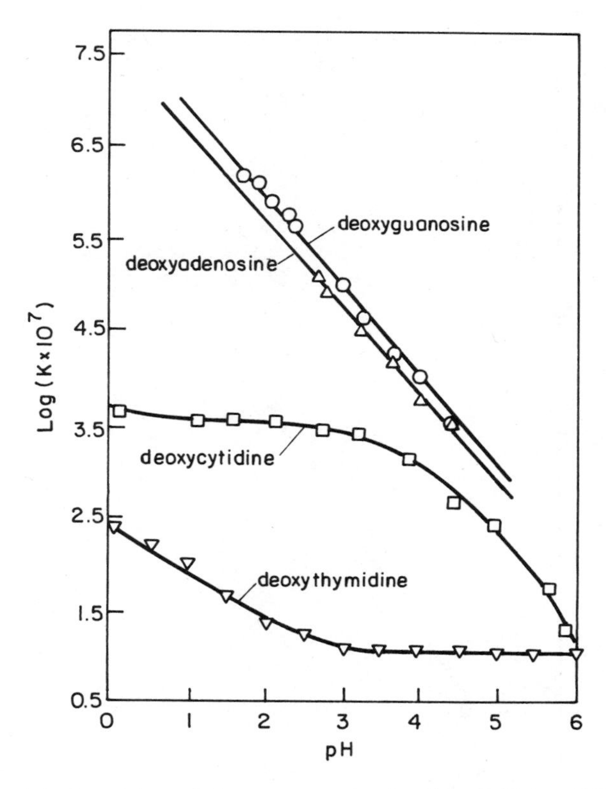

Figure III-2. Plot of the logarithms of the rate constants (sec^{-1}) at 95°C vs. pH for the hydrolysis of the four common, naturally occurring deoxyribonucleosides. Adapted from Shapiro and Danzig (1972) *Biochemistry* **11,** 23.

perature-dependent in the pH-independent plateau area (Figure III-3) and thus likely to be temperature-dependent at any pH.

In contrast to deoxypurine nucleosides, the glycosyl bond of ribopurine nucleosides is three orders of magnitude more stable in acid (Table III-1). The $t_{1/2}$ (pH 1, 37°C) of Guo is 200 hr while that of dGuo is 0.23 hr. Comparable half-lives of Ado and dAdo are 530 and 0.45 hr, respectively. Ribopyrimidines are also more stable than deoxypyrimidines. Although several theories have been proposed, there is no definitive explanation of how a small change in the sugar can so dramatically affect cleavage of the N–C bond. A summary of kinetic parameters of acid hydrolysis of nucleosides and nucleotides is given in Table III-2.

There are several modifications of the purine and pyrimidine ring that can labilize the glycosyl bond. In all but one case, these reactions form quaternary bases that have an intrinsic instability. Derivatives studied in terms of increased chemical lability of the glycosyl bond are 3- and 7-alkyldeoxypurine nucleosides; O^2-alkylcytidine, O^2-aldyldeoxycytidine, and O^2-alkyl-

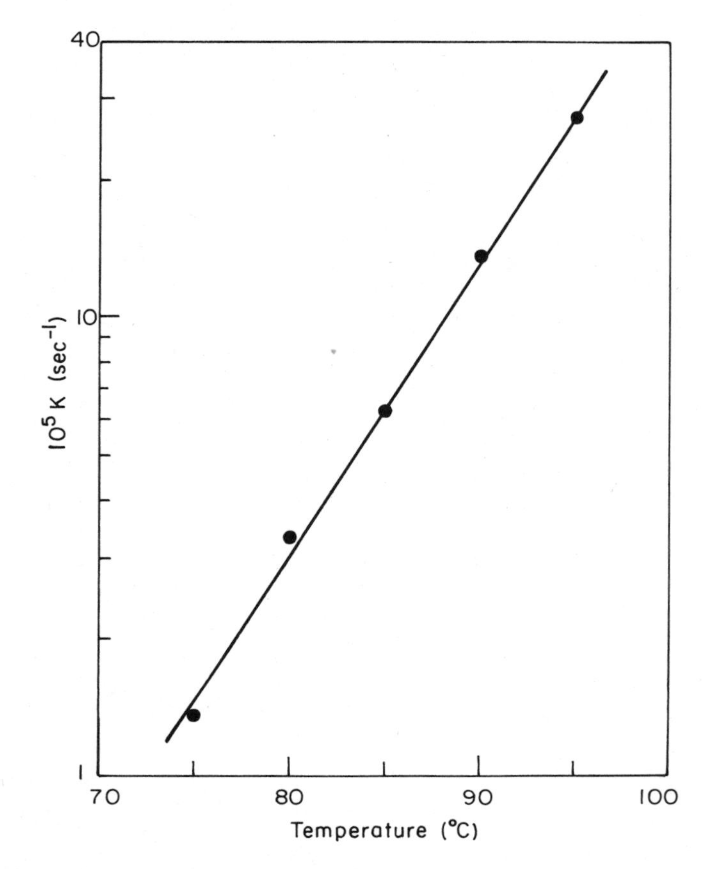

Figure III-3. Hydrolysis of deoxycytidine. Variation with temperature in pH-independent plateau area. See Figure III-2 for plot of the rate constant for depyrimindation vs. pH. Data from Shapiro and Danzig (1972) *Biochemistry* **11,** 23.

thymidine (Tables III-1, III-3). N^6-methyl A and 5-methyl C, normal constituents of DNA, are released at rates similar to those of unmodified bases.

A relevant and important question is whether, and to what extent, glycosyl bond cleavage occurs in nucleic acids under physiological conditions. The stability of the few nucleosides and derivatives measured under such conditions indicates that the nucleic acid structure greatly decreases depurination (Table III-1). Neutral depyrimidination in DNA has only been measured at high temperature, due to its extremely low rate, estimated to be 20 times slower than depurination. Nevertheless, both free cytosine and thymine were found to be released from DNA (pH 7, 95°C) (Figure III-4) and calculation of the expected rate at neutrality has indicated that both depurination and depyrimidination would occur intracellularly to a significant extent (Table III-4).

Table III-1. Relative Stability of Glycosyl Bonds of Purine Nucleosides, Nucleotides, and Alkyl or Deaminated Derivatives[a]

Compound	$t_{1/2}$ (hr) at 37°C			
	pH 0	pH 1	pH 3.35	pH 7
Nucleosides				
Guanosine	1.9	200		
3-Methylguanosine		$\ll$0.1[b]		
7-Methylguanosine	7.1			
7-Ethylguanosine	9.9			
Deoxyguanosine		0.23	25	9.1×10^4
Deoxyguanosine (in DNA)				$\simeq 1.5 \times 10^7$
7-Methyldeoxyguanosine				6.5
7-Methyldeoxyguanosine (in DNA)				$\simeq$155
7-Ethyldeoxyguanosine				5.6
1,7-Diethyldeoxyguanosine				7
O^6-Methyldeoxyguanosine		$\simeq$0.1[c]		
O^6-Methyldeoxyguanosine (in DNA)		<8		
O^6-Ethyldeoxyguanosine (in DNA)				$\gg$240
Adenosine	7.9[d]	530		
1-Methyladenosine	12.6[d]			
N^6-Methyladenosine	7.9[d]			
3-Methyladenosine			17[e]	
Deoxyadenosine		0.45	830	
Deoxyadenosine (in DNA)				$\simeq 1.5 \times 10^7$
3-Methyldeoxyadenosine			0.04[e]	0.5
3-Methyldeoxyadenosine (in DNA)				$\simeq$26
7-Methyldeoxyadenosine (in DNA)				$\simeq$3
Nucleotides				
Deoxyguanosine-5′-phosphate		1.1		
7-Methyldeoxyguanosine-5′-phosphate				16
7-Ethyldeoxyguanosine-5′-phosphate				19
7-[Hydroxyethylthioethyl]-deoxy-guanosine-5′-phosphate				8
Deoxyxanthosine-5′-phosphate			64	
Deoxyadenosine-5′-phosphate		0.62		
Deoxyinosine-5′-phosphate			2	

[a] Data have been calculated in many cases from the published rate constants. See Figure III-2 for the pH dependence of the rate constants for deoxyadenosine and deoxyguanosine. pH 0 is 1 N HCl, pH 1 is 0.1 N HCl, pH 3.35 and pH 7 are aqueous buffers. Data for depurination of DNA in the pH 7 column include pH 6.8–7.4.
[b] 20°C.
[c] 23°C.
[d] 41°C.
[e] 25°C.

2. *Deamination*

Hydrolytic deamination also occurs under physiological conditions as judged primarily by indirect methods that involve measuring mutagenesis before and after heating. In a double-stranded nucleic acid, deamination is relatively rare but certainly of biological significance. Deamination of C → U,

Table III-2. Kinetic Parameters of Acid Hydrolysis of Nucleosides and Their Monophosphates at pH 1[a]

Original compound	k_1 (sec^{-1}) at 37°C	ΔH_a (kcal/mole)
2′-Deoxyadenosine	4.3×10^{-4}	6.9
2′-Deoxyguanosine	8.3×10^{-4}	8.1
2′-Deoxycytidine	1.1×10^{-7}	—
2′-Deoxyuridine	10^{-7}	—
Adenosine	3.6×10^{-7}	6.0
Guanosine	9.36×10^{-7}	11.5
Cytidine	10^{-9}	—
Uridine	10^{-9}	—
2′-Deoxyadenosine-5′-phosphate	3.1×10^{-4}	—
Adenosine-5′-phosphate	3.8×10^{-7}	—
Adenosine-2′(3′)-phosphate	3.3×10^{-7}	—
2′-Deoxyguanosine-5′-phosphate	1.8×10^{-4}	—
Guanosine-2′(3′)-phosphate	6.6×10^{-7}	—
2′-Deoxycytidine-5′-phosphate	2.0×10^{-8}	—
Thymidine-5′-phosphate	2.0×10^{-8}	—

[a] From *Organic Chemistry of Nucleic Acids,* Part B (N. K. Kochetkov and E. I. Budovskii, eds.), Plenum Press (1972). For derivatives of adenine, guanine, and thymine, log k_1 is a linear function of pH; for cytosine derivatives, log k_1 is independent of pH within the range pH 1–4.

Table III-3. Relative Stability of Glycosyl Bonds of Pyrimidine Nucleosides and Alkyl Derivatives[a]

Compound	$t_{1/2}$ (hr) pH 0, 100°C	pH 1, 37°C	pH 7, 95°C
Deoxyuridine	1.7	1.9×10^3	64
O^2-Ethyldeoxyuridine			1
Deoxythymidine	1		160
O^2-Ethyldeoxythymidine			2.7
Uridine	⩾50	2×10^5	
O^2-Ethyluridine			100
Cytidine		2×10^5	
O^2-Ethylcytidine			5
Deoxycytidine	1.8	1.7×10^3	
O^2-Ethyldeoxycytidine			<0.1

[a] Hydrolysis of the glycosyl bond of deoxypyrimidines is independent of pH over the range pH 2–8 (Figure III-2). pH 0 is 1 N HCl, pH 1 is 0.1 N HCl, and pH 7 is 0.1 M phosphate buffer.

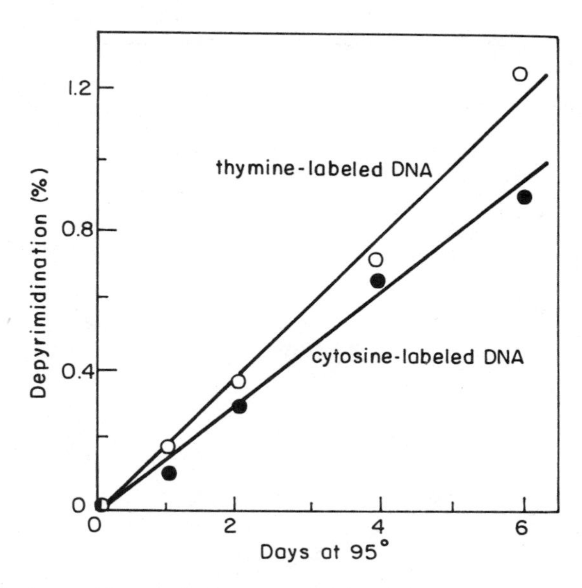

Figure III-4. Rate of release of free pyrimidines from ^{14}C-labeled *E. coli* DNA at pH 7.4. ^{14}C-Cytosine-labeled (●); ^{14}C-thymine-labeled (○). From Lindahl and Karlström (1973) *Biochemistry* **12,** 5151.

Table III-4. Thermodynamic Characteristics of Nucleosides Leading to Mutation

	Events/day/10^{10} bases (rat liver cell) Strandedness	
	Double[a]	Single[a]
Depurination	$\simeq 10^4$	$\simeq 4 \times 10^4$
Depyrimidination	$\simeq 5 \times 10^2$	$\simeq 2 \times 10^3$
Deamination	$\simeq 1.7 \times 10^2$	$\simeq 4.3 \times 10^4$
	Equilibrium constant ($\text{mole}^{-1}\ \text{sec}^{-1}$)	
	Double	Single[b] (calculated from monomer)
All tautomeric shifts[c]	not determined	$\simeq 10^{-4}$
Total ionization[d]	not determined	$\simeq 10^{-2}$–10^{-5}
Base rotation[e]	not determined	$\simeq 10^{-1}$–10^{-2}

[a] Adapted from Lindahl (1979) *Prog. Nucleic Acid Res. Mol. Biol.* **22,** 135, and Hartman (1980) *Environ. Mutagenesis* **2,** 3. Single strand represents about 1–3% of double strand.
[b] Data extrapolated from physical constants for bases, nucleosides, and nucleotides, including model compounds.
[c] Topal and Fresco (1976) *Nature (London)* **263,** 285, 289 and references therein.
[d] From published pK values by Dunn and Hall; Singer (1975) In *Handbook of Biochemistry and Molecular Biology,* 3rd ed., Vol. 1, *Nucleic Acids,* CRC Press.
[e] Calculated by Haschemeyer and Rich (1967) *J. Mol. Biol.* **27,** 369.

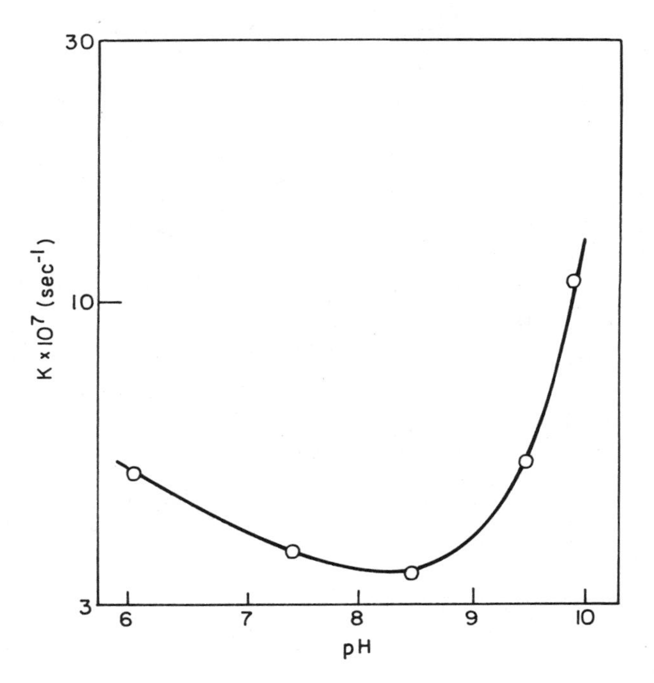

Figure III-5. pH dependence of the rate of deamination of cytosine residues in *E. coli* DNA at 100°C. From Lindahl and Nyberg (1974) *Biochemistry* **13,** 3405.

or A → HX, if unrepaired, represents direct mutational events if expressed. Although 100°C was used to chemically determine deamination of C in *E. coli* DNA, there is little doubt that the rate constant can be extrapolated to 37°C (Figure III-5). Deamination of A is considerably slower than that of C, estimates being 1–2% the rate of C deamination. There is, however, efficient repair of these deamination products (discussed in Chapter VIII).

Deamination of the minor base 5-methylcytosine to thymine results in transition mutations since thymine is a normal DNA base, thus unrepaired. In *E. coli* 5-methylcytosine residues are hot spots for spontaneous mutations.

3. *Phosphodiester Cleavage*

Phosphodiester cleavage by hydrolysis occurs in RNA, even in weak alkali, but not in DNA. This is not a matter of strandedness, but rather is due to the presence of the C2′-OH group of the ribose, which favors a trans-esterifiction under OH^- catalysis (the C3–C5 diester being transformed to a cyclic C3–C5 diester intermediate).

For DNA there is limited evidence for a direct hydrolytic path of chain breakage. Breaks generally occur as a consequence of depurination/depyrimidination. Hydrolytic cleavage of a base results in deoxyribose residues that exist, in part, in the aldehyde form, which through β-elimination causes the cleavage of the phosphodiester bond. The rate constant for a hydrolytic break of the phosphodiester bond at an apurinic site (pH 7.4, 37°C) is about 10^4 faster than depurination. Thus, all apurinic (AP) sites could eventually be sites of breakage, even in the absence of endonucleases. However, the estimated rate constants (pH 7.4, 37°C) in double-stranded DNA are on the order of 10^{-11}/sec^{-1} for hydrolytic chain breakage at AP sites, or an average lifetime of 400 hr. It is therefore unlikely that nonenzymatic hydrolysis of phosphodiesters in DNA is of physiological importance.

C. REVERSIBLE EQUILIBRIUM REACTIONS

1. *Tautomerism*

All nucleic acid bases can exist in at least two tautomeric forms. The most stable form has been calculated from a consideration of experimental data for the energy of isolated bonds and the resonance energy. These theoretical data and others indicate that the keto form is more stable than the enol, and the amino form is more stable than the imino (Table III-5). From these data, it would be expected that U tautomers would occur at a much higher frequency than A tautomers. Various spectral methods confirm that the thermodynamically favored tautomers are predominant both in solution and in the vapor phase. Direct evidence for the rare tautomer of common bases is not available because spectroscopic methods can only detect a few percent of this tautomer under ideal conditions. Thus, K_t values for monomers must be derived from the pK values of fixed tautomers. These are on the order of 10^{-4}–10^{-5} (Table III-4).

It is interesting to note that the introduction of an electronegative substituent such as bromine in the 5 position of uridine shifts the equilibrium toward the enol form, which may be relevant to the observations that 5-BrU

Table III-5. Calculated Energy of Transformation of Tautomeric Forms of Some Nucleic Acid Bases[a]

Base	Type of tautomeric equilibrium	ΔH(eV)
Adenine	Amine–imine	1.477
Cytosine	Amine–imine	0.550
Guanine	Keto–enol	0.185
Thymine	Keto–enol	0.345
Uracil	4-Keto–4-hydroxy	0.329

[a] From *Organic Chemistry of Nucleic Acids,* Part A (N.K. Kochetkov and E. I. Budovskii, eds.), Plenum Press (1972).

is mutagenic when incorporated into prokaryotes, even though it basepairs normally. Another biologically relevant effect of the presence of an unfavored tautomer is the postulated G · U wobble pair, which involves pairing G(enol) with U(keto). The only direct evidence for the wobble comes from tRNA where G · U pairing has been shown from the three-dimensional structure of $tRNA^{Phe}$.

In contrast to the lack of direct evidence for tautomers of common bases, at the polymer level the rare tautomers can be detected using UV spectroscopy. Hydrogen-bonding schemes have been proposed by Fresco and colleagues for A · C and I · U base mispairs in helices dominated by A · U or I · C pairs. Each mispair contains one of its bases in the rare tautomeric form. In Figure III-6 the rare form is on the left side and the normal base, which has equivalent basepairing, is on the right side. Which of the bases in a mispair is in the unfavored form depends upon the stacking interactions of these bases with their nearest-neighbor Watson–Crick pairs. In this case the enhancement of energetically unfavored tautomers is at the expense of the stacking interactions. Similar effects of neighbor bases on mispairing are shown for 2-aminopurine, which, in transcription, acts more like G when in a C polymer than in an A polymer. Even more distant neighbors may also exert an effect.

The substitution of the 4-amino group in cytidine by hydroxylamine or methoxyamine is an example of a reaction causing a tautomeric shift: the imino tautomer is predominant in water solution and is the only form observed in nonpolar solutions. Model experiments on possible basepairing have also been reported. 1-Methyl-N^4-methoxycytosine forms 1 : 1 complexes with a 9-substituted adenine, thus behaving like U. No interaction between 1-substituted N^4-methoxycytosine and 9-ethylguanine was observed in nonpolar solution, which is in line with the observed tautomerism favoring the imino form. The association constants of 1-methyl-N^4-methoxycytosine and its 5-methyl derivative with adenosine in chloroform solutions are identical. This is contrary to expectation, since the 5-methyl substituent should, based on analogous experiments, sterically hinder rotation of the exocyclic N^4-methoxyamino group, so that it is constrained to the *syn* form with respect to the ring N(3). This would prevent Watson–Crick basepairing (although monomer interactions are not restricted to Watson–Crick pairing). Nevertheless, experimentally, Watson–Crick pairing occurs in poly(U, N^4-methoxy C) complexed with poly(A), as shown by mixing curves (Figure III-7). Unlike (U, C) copolymers, poly(U, N^4-methoxy C) annealed with poly(A) on a 1 : 1 basis, indicating that the modified C was held within the helix. Watson–Crick basepairing was shown by cooperative melting of the annealed complex.

An example of a "frozen" tautomer of a purine base is O^6-alkylguanine. In the enol form, this modified base can form a wobble pair with U, which would be a mutagenic event. However, there is no evidence for any complex formation between O^6-methyl G and U, C, or A in chloroform at the monomer level.

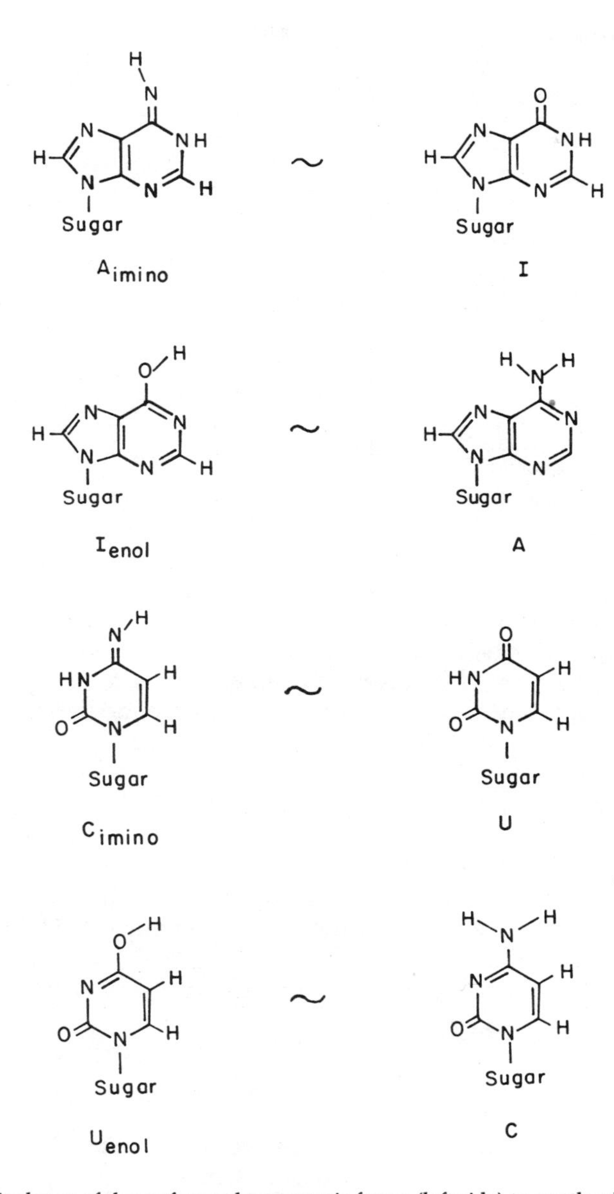

Figure III-6. Equivalence of the unfavored tautomeric forms (left side) to another base (right side) in the favored form. Adapted from Fresco, Broitman, and Lane (1980) In *Mechanistic Studies of DNA Replication and Genetic Recombination* (B. Alberts, ed.), Academic Press, pp. 753–768.

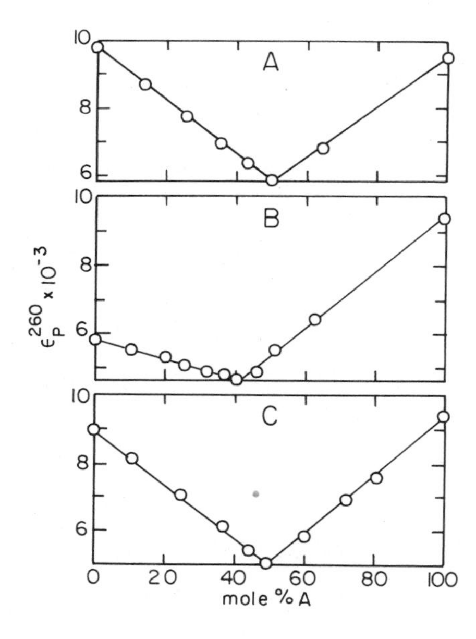

Figure III-7. Mixing curves of copolymers with poly(A) in SSC at 25°C. Panel A, poly(U) plus poly(A); panel B, poly(U, 35% C) plus poly(A); panel C, poly(U, 39% mo⁴C) plus poly(A). From Spengler and Singer (1981) *Biochemistry* **20,** 7290.

Base tautomerism, which does occur, is difficult to measure directly and unequivocally, but this property of N^4-hydroxycytidine has been used in elegant experiments on site-directed mutagenesis (discussed in Chapter IX). Polymerases can utilize N^4-hydroxy CTP in the absence of CTP. Once incorporated, this base can mispair when it is in the imino form. Thus, C → T transitions have been recovered *in vivo.*

2. Ionization

Ionization is a function of pH, in contrast to tautomerism, which is an innate property of neutral molecules. Under physiological conditions the degree of ionization, i.e., the ratio of the charged to uncharged species, ranges from 10^{-2} to 10^{-5}, depending on the pK of the nucleoside. The resultant protonation or deprotonation can cause mispairing in the same way as tautomerism.

3. Syn–Anti Rotation

Rotation of a base around the glycosyl bond increases the ways in which changed base–base interaction can occur (Chapter VII, Figure VII-1). The purine–purine base pair in which one purine is in the rare tautomeric form and the other is in the *syn* conformation has been postulated by Topal and Fresco to explain some spontaneous transversions. Purine nucleosides have long been considered capable of existing in the *syn* conformation. Recently, certain pyrimidine nucleosides [m^6dU and pseudouridine] have also been shown to exist as the *syn* rotamer. Although in crystals the nucleosides and nucleotides are usually found in either *anti* or *syn* conformations, there is one report that crystals can contain two molecules in the asymmetric cell unit, one *syn*, the other *anti*. A new finding of the significance of rotamers involves alternating d(G, C), which forms a double-stranded left-handed helix in which G is *syn* (discussed in Chapter VII), in striking contrast to the more common right-handed DNA helix with bases in the *anti* orientation.

In solution, when there is freedom of rotation, the evaluation of the population of *syn* and *anti* rotamers of nucleosides and nucleotides varies from more than 50% of *syn* to 70–100% of *anti*. This wide variation in rotamers can be attributed to at least two factors. There are differences in the experimental techniques and also in the theoretical models used for evaluation of experimental data. In addition, the energy differences between the two rotamers may be very low, so that minor experimental conditions would have a large effect on the forms observed. Even in the case of the sterically constrained 8-bromoadenosine, this energy difference is on the order of only 1 kcal–mole. This means that in enzymatic reactions the proper conformation can be easily adopted and maintained, particularly in a helix.

It has been suggested that since a modified nucleotide such as m^5C promotes the formation of left-handed helix in alternating poly(dG-m^5dC), a possible consequence of chemical reactions could be rotation of the base around the glycosyl bond. A further consequence would be an increased availability for reaction of normally unavailable or buried sites on nucleotides. Although these proposals are based on the behavior of synthetic oligonucleotides, such DNA structures have been postulated to play a role in mutagenesis.

D. FIDELITY IN NUCLEIC ACID SYNTHESIS

The central postulate of the very specific formation of Watson–Crick basepairs is generally accepted. One would expect that infidelity, or mutagenesis resulting from hydrolytic or reversible reactions, would also be explained by the parameters for basepair formation. The fidelity of information transfer relies primarily on the ability to form a specific basepair with the appropriate number of hydrogen bonds. However, tautomerism, sugar conformation, base orientation, steric hindrance, appropriate enzyme activity, and the participation of neighboring groups can each have a distinct effect; either to increase or decrease fidelity.

In Section E, the effect on fidelity of mutagen–carcinogen modifications of nucleic acids is discussed. In this Section, nucleic acids are assumed to contain only the intrinsic modifications occurring as a result of normal physiological conditions (Sections B, C). Without a polymerizing enzyme, it has been calculated from the free energy difference between complementary and noncomplementary nucleotides that mismatches could occur with an error frequency of one mispaired nucleotide out of every 10 to 100 nucleotides incorporated. This "error rate" could not possibly sustain the genetic identity of a DNA, which actually occurs with an experimentally determined *in vitro* fidelity of one error per 10^7 nucleotides and, *in vivo,* in the range of 10^{-7}–10^{-11} misincorporations. The latter data are derived from a study of forward mutations in genomes as diverse as *E. coli* and *Drosophila.* In contrast, fidelity of DNA polymerases copying synthetic templates, summarized by Loeb and Kunkel (Table III-6), is lower regardless of whether the polymerase is of prokaryotic or eukaryotic origin. It is speculated that synthetic polynucleotides may allow more errors than DNA due to their structure, which permits slippage of the primer along the template, or because of secondary structure and stacking interactions, which again differ from natural DNA. In spite of these arguments, much data have been obtained that are relevant to DNA synthesis. In general, prokaryotic DNA polymerases are more accurate than eukaryotic polymerases, although this may be a function of the lack of proofreading in many eukaryotic DNA polymerases.

A very useful and highly sensitive experimental system for studying fidelity, used in a number of laboratories, is the replication of the double-stranded form of bacteriophage φX174 to form single-stranded φX174. The number of revertants produced from phage mutated at specific loci indicate that fidelity of *E. coli* Pol I, Pol III, and bacteriophage T4 polymerases is similar to *in vivo* fidelity (Table III-7). Clearly, eukaryotic polymerases or reverse transcriptase (AMV) lack this great specificity for base selection. Since eu-

Table III-6. Fidelity of Purified DNA Polymerases with Synthetic Polynucleotide Templates[a]

Enzyme source	Template	Incorrect substrate	Error rate	Reference[b]
Bacteria and Bacteriophage				
E. coli Pol I	poly[d(A-T)]	G	1/80,000	1
	poly[d(A-T)]	C	1/8,000	1
	poly(dA) · oligo(dT)	C	1/3,200	11
E. coli Pol II	poly(dA) · oligo(dT)	C	1/13,300	11
E. coli Pol III (core)	poly(dA) · oligo(dT)	C	1/16,000	11
Pol III*	poly(dA) · oligo(dT)	C	1/4,050	7
Bacteriophage T4, wild type	poly[d(A-T)]	G	1/42,300	10
	poly(dC) · oligo(dG)	T	1/96,600	6
Bacteriophage T4, mutator L56	poly(dC) · oligo(dG)	T	1/27,700	6
Eukaryotic cells				
DNA polymerase-α				
Regenerating rat liver	poly[d(A-T)]	G	1/100,000	13
Normal rat liver	poly[d(A-T)]	G	1/16,500	3
Rat hepatoma	poly[d(A-T)]	G	1/3,360	5
Human placenta	poly[d(A-T)]	G	1/12,000	14
Calf thymus	poly[d(A-T)]	G	1/4,500	10
Human lymphocytes	poly[d(A-T)]	G	1/5,330	5
HeLa cell	poly[d(A-T)]	G	1/4,830	5

Table III-6. (Continued)

Enzyme source	Template	Incorrect substrate	Error rate	Reference[b]
DNA polymerase-β				
Calf thymus	poly(dC) · oligo(dG)	T	1/8,000	4
		C	1/8,000	4
		A	1/1,400	4
	poly(dA) · oligo(dT)	C	1/180,000	4
	poly[d(A-T)]	G	1/11,800	5
Calf liver	poly[d(A-T)]	G	1/15,200	5
Rat hepatoma	poly[d(A-T)]	G	1/4,500	10
Rat liver	poly[d(A-T)]	G	1/18,900	3
Guinea pig liver	poly[d(A-T)]	G	1/44,200	5
Human placenta	poly[d(A-T)]	G	1/20,000	14
DNA polymerase-γ				
Human placenta	poly(dI) · poly(dC)	T	1/1,440	8
Fetal calf liver	poly[d(A-T)]	G	1/7,250	9
HeLa cell	poly[d(A-T)]	G	1/3,120	9
Animal tumor viruses				
Avian myeloblastosis virus	poly(rA) · oligo(dT)	C	1/700	2
	poly[d(A-T)]	G	1/1,400	16
	poly(dG) · poly(dC)	A	1/670	2
Avian leukosis virus	poly(dA) · oligo(dT)	C	1/235	17
Rauscher leukemic virus	poly(dA) · oligo(dT)	C	1/525	15
Spleen necrosis virus	poly(dA) · oligo(dT)	G	≃1/300	12

[a] From Loeb and Kunkel (1982) *Annu. Rev. Biochem.* **51,** 429.
[b] References:
1. Agarwal, Dube, and Loeb (1979) *J. Biol. Chem.* **254,** 101.
2. Battula and Loeb (1974) *J. Biol. Chem.* **249,** 4086.
3. Chan and Becker (1979) *Proc. Natl. Acad. Sci. USA* **76,** 814.
4. Chang (1973) *J. Biol. Chem.* **248,** 6983.
5. Dube, Kunkel, Seal, and Loeb (1979) *Biochim. Biophys. Acta* **561,** 369.
6. Hall and Lehman (1968) *J. Mol. Biol.* **36,** 321.
7. Jimenez-Sanchez (1976) *Mol. Gen. Genet.* **145,** 113.
8. Krauss and Linn (1980) *Biochemistry* **19,** 220.
9. Kunkel, unpublished.
10. Kunkel, Meyer, and Loeb (1979) *Proc. Natl. Acad. Sci. USA* **76,** 6331.
11. Miyaki, Murata, Osabe, and Ono (1977) *Biochem. Biophys. Res. Commun.* **77,** 854.
12. Mizutani and Temin (1976) *Biochemistry* **15,** 1510.
13. Salisbury, O'Connor, and Saffhill (1978) *Biochim. Biophys. Acta* **517,** 181.
14. Seal, Shearman, and Loeb (1979) *J. Biol. Chem.* **254,** 5229.
15. Sirover and Loeb (1974) *Biochem. Biophys. Res. Commun.* **61,** 410.
16. Sirover and Loeb (1977) *J. Biol. Chem.* **252,** 3605.
17. Weymouth and Loeb (1977) *Biochim. Biophys. Acta* **478,** 305.

karyotes obviously are able, *in vivo,* to maintain genetic stability, such experimental infidelity may result from a lack of factors present in cells, including proofreading, or they may reflect a change in the secondary or tertiary structure of the polymerase upon isolation.

In view of the significant number of dynamic events (tautomerism, base rotation, protonation) as well as irreversible hydrolytic events (base cleavage, deamination, chain breakage) occurring in DNA, the fidelity of DNA synthesis

Table III-7. DNA Polymerase Fidelity with ϕX174 am3 DNA[a]

DNA polymerase	Incorrect deoxynucleotide(s)	Error rate	Reference[b]
E. coli Pol I	C	1/680,000 → 1/2,000,000	2, 3, 5
	CαS[c]	1/100,000	2
	A	≤1/6,300,000	3
	G	1/1,080,000	5
E. coli Pol III	C	<1/12,200,000	6
	A	<1/12,200,000	6
Bacteriophage T4, wild type	C	$<1/10^7$	2, 5
	A	$<1/10^7$	5
	CαS[c]	1/20,000	2
Avian myeloblastosis virus	C, A, G	1/329 → 1/17,000	1, 2, 5
Calf thymus Pol-α	C, A, G	1/23,800 → 1/30,500	4
Mouse myeloma Pol-α	C, A, G	1/47,500	4
Rat hepatoma Pol-β	C, A, G	1/6,660	4
Mouse myeloma Pol-β	C, A, G	1/2,930 → 1/4,660	4
HeLa cell Pol-γ	C, A, G	1/6,660 → 1/8,070	4

[a] From Loeb and Kunkel (1982) *Annu. Rev. Biochem.* **51**, 429.
[b] References:
1. Gopinathan, Weymouth, Kunkel, and Loeb (1979) *Nature (London)* **278**, 857.
2. Kunkel, Eckstein, Mildvan, Koplitz, and Loeb (1981) *Proc. Natl. Acad. Sci. USA* **78**, 6734.
3. Kunkel and Loeb (1980) *J. Biol. Chem.* **255**, 9961.
4. Kunkel and Loeb (1981) *Science* **213**, 765.
5. Kunkel, Schaaper, Beckman, and Loeb (1981) *J. Biol. Chem.* **256**, 9883.
6. Loeb, Kunkel, and Schaaper (1980) *ICN–UCLA Symp. Mol. Cell Biol.* **19**, 735.

[c] Sulfur substituted for oxygen at the α phosphorus of dCTP. The incorporated CαS is not hydrolyzed by the exonucleases associated with *E. coli* Pol I or T4 DNA polymerase.

is actually much higher than would be predicted. Therefore, the associated proofreading exonucleolytic activity of polymerases must contribute greatly in error prevention. *In vivo,* there is evidence for a second type of proofreading, namely mismatch-specific endonucleases, which, in bacteria, excise noncomplementary bases postsynthetically. Errors produced by nonspecific base insertion opposite apurinic–apyrimidinic sites, for example, may be repaired in this way. Nevertheless, errors of a high magnitude can be induced by various perturbations of the experimental systems. Polymerases themselves can be error-prone, as found in enzymes isolated from the liver of carcinogen-treated animals or in leukemic lymphocytes. A number of metals normally present in cells decrease fidelity, probably by interaction with the template. Modification of the triphosphates may lead to errors if the modified NTP is an analog that can basepair to give a transition (e.g., 2-aminopurine NTP). Changes of intracellular nucleotide pools, such as occur in diseases due to the lack of specific kinases, are likely to lead to error. In model experiments, biasing specific NTPs relative to other NTPs in the pool greatly increased the reversion frequency of ϕX174. An exampple is shown in Table III-8. The

Table III-8. Effect of Nucleotide Pool Bias on Fidelity of φX174 Synthesis[a]

Relative concentration of biased nucleotide (dCTP)	Reversion frequency ($\times 10^{-6}$)	Estimated error rate
—	0.50 ± 0.50	
5×	1.52 ± 0.70	1/128,000
10×	3.37 ± 1.15	1/57,900
25×	6.22 ± 1.90	1/31,400
50×	13.40 ± 4.22	1/14,600

[a] Data from Kunkel and Loeb (1980) *J. Biol. Chem.* **255,** 9961. Reactions were performed using 10 μM unbiased dNTPs, *E. coli* DNA polymerase I, and 5 mM $MgCl_2$. The method of calculation of the estimated error rate is given in the reference.

reverse type of experiment, namely greatly decreasing or eliminating one NTP, also forces errors since polymerases can insert the noncomplementary base, but synthesis is greatly decreased. Competition studies between the K_m of the incorrect substrate and the correct substrate indicated a ratio of 10^{-2}–10^{-3}.

In summary, fidelity in DNA synthesis relies to a great extent on the ability of DNA polymerases to distinguish NTPs and to correct errors, rather than solely on the hydrogen-bonding properties of nucleotides. Ultimately, however, genetic information is replicated through basepairing compatible with the stereochemical constraints of a Watson–Crick helix. When a mispairing base fitting the geometry of the helix can be stabilized in the helix through stacking or alternative hydrogen-bonding, strengthened by nearest-neighbor environment, both transitions and transversions may occur. It should be emphasized that mispairing is greatly dependent on the precise sequence of nucleotides. In this way, unfavored tautomers or increased protonation contribute to mutation.

E. TRANSCRIPTION AND REPLICATION OF MODIFIED TEMPLATES

The precise mechanism by which a mutagen or carcinogen exerts its biological effect has not been defined in any system, inasmuch as there can be both direct and indirect effects at every stage in replication. However, several types of experiments, in which one component is modified, are useful as models of point mutation. These include translation of messengers, measurement of codon–anticodon interaction, hybridization of polynucleotides, and transcription or replication using natural and synthetic templates. When the modified nucleoside can form at least two hydrogen bonds, templates and messengers are active. In hybridization studies, secondary structure or conformation can prevent this bonding, even though it is theoretically possible. Some examples are polyformycin (adenosine with the sugar attached to a carbon replacing a nitrogen at the 9 position), which does not complex with poly(U), a fact attributed to the base being in the *syn* conformation; and poly(2-thiouridine), which does not complex with poly(A), a fact attributed to its high stacking force, which leads to a stable self-helical molecule. Both these modified bases, as NTPs, can be utilized by polymerases. When copolymerized with other bases and transcribed, they exhibit the expected base-pairing. Thus, when studying the effects of carcinogen adducts on nucleic acid synthesis, it is important to use polymers in which the modified base does not markedly change the secondary structure.

Relatively few nucleic acid derivatives have been studied in terms of their

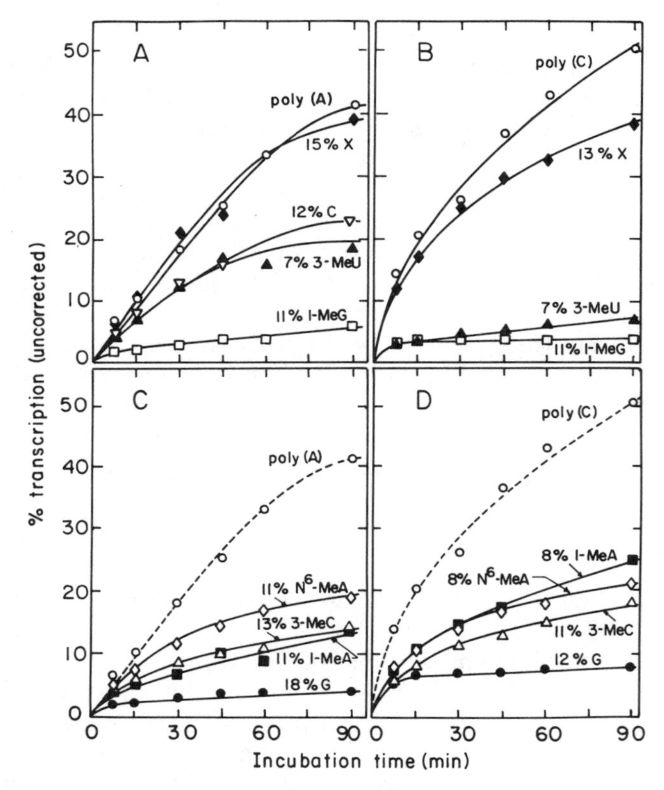

Figure III-8. Rate of transcription of homoribopolynucleotides as compared with copolymers. All polymers were synthesized using polynucleotide phosphorylase. The percent and type of nucleotide, other than the major nucleotide (or carrier), are shown in the figure. Percent transcription is the absolute amount of [^{3}H]-UMP incorporated opposite poly(A) homo- or copolymers (panels A, C) and of [^{3}H]-GMP opposite poly(C) homo- or copolymers (panels B, D). All four triphosphates were present. From Kröger and Singer (1979) *Biochemistry* **18**, 3493.

ability to direct transcription. Mutagenic modification of a polynucleotide, with very few exceptions, leads to at least two products and usually more (see Chapters IV, VI). Particularly in the case of the large aromatic metabolites, not all products have been identified. It does appear that transcription is inhibited by the adducts formed by metabolites of acetylaminofluorene and benzo[*a*]pyrene. In these instances, transcription is terminated at, or near, the modified base.

Ribopolynucleotides synthesized with a single type of modified base incorporated to a level of 1–25% have been used as templates for *in vitro* transcription or translation. Regardless of whether or not errors result, the rate and extent of polymer synthesis are generally inhibited. Figure III-8 shows that not only modifications blocking an essential hydrogen bond (N-3 C, N-1 A, N-3 U, N-1 G) decrease transcription, but also the presence of G slows transcription dramatically, presumably due to changes in secondary structure. In Figure III-9, it can also be seen that inhibition is a function of the amount of modified base. While these figures illustrate transcription using DNA-dependent RNA polymerase (in the presence of Mn^{2+}) the same kinetic effects are found, at even lower levels of modified bases, when deoxypolynucleotides are the template for DNA polymerase I or RNA-dependent RNA polymerase. Thus, it is likely that all *in vivo* tests of misincorporation resulting from base modification give much higher error rates than *in vivo* where only a few

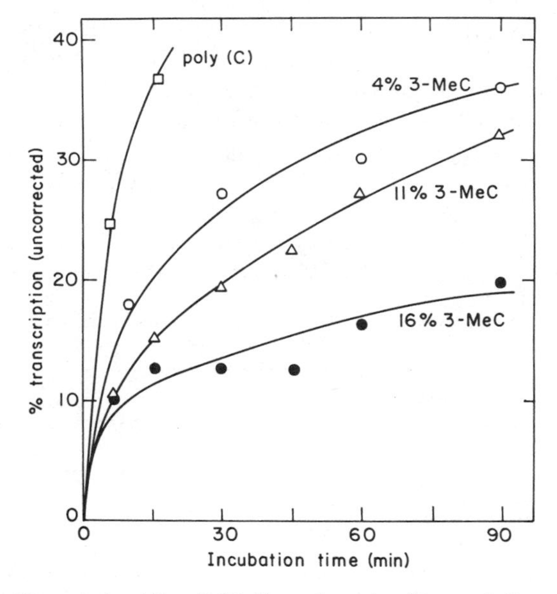

Figure III-9. Effect of 3-methylcytidine (3-MeC) on the rate of transcription of poly(rC) by using [^{3}H]-GTP and unlabeled ATP, UTP, and CTP. Percent transcription is the absolute amount of [^{3}H]-GMP incorporated, uncorrected for the amount and transcriptional effect of 3-MeC. From Kröger and Singer (1979) *Biochemistry* **18,** 3493.

molecules of modified DNA will be completely replicated, given the kinetics observed *in vitro*.

Nevertheless, *in vitro* copying of polynucleotides is an important clue as to the type and extent of errors produced by specific modification.

A considerable amount of data on the effects of mutagens and carcinogens adding small aliphatic groups to nucleosides has been assessed by transcription of polyribonucleotides with prokaryotic DNA-dependent RNA polymerases. Whether Mn^{2+} or the normal cation, Mg^{2+}, is used, no qualitative differences have been observed (Table III-9). An example of quantitative differences is shown in the table. In the presence of Mn^{2+}, m^3dC or m^3rC both incorporate C preferentially, while when Mg^{2+} is used, U incorporation is favored. The data in Table III-9 are obtained with an *E. coli* polymerase. Experiments using a eukaryotic DNA-dependent RNA polymerase from calf thymus to transcribe deoxypolymers are generally confirmatory, but here, too, there are quantitative differences from results using the prokaryotic polymerase.

Although it would be preferable for model studies to use defined polydeoxynucleotides or site-specific insertion of the modified derivative in natural DNAs, the problems and labor to synthesize such models are great compared to the relative ease of preparing defined ribopolynucleotides. There is also an advantage to using ribopolynucleotides and an RNA polymerase: the lack of proofreading and repair maximizes any errors so that polymers with a very small proportion of modified nucleoside can be tested.

Apparently the only defined deoxypolynucleotides prepared synthetically for studies of infidelity in synthesis with DNA polymerases contain the analog, 2-aminopurine, O^6-methyl G, O^2-methyl T, or O^4-methyl T. Two deoxypolynucleotides have been chemically modified to produce a single derivative, as judged by HPLC analysis. These are poly[d(C, 3-methyl C)] and

Table III-9. Ambiguities Resulting from Ethenocytosine or 3-Methylcytosine in Poly(rC) or Poly(dC) Transcribed by E. coli DNA-Dependent RNA Polymerase[a]

	Nearest-neighbor sequence analysis ^{32}P radioactivity (%)							
	2 mM Mn^{2+}				10 mM Mg^{2+}			
Template[b]	CpG	ApG	UpG	Total	CpG	ApG	UpG	Total
Poly(rC_{93}, εrC_7)	0.3	2.2	3.2	*5.7*	0.2	1.1	2.8	*4.1*
Poly(rC_{83}, m^3rC_{17})	5.9	3.6	2.2	*11.7*	3.6	2.8	6.4	*12.8*
Poly(dC_{93}, εdC_8)	0.3	2.6	3.0	*5.9*	0.2	0.9	3.0	*4.1*
Poly(dC_{84}, m^3dC_{16})	7.1	3.8	3.9	*14.8*	4.8	1.4	5.0	*11.2*

[a] The Mn^{2+} and Mg^{2+} concentrations were optimal for poly(rC) and poly(dC), respectively. All four ribotriphosphates were present with GTP $\alpha^{32}P$-labeled. Data from Kuśmierek and Singer (1982) *Biochemistry* **21,** 5723.

[b] Abbreviations: εrC, 3,N^4-ethenocytidine; m^3rC, 3-methylcytidine. εdC and m^3dC are the deoxy derivatives.

Table III-10. Summary of Effect of Nucleoside Modification on Transcription of Polynucleotide Templates[a]

Basepairing unchanged	Ambiguous behavior[b]	Changed pairing only
N^4-Acetyl C	N^4-Hydroxy C	N^4-Methoxy C
N^6-Methyl A	N^6-Methoxy A	Iso-C
N^6-Hydroxyethyl A	N^4-Methyl C	(2-Amino-2-deoxy U)
N^6-Isopentenyl A	N^2-Methyl G	Xanthosine
5-Halo, methyl, hydroxyl U or C	O^2-Alkyl U	
2′-*O*-methyl A or C or U	O^4-Alkyl U	
	O^2 Methyl T[c,e]	
	O^4 Methyl T[c,e]	
7-Methyl G	O^6-Alkyl G[c]	
2-Thio U or C	1-Methyl A	
4-Thio U	3-Methyl C[c,e]	
Iso-A (3-ribosyladenine)	3-Methyl U	
	1,N^6-Etheno A	
	3,N^4-Etheno C[c,d,e]	
	2-Aminopurine[e]	

[a] Many of these data were published in a different form by Singer and Kröger [*Prog. Nucleic Acid Res. Mol. Biol.* **23,** 151 (1979)]. Additional data are in references in the text. Except where noted (by footnotes *c, e*), ribopolynucleotides were used.

[b] The term ambiguous behavior is used for two classes of modified bases. The first class can form hydrogen bonds with two different base residues (N^4-hydroxy C, N^6-methoxy A, 2-aminopurine). All the remaining modifications lead to misincorporation of more than two bases, usually without stable hydrogen bonds. This latter type of ambiguity is not completely random and is consistent for each modification. Experiments in which the effect of the presence of Mn^{2+} versus Mg^{2+} was studied indicate that this divalent ion does not change the pattern of misincorporation.

[c] Misincorporation of ribonucleotides also from deoxypolynucleotide templates containing a single modification.

[d] Reaction of chloroacetaldehyde with poly(rC) or poly(dC) leads to two products. Transcription of these modified polymers without prior dehydration to a single product (εC) does not represent the specific misincorporations resulting from εC. Misincorporations resulting from the presence of the hydrated intermediate differ markedly from those of εC [Kuśmierek and Singer (1982) *Biochemistry* **21,** 5723].

[e] Misincorporation of deoxynucleotides from deoxypolynucleotide template.

poly[d(C, 3,N^4-etheno C)]. Results obtained with these polymers as templates for *E. coli* DNA polymerase I will be discussed separately.

Table III-10 summarizes data for both ribopolymers and deoxyribopolymers as templates for DNA-dependent RNA polymerases. The data are arranged in three categories: modifications having no discernible effect on basepairing, modifactions causing ambiguous incorporation of more than one nucleoside, and modifications completely changing basepairing through a change in hydrogen-bonding.

Modifications that are on the non-Watson–Crick side or on the sugar (such as C-5 pyrimidines, 2′-*O*, N-7 of G and iso A) have no effect on basepairing. The well-documented mutagenicity of 5-Br dU incorporated into bacteria is, as stated in Section C, probably due to enhanced keto–enol tautomerism resulting in increased C-like behavior. In terms of efficiency, this effect is too small to be detected by *in vitro* tests. All three pyrimidine oxygens can be changed to sulfur without significant distortion of hydrogen-bonding. Modification of exocyclic amino groups can lead to any of the three categories in Table III-10, but if the substituent in the polymer is *anti* to the Watson–Crick side, normal pairing occurs. In transcription, this is the case for N^4-acetyl C and the three N^6 modified derivatives of A.

Derivatives show ambiguous behavior when an essential Watson–Crick site is blocked (N-1 of purines or N-3 of pyrimidines) (Figure III-10) and no more than a single hydrogen bond can be formed. Transcription, however, is not terminated at these sites. This is also true of the etheno derivatives, 3,N^4-ethenocytidine (εC) and 1,N^6-ethenoadenosine (εA) (Chapter IV, Figure IV-29), in which a bulky, planar substituent blocks two Watson–Crick sites. The inability to form a specific basepair then leads to misincorporation. These misincorporations are not completely ambiguous since no incorporation of G occurs with εC-containing templates. Figure III-11 illustrates the possible orientation of εC and misincorporated bases. It is postulated that while stable hydrogen bonds may not be formed, proper positioning of the two bases without steric hindrance is necessary.

Another reason for amibuity is rotation of a substituent on an exocyclic amino group into the Watson–Crick side, which then sterically interferes with

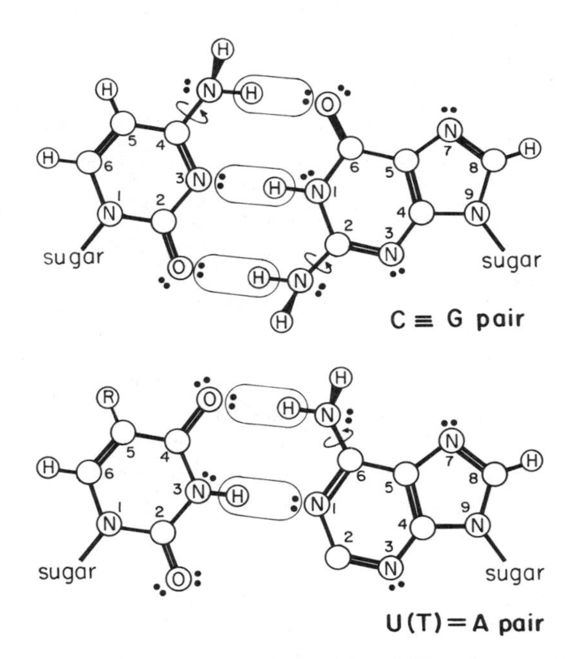

Figure III-10. Participation of electron pairs in basepairing. The ovals represent hydrogen bonds in a Watson–Crick structure. Solid pairs of dots represent electron pairs. Electron pairs inside ovals, when involved in hydrogen-bonding, are unreactive. Others may react when in double-stranded structures, if physically accessible. Note that the oxygens all have a free electron pair, reactive in both single-stranded and double-stranded nucleic acids. The three exocyclic amino groups may rotate, as indicated by arrows, so that the two hydrogens are equivalent. A substituent may, therefore, interfere with normal hydrogen-bonding if rotated into the basepairing side, as well as if it is on the N-1 of purines and N-3 of pyrimidines.

basepairing and eliminates a possible hydrogen bond (Figure III-10). It can be inferred from transcription data that N^4-hydroxy C not only acts like U due to the favored tautomer and *anti* conformation, but the hydroxyamino group can rotate *syn,* which results in some ambiguity. N^6-Methoxy A can, as a result of tautomerism, act like G as well as like A, but not like U or C. The experimental results indicate that there is a shift in the tautomeric equilibrium, but in this case the methoxyamino substituent lies *anti* in transcription.

The methyl group in both N^4-methyl C and N^2-methyl G prefers to be *anti,* but also rotates. The *O*-alkyl derivatives studied (O^6-alkyl G, O^2-alkyl U/T, O^4-alkyl U/T) appear to have the substituent *syn,* on the basis of transcription data, but the changed electron distribution can also be responsible. The crystal structure of O^6-methyl G also indicates that the methyl group is *syn*. However, although N^6-methyl A has the methyl group preferentially *syn* in the monomer, when incorporated into polymers it is *anti* as indicated by the ability of N^6-methyl A to substitute for A in *in vitro* synthesis (Table III-11). Therefore, rotational freedom of exocyclic substituents in polymers may be quite limited. As shown in Figure III-12, the size of various substituents on the N^6 of A would, if oriented *syn,* completely prevent formation of any hydrogen bond. Nevertheless, in transcription when all four NTPs are present these derivatives all behave like A only.

Three derivatives (right column of Table III-10) show changed basepairing that can be explained on a chemical basis. N^4-Methoxy C, which is formed by reaction with methoxyamine, is in the imino form and thus substitutes for U (Section C-1). The behavior of the other two derivatives is also predictable, since iso C, a model compound, and xanthosine (deaminated G) have the hydrogen-bonding equivalent of U.

The extent of total changed or mis-incorporation in transcription is approximately equal to the amount of modified nucleoside in the polymer. Thus,

Table III-11. In Vitro Synthesis of Poly[d(m⁶A, A-T)] with E. coli DNA Polymerase I[a]

Input ratio (dATP : m^{6}dATP)	Incorporation (into polymer) ratio (dA : m^{6}dA)	T_m (°C)[b]
100 : 0	100 : 0	61.5
90 : 10	89 : 11	61.0
75 : 25	86 : 14	59.5
62 : 38	58 : 42	55.0
50 : 50	47 : 53	53.0
0 : 100	0 : 100	44.5

[a] From Engel and von Hippel (1978) *J. Biol. Chem.* **253,** 935.
[b] In 0.1 M NaCl 0.01 M sodium phosphate, 10^{-4} M Na_2EDTA, pH 6.8 buffer.

(1)

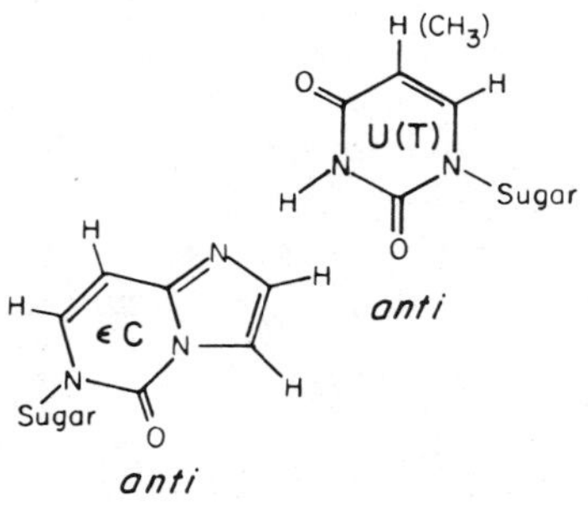

U(T) *anti*

εC *anti*

(2)

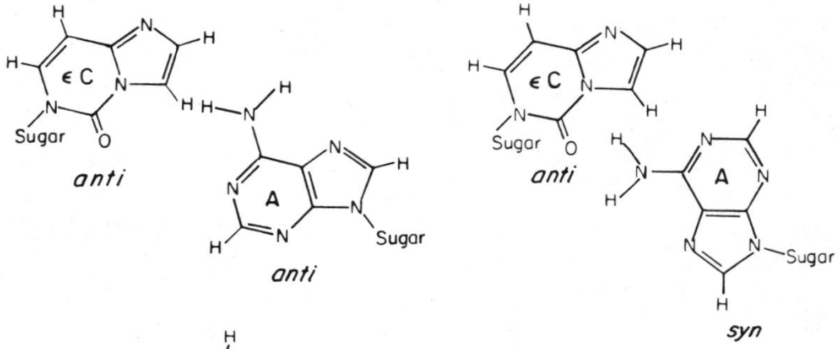

εC *anti* — A *anti*

εC *anti* — A *syn*

(3)

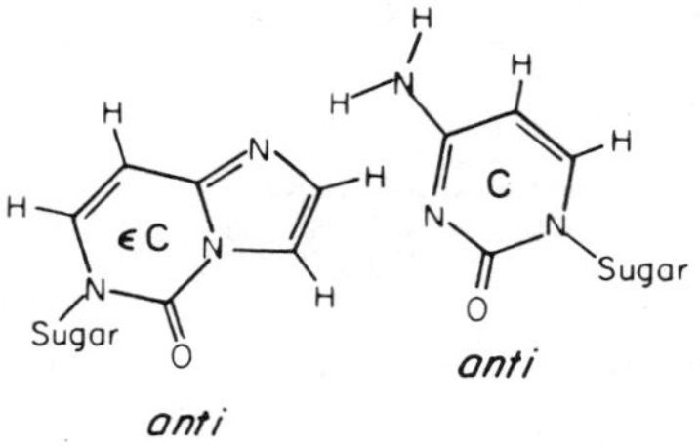

εC *anti* — C *anti*

(4)

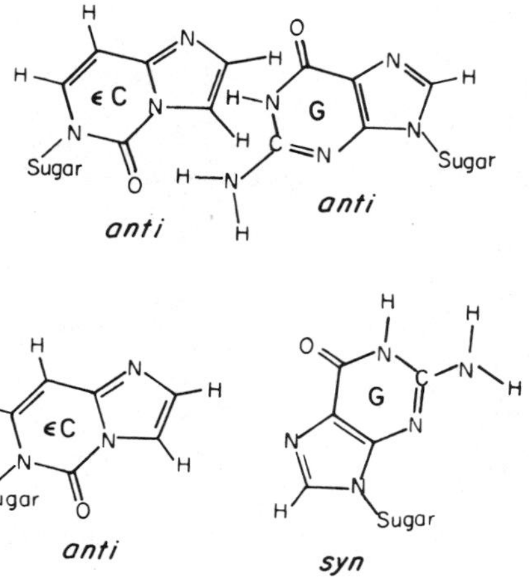

εC *anti* — G *anti*

εC *anti* — G *syn*

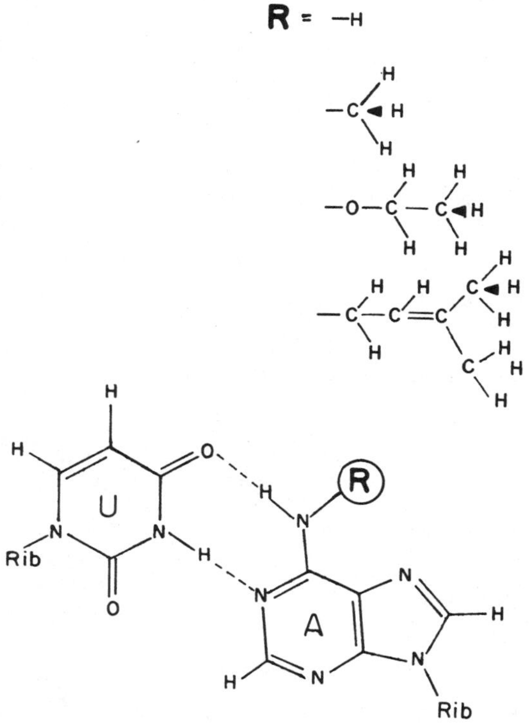

Figure III-12. Hydrogen-bonding of N^6-subsituted A with U. R = substituents as shown above the basepair. The substituents are (top to bottom): hydrogen (A), methyl (m^6A), hydroxyethyl (he^6A), isopentenyl (i^6A). Rotation of any substituent except the hydrogens would probably prevent the formulation of any hydrogen bonds by shielding (m^6A) or gross steric interference (he^6A, i^6A). From Singer and Spengler (1981) *Biochemistry* **20,** 1127.

the misincorporations observed are not rare events. Nearest-neighbor analysis (Figure III-13) is an extremely sensitive tool and can detect minor misincorporations readily since transcription by DNA-dependent RNA polymerases is not appreciably blocked by small alkyl bases.

Prokaryotic DNA polymerases are greatly inhibited by modified bases in deoxypolynucleotide templates and, particularly, *E. coli* DNA polymerase I

Figure III-11. Effect of 3,N^4-etheno C (εC) on transcription of ribo- and deoxyribopolynucleotides using DNA-dependent RNA polymerase in the presence of either Mn^{2+} or Mg^{2+}. εC in copolymers directs the incorporations of U/T > A >>> C. These interactions without hydrogen bonds included are shown in (1) εC · U/T; (2)εC · A (*anti* and *syn*); and (3) εC · C. A single hydrogen bond can be visualized in (1), (2) if A is *syn*, and (3). No incorporation of G is detected when εC-containing templates are used. The steric hindrance preventing interaction is shown in (4). When G is rotated *syn*, proper positioning of the sugar is prevented and no hydrogen bonds are possible (4).

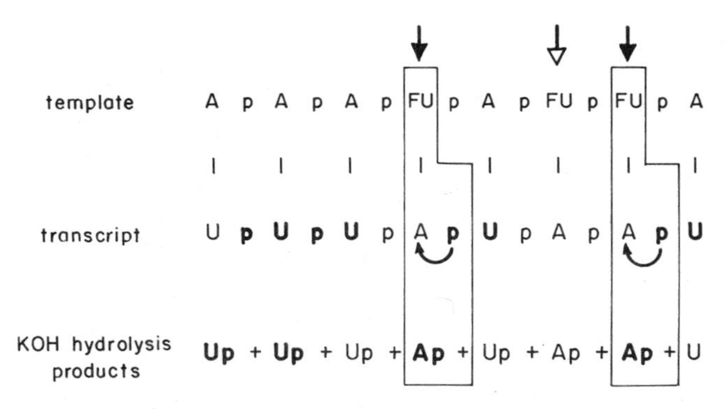

Figure III-13. Principle of nearest-neighbor analysis of transcription products using α^{32}P-labeled nucleoside triphosphates. A copolymer of A and FU is used as an example. In the transcript, pU, in bold type, denotes the $5'^{32}$P-labeled nucleotide; pA, in light type, is unlabeled. Bold type in the hydrolysis products shows where the ^{32}P label has transferred. When the product of transcription is a deoxypolynucleotide, spleen phosphodiesterase is used for hydrolysis to 2′(3′) nucleotides. Solid arrows indicate the conditions under which the radioactivity is transferred to Ap, while the open arrow indicates that no transfer occurs when two FUs are adjacent. Thus, polynucleotides with clustered modified nucleotides will not give quantitative data by using nearest-neighbor analysis.

contains an editing function so that few errors are made or detected. There are several studies utilizing chemically modified deoxy templates but, in most cases, the products of reaction were not analyzed. It is therefore difficult to interpret these data in terms of which modification(s) leads to misincorporation. There are three cases where the extent and nature of the modification were analyzed for. These are 3,N^4-etheno dC and its hydrated intermediate and 3-methyl dC in poly(dC). All block Watson–Crick sites and all cause differing ambiguity in transcription. However, when studying DNA synthesis, different results are obtained. As illustrated in Figure III-14, the hydrated intermediate of chloroacetaldehyde-treated poly(dC) does not miscode, while the stable εdC leads to TMP misincorporation: 1/20 εdC. m^{3}dC also causes TMP misincorporation but to a lesser extent: 1/80 m^{3}dC. Thus, both qualitatively and quantitatively, replication errors resulting from carcinogen-modified bases are less frequent than errors in transcription of the same deoxypolynucleotides (Table III-9).

One of the important sites of modification of nucleic acids by carcinogens are the phosphodiesters. In ribopolymers phosphotriesters are unstable, but in deoxypolymers they are extremely stable and resistant to enzymatic hydrolysis. Decadeoxynucleotides containing an ethyl triester in either an *R* or an *S* configuration have been prepared and used as templates for *E. coli* DNA polymerase I. Although the rates of synthesis of the complementary deoxyoligomers were decreased compared to an oligomer lacking the triester, the polymerase was able to "read through" the triester. The two isomers, occurring also in poly(dA-dT) containing triesters, differ in the rates of synthesis

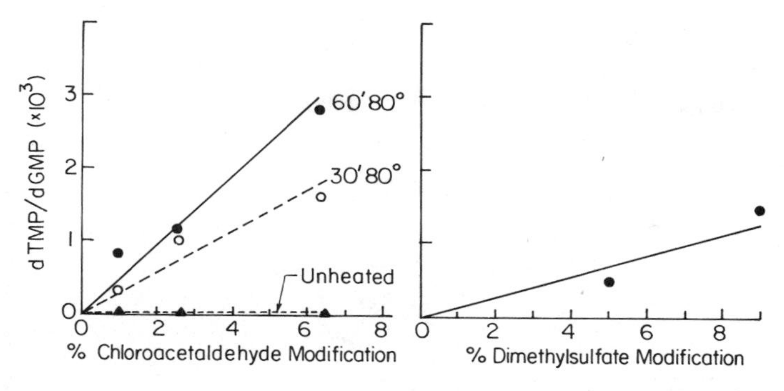

Figure III-14. Misincorporation of dTMP by *E. coli* DNA polymerase I using as templates poly(dC) modified by chloroacetaldehyde (left) or dimethylsulfate (right). The modification in the unheated chloroacetaldehyde sample is >95% hydrated 3,N^4-etheno dC. Heating at 80°C in 0.01 M pH 7.2 Tris quantitatively converts the hydrated form to 3,N^4-etheno dC (see Chapter IV, Section F). The only modified base, found by HPLC, in dimethylsulfate-treated poly(dC) is m^3dC. From Singer *et al.* (1983) *Proc. Natl. Acad. Sci. USA* **80,** 969–972.

(Figure III-15). This suggests that the orientation of the ethyl group relative to the rest of the template backbone is an important factor in inhibition of synthesis, possibly causing unfavorable steric interactions with the polymerase. It is interesting to note that the replacement of a negative charge by an ethyl group does not block DNA synthesis, although the strength of the hydrogen-bonding probably differs from that of normal nucleotides. This later supposition is based on earlier definitive studies on the effect of triesters on codon–anticodon binding.

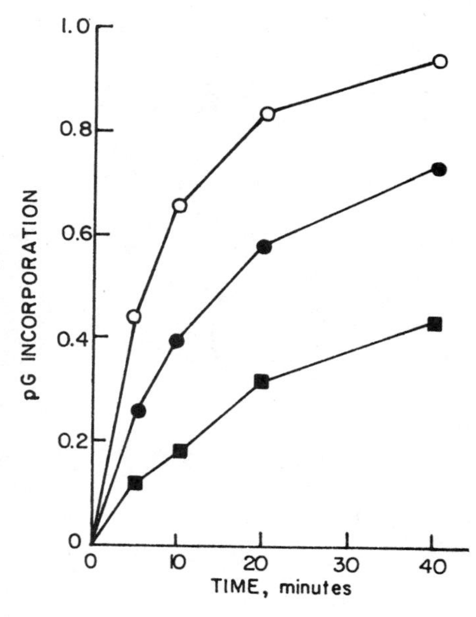

Figure III-15. DNA polymerase I-catalyzed polymerization directed by d-CpCpApApGpApTpTpGpG(pA)$_9$(○);d-CpCpApApGp(Et)ApTpTpGpG(pA)$_{12}$, isomer I (●); and d-CpCpApApGp(Et)ApTpTpGpG(pA)$_{11}$, isomer II (■). Incorporation of d[^{32}P]-GMP into oligomeric material was assayed on PEI cellulose. For polymerization to occur, the polymerase must copy through the ethylphosphotriester. The extent of polymerization using the unmodified template levels off after approximately one dGTP residue is incorporated. This suggests that the 3 → 5 exonucleolytic activity of the polymerase prevents complete copying of the 5′ end. Adapted from Miller *et al.* (1982) *Biochemistry* **21,** 5468.

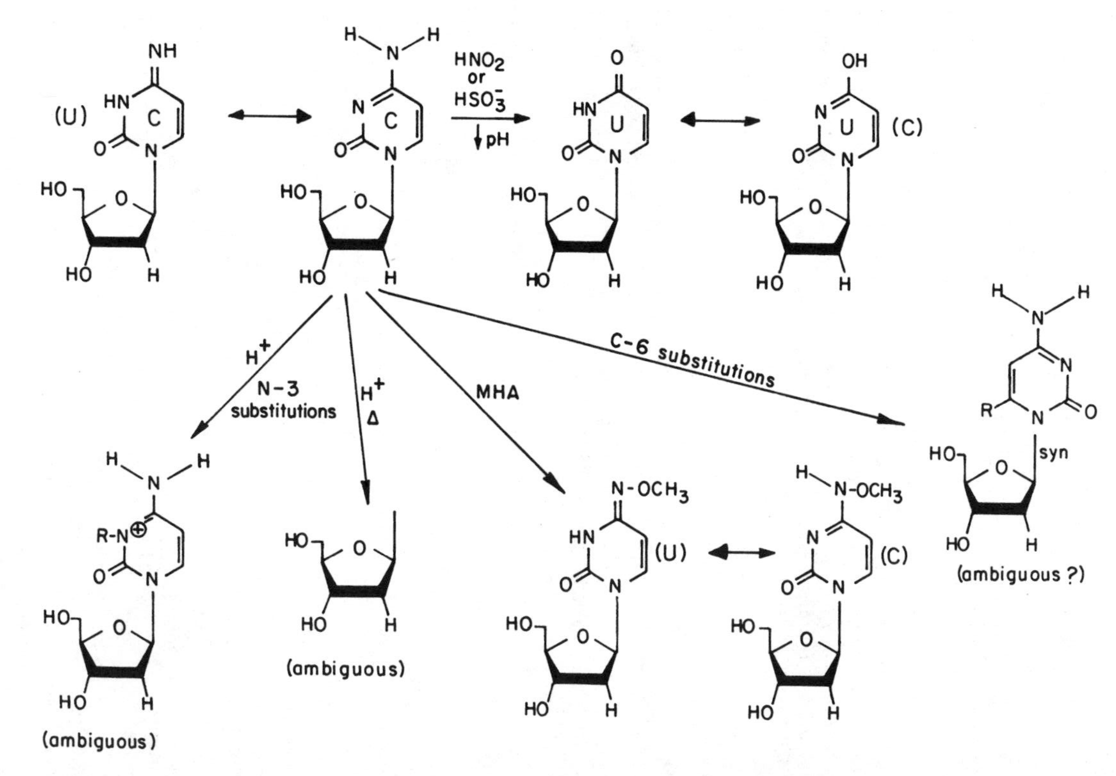

Figure III-16. Various reactions of cytidine leading to changes in base pairing. The figure illustrates the effects of tautomeric shift, deamination, base rotation, depyrimidination, change in ionization, and substitution of the amino group or modification of the N-3. Experimentally determined effects on transcription are shown in parentheses. MHA is *O*-methylhydroxylamine, also termed methoxyamine. From Singer and Kuśmierek (1982) *Annu. Rev. Biochem.* **51,** 655.

Based on polynucleotide studies, some general conclusions can be drawn regarding possible mutational effects on modified nucleosides.

The bulky aromatic substituents, regardless of their position on a nucleoside, will interfere with replication simply because of size. However, the rapid depurination of some of these products may exert a mutagenic effect through error-prone repair of the apurinic site. This is also true for other easily depurinated derivatives such as N-3 and N-7 alkylpurines.

Smaller alkyl derivatives, with substituents affecting hydrogen-bonding, if unrepaired, are likely to cause misincorporation without necessarily forming hydrogen bonds if they can be held in the helix by stacking or other interactions.

Mutation by most nonalkylating agents can be ascribed to a change in a base by deamination or a change in the preferred tautomer in the modified derivative.

Finally, Figure III-16 illustrates, with cytidine, how all the reactions in this chapter can affect replication, although the data on which the figure is based are mostly derived from transcription experiments. Starting from cytidine (top, second from left), tautomerism, whether intrinsic or chemically caused, leads to U as does deamination. Deaminated C, which is U, also can exhibit tautomerism and thus would act like C. On the bottom line, from the left, are shown a change in ionization, depyrimidination, a change to the imino tautomer resulting from chemical modification (and the less-favored tautomer), and rotation of the base around the glycosyl bond. Ambiguity in transcription results from three of these reactions.

NOTE ADDED IN PROOF

O^2-Methylthymidine and O^4-methylthymidine (as the triphosphates) can be substituted for thymidine in DNA polymerase I (Pol I) catalyzed synthesis of the alternating polymer d(A-T). The parameters as determined by thermal denaturation are the same as pol d(A-T). The incorporated O-methylthymidines cause misincorporation of G, both in transcription and replication.

SELECTED REFERENCES

Bodell, W. J., and Singer, B. (1979) Influence of hydrogen bonding in DNA and polynucleotides on reaction of nitrogens and oxygens toward ethylnitrosourea. *Biochemistry* **18,** 2860–2863.

Cullis, P. M., and Wolfenden, R. (1981) Affinities of nucleic acid bases for solvent water. *Biochemistry* **20,** 3024–3028.

Drake, J. W. (1969) Comparative rates of spontaneous mutation. *Nature (London)* **221,** 1132.

Dukar, N. J., Hart, D. M., and Grant, C. L. (1982) Stability of the DNA apyrimidinic site. *Mutat. Res.* **103,** 101–106.

Fersht, A. R., and Knill-Jones, J. W. (1981) DNA polymerase accuracy and spontaneous mutation rates: Frequencies of purine · purine, purine · pyrimidine, and pyrimidine · pyrimidine mismatches during DNA replication. *Proc. Natl. Acad. Sci. USA* **78,** 4251–4255.

Fersht, A. R., Shi, J.-P., and Tsui, W.-C. (1983) Kinetics of base misinsertion by DNA polymerase I of *Escherichia coli*. *J. Mol. Biol.* **165,** 655–667.

Fresco, J. R., Broitman, S., and Lane, A.-E. (1980) Base mispairing and nearest neighbor effects in transition mutations. In *Mechanistic Studies of DNA Replication and Genetic Recombination* (B. Alberts, ed.), Academic Press, New York, pp. 753–768.

Grunberger, D., and Weinstein, I. B. (1979) Biochemical effects of the modification of nucleic acids by certain polycyclic aromatic carcinogens. *Prog. Nucleic Acid Res. Mol. Biol.* **23,** 106–149.

Katritzky, A. R., and Waring, A. J. (1962) Tautomeric azines. Part I. The tautomerism of 1-methyluracil and 5-bromo-1-methyluracil. *J. Chem. Soc.*, 1540–1544.

Kochetkov, N. K., Budovskii, E. I., Sverdlov, E. D., Simukova, N. A., Turchinskii, M. F., and Shibaev, V. N. (1971) The secondary structure of nucleic acids. In *Organic Chemistry of Nucleic Acids,* Part A (N. K. Kochetkov and E. I. Budovskii, eds.), Plenum Press, New York, pp. 183–268.

Kochetkov, N. K., Budovskii, E. I., Sverdlov, E. D., Simukova, N. A., Turchinskii, M. F., and Shibaev, V. N. (1972) Cleavage of phosphoester bonds and some other reactions of phosphate groups of nucleic acids and their components. In *Organic Chemistry of Nucleic Acids,* Part B (N. K. Kochetkov and E. I, Budovskii, eds.), Plenum Press, New York, pp. 477–532.

Kochetkov, N. K., Budovskii, E. I., Sverdlov, E. D., Simukova, N. A., Turchinskii, M. F., and Shibaev, V. N. (1972) Hydrolysis of *N*-glycosidic bonds in nucleosides, nucleotides, and their derivatives. In *Organic Chemistry of Nucleic Acids,* Part B (N. K. Kochetkov and E. I. Budovskii, eds.), Plenum Press, New York, pp. 425–448.

Kulikowski, T., and Shugar, D. (1979) Methylation and tautomerism of 1-substituted 5-fluorocytosines. *Acta Biochim. Pol.* **26,** 145–160.

Loeb, L. A., and Kunkel, T. A. (1982) Fidelity of DNA synthesis. *Annu. Rev. Biochem.* **51,** 429–457.

Lomant, A. J., and Fresco, J. R. (1975) Structural and energetic consequences of noncomplementary base oppositions in nucleic acid helices. *Prog. Nucleic Acid Res. Mol. Biol.* **15,** 185–218.

Nelson, J. W., Martin, F. H., and Tinoco, I. (1981) DNA and RNA oligomer thermodynamics: The effect of mismatched bases on double-helix stability. *Biopolymers* **20,** 2509–2531.

Niemczura, W. P., Hruska, F. E., Sadana, K. L., and Loewen, P. C. (1981) Proton magnetic resonance study of nucleosides, nucleotides and dideoxynucleoside monophosphates containing a *syn* pyrimidine base. *Biopolymers* **20,** 1671–1690.

Patel, D. J., Kozlowski, S. A., Marky, L. A., Rice, J. A., Broka, C., Dallas, J., Itakura, K., and Breslauer, K. I. (1982) Structure, dynamics, and energetics of deoxyguanosine-thymidine wobble base pair formation in the self-complementary d(CCTGAATTCGCG) duplex in solution. *Biochemistry* **21,** 437–444.

Pullman, B. (1979) The macromolecular electrostatic effect in biochemical reactivity of the nucleic acids. In *Catalysis in Chemistry and Biochemistry: Theory and Experiment* (B. Pullman, ed.), Reidel, Dordrecht, pp. 1–10.

Rich, A., and RajBhandary, U. L. (1976) Transfer RNA: Molecular structure, sequence, and properties. *Annu. Rev. Biochem.* **45,** 805–860.

Shapiro, R. (1968) Chemistry of guanine and its biologically significant derivatives. *Prog. Nucleic Acid Res. Mol. Biol.* **8,** 73–112.

Singer, B., and Kröger, M. (1979) Participation of modified nucleosides in translation and transcription. *Prog. Nucleic Acid Res. Mol. Biol.* **23,** 151–194.

Singer, B., and Kuśmierek, J. T. (1982) Chemical mutagenesis. *Annu. Rev. Biochem.* **51,** 655–693.

Singer, B., Kröger, M., and Carrano, M. (1978) O^2- and O^4-alkyl pyrimidine nucleosides: Stability of the glycosyl bond and of the alkyl group as a function of pH. *Biochemistry* **17,** 1246–1250.

Spengler, S., and Singer, B. (1981) The effect of tautomeric shift on mutation: N^4-Methoxycytidine forms hydrogen bonds with adenosine in polymers. *Biochemistry* **20,** 7290–7294.

Strauss, B., Rabkin, S., Sagher, D., and Moore, P. (1982) The role of DNA polymerase in base substitution mutagenesis on non-instructional templates. *Biochimie* **64,** 829–832.

Topal, M. D., Hutchinson, C. A., III, Baker, M. S., and Harris, C. (1980) Studies on the effects of chemical mutagens on the fidelity of DNA replication. In *Mechanistic Studies of DNA Replication and Genetic Recombination* (B. Alberts, ed.), Academic Press, New York, pp. 725–734.

IV

Reactions of Directly Acting Agents with Nucleic Acids

A. Introduction

Mutagens and carcinogens can be divided into those called "direct" because they are reactive with macromolecules without enzymatic activation, and those that require metabolic activation. In this chapter, only directly acting agents will be discussed, with one exception. Dialkylnitrosamines must undergo only a simple enzymatic step to be reactive, and for this reason they are included in the section on *N*-nitroso compounds rather than in Chapters V and VI, which deal with metabolically activated agents.

A further division of directly acting compounds can be made into those that do not alkylate but change basepairing by deamination or a shift in tautomeric equilibria (Section B), and those that substitute an alkyl residue for a proton (Sections C, D).

For virtually all the chemical reactions described in this chapter, model experiments on modification of nucleosides or homopolynucleotides preceded those with RNA, DNA, or *in vivo*. It appears that no reaction with alkylating agents occurs *in vivo* that does not occur in *in vitro* model systems. However, some reactions found in models are undetectable *in vivo*, probably due to the very low extent of reaction. Also, the use of high concentrations of most of the nonalkylating agents is required for modification in model systems, while the low concentrations used *in vivo* can lead to unexpected results. These are usually due to indirect effects, some of which are attributed to the formation of free radicals or oxidative degradation of the reagent.

B. Simple Nonalkylating Agents

This class of reagents (Figure IV-1) consists of those mutagens that can directly substitute exocyclic amino groups of nucleosides, either through oxidative mechanisms or by direct electrophilic attack. Their predominant effect

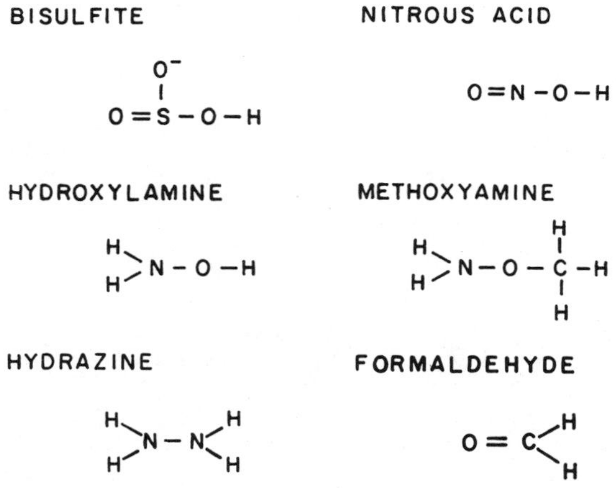

BISULFITE

$O{=}S(O^-){-}O{-}H$

NITROUS ACID

$O{=}N{-}O{-}H$

HYDROXYLAMINE

$H_2N{-}O{-}H$

METHOXYAMINE

$H_2N{-}O{-}CH_3$

HYDRAZINE

$H_2N{-}NH_2$

FORMALDEHYDE

$O{=}CH_2$

Figure IV-1. Structural formulas of simple nonalkylating mutagens and/or carcinogens.

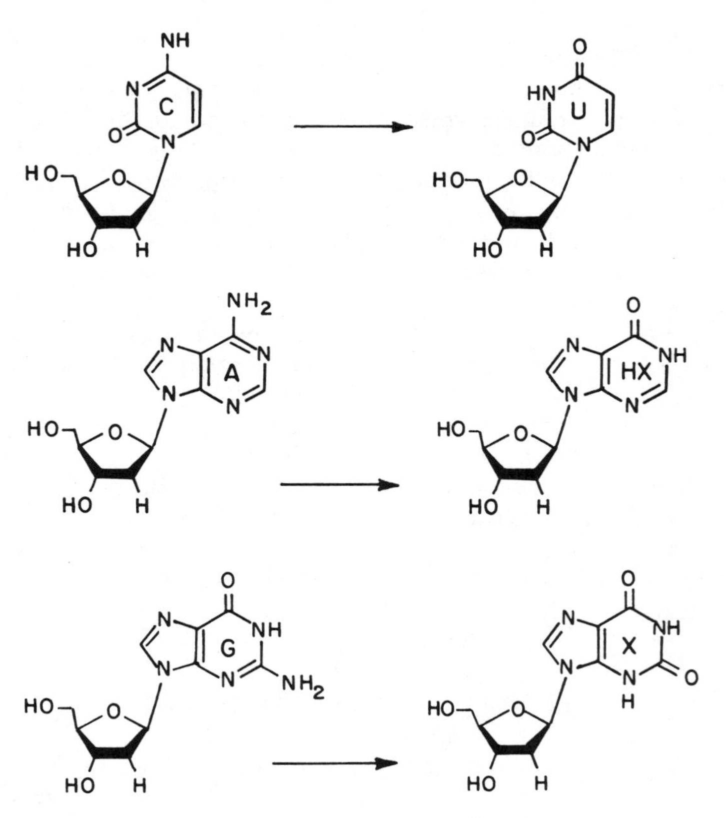

Figure IV-2. Deamination of nucleosides by nitrous acid.

is to change basepairing by deamination or a shift in the tautomeric equilibrium. In addition, bisulfite, hydroxylamine, methoxyamine, and hydrazine can add to the 5,6 double bond of C or (except for methoxyamine) to the 5,6 double bond of U. All react either primarily or solely on single-stranded polynucleotides.

1. Nitrous Acid

Nitrous acid, the classical oxidative deamination reagent ($-NH_2 \rightarrow -OH$), was once considered the "perfect" mutagen. Deamination of C → U, A → HX, or G → X (Figure IV-2) changes basepairing in a predictable manner. However, it was recognized very early that crosslinks, considered to be inactivating, were formed in HNO_2-treated *Bacillus subtilis* transforming DNA, and crosslinked diguanyl and guanyladenine were later isolated from DNA (Figure IV-3). Based on a comparison of data for HNO_2 mutagenesis as a function of deamination, there also appear to be additional unelucidated reactions occurring in DNA.

The kinetics of reaction with nucleosides or nucleotides indicate that the rate of deamination rises rapidly with a decrease in pH and that the protonated form of the nucleoside is attacked more rapidly than the neutral form. At pH 4.5 the ratio between the velocities of reactions of guanosine and cytidine,

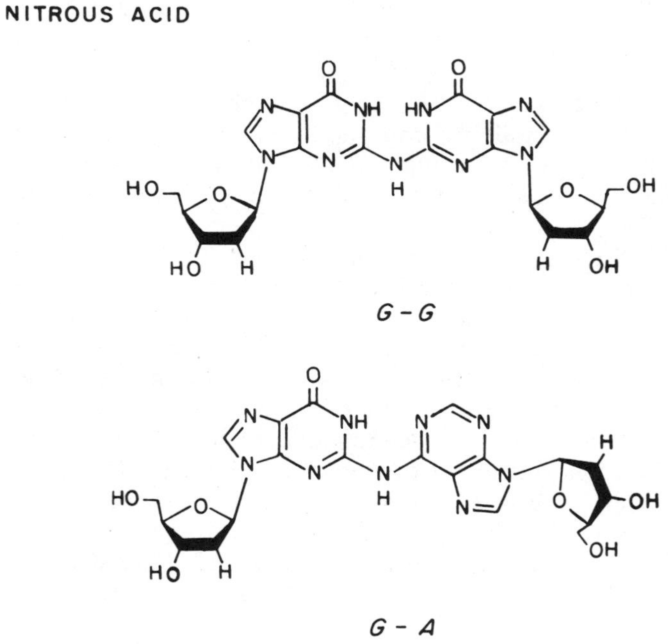

NITROUS ACID

G - G

G - A

Figure IV-3. Crosslinks between nucleic acid bases formed by nitrous acid.

Table IV-1. Relative Rates of Deamination of Bases with Nitrous Acid[a]

Substrate	G	A	C	Conditions
Nucleoside	6.5	2.2	1.0	pH 4.5, 21.5°C
tRNA *(E. coli)*	2.1	1.4	1.0	pH 4.3
TMV RNA	2.1	1.9	1.0	pH 4.3, 21.5°C
TMV	0.01	0.53	1.0	pH 4.2, 21.0°C
DNA (calf thymus)	2.2	0.44	1.0	pH 4.2, 20°C
DNA (heat-denatured)	1.3	0.63	1.0	pH 4.2, 20°C

[a] From Shapiro and Pohl (1968) *Biochemistry* 7, 448.

when both are primarily in the neutral form, is 6.5; but at pH 3.75, when guanosine is mainly in the neutral form and cytidine is protonated, it is 3.6. At all pHs when reaction is rapid (pH > 5), relative reaction rates of nucleosides are G > A > C. In polynucleotides, where secondary structure plays an important role, the relative reactivity of guanosine is greatly decreased (Table IV-1), as are the rate and extent of deamination.

The conditions required for generating HNO_2 from a precursor such as sodium nitrate or nitrite are inappropriate for mutation studies with cells. However, the optimal pH for $NO_2^- \rightarrow HNO_2$ is known to exist in the human stomach. The ubiquity of nitrate makes it likely that wherever the pH is acid some oxidative deamination may occur in addition to hydrolytic deamination.

2. *Bisulfite*

HSO_3^- is produced endogenously in mammals as an intermediate in the catabolism of sulfur-containing amino acids. Environmental exposure to sulfur dioxide is also a source of bisulfite. In water, sulfur dioxide becomes hydrated to form H_2SO_3, which, at neutral pH, dissociates to a mixture of bisulfite and sulfite.

Bisulfite undergoes a number of reactions with nucleic acids, but the observed mutations are likely to be the result of the deamination of C $\rightarrow$ U. Both Shapiro and Hayatsu studied the reactions of bisulfite with nucleosides or DNA and agree that only C is modified, but there are multiple products as a consequence of the mechanism by which C is deaminated (Figure IV-4). These include, as intermediates, the addition product of bisulfite to the 5,6 double bond, which leads by deamination to the analogous dihydrouracil derivative, and, following the slow removal of bisulfite, to the final product, uracil. In the presence of an appropriate amine, transamination can occur, yielding an N^4-substituted C. This reaction proceeds effectively at neutral pH, with lower concentrations of bisulfite than needed for deamination. Protein–nucleic acid crosslinks can be formed by such a transamination reaction involving the ε-amino group of lysine (or a terminal $-NH_2$) and the 4-NH_2 of cytidine. Such crosslinks have been reported to occur in bisulfite-treated

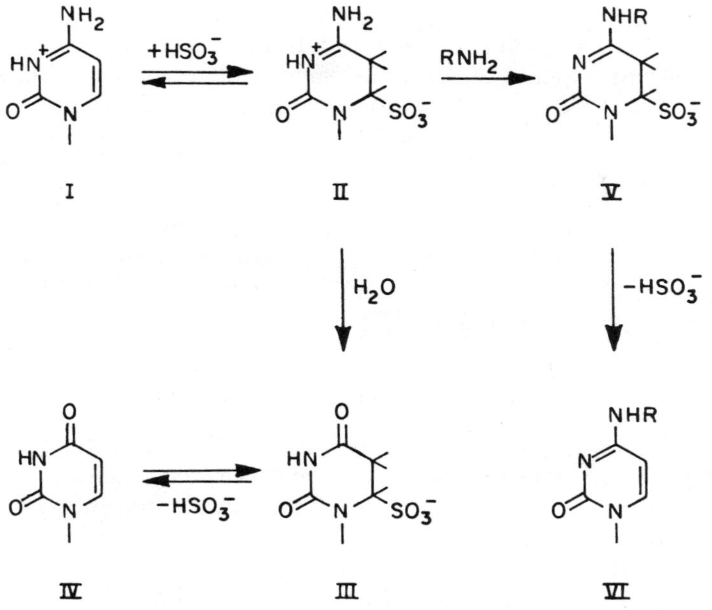

Figure IV-4. Deamination and transamination of cytidine. I–IV show the overall mechanism of bisulfite deamination. Transamination of the addition product II occurs when certain amines compete with water to displace the amino group (V) followed by regeneration of the double bond (VI). Adapted from Shapiro (1982) In *Induced Mutagenesis: Molecular Mechanisms and Their Implications for Environmental Protection* (Lawrence, Prakash, and Sherman, eds.), Plenum Press, pp. 35–60.

S_D phage and turnip yellow mosaic virus. The reaction with cytidine *in situ* can stop at the intermediate product, 5,6-dihydro-6-sulfopyrimidine, presumably as the result of shielding of the amino group of cytidine by protein.

5-Methylcytosine (m^5C), found in DNA, is also deaminated by bisulfite, but at a rate two orders of magnitude slower than C. In contrast to deamination of C → U, deamination of m^5C in DNA can lead to T, which, not being repaired (as is U by a uracil glycosylase), consequently would cause GC → AT transition. m^5C deamination could be the cause of some "hotspots" for spontaneous transition mutations, although it is doubtful if the formation of T is the only mutational event. In T-even bacteriophages, where 5-hydroxymethylcytidine (hm^5C) and its glucosylated derivative replace cytidine, reaction with bisulfite does not result in deamination, but rather gives 5-hydroxymethylenesulfonate as a product.

Except for this reaction with hm^5C, bisulfite reaction with C has been detected only in single-stranded nucleic acids. This general specificity for a non-hydrogen-bonded base suggests that the C-6 of the 5,6 double bond, to which bisulfite must add as the first step in deamination, may be sterically unavailable to the bulky SO_3^- ion. Because of its single-strand specificity, bisulfite has been used successfully to determine secondary structure and as a site-specific mutagen. Cytidine was deaminated in a selected single-stranded section of SV40 DNA, and the gap was filled using DNA polymerase I. A GC → AT transition was detected following *in vivo* replication.

Deamination of cytosine requires high concentrations of bisulfite (1 M) at pH 5.6. At a lower concentration (0.1 M), no mutations were found in *E. coli*. However, various genetic effects have been observed in neutral pH at even lower concentrations, and these are assumed to be caused by side reactions or reactions with proteins, such as sulfitolysis of disulfide bonds. One further possibility is that free radicals are formed that can inactivate enzymes.

3. *Hydroxylamine, Methoxyamine*

Hydroxylamine (HA), a presumed intermediate in nitrate reduction *in vivo*, and methoxyamine (MHA) are among the few mutagens that can directly cause a change in basepairing by favoring a tautomeric shift. Although HA can react with A, C, and U, the reaction with U, which degrades the nucleoside to ribosylurea (Figure IV-5), is very slow, and significant only at pH 10. The guanine ring does not appear to be reactive. Alkyl substituents at C-5 and C-6 of the uracil nucleus prevent formation of an adduct, probably due to a sharp increase in the energy of electron delocalization of the C5–C6 double bond. For this reason thymine is unreactive. MHA does not react with U or T, which is attributed to the absence of intramolecular stabilization.

The reaction of cytidine at pH 5–6 with HA or MHA leads to replacement of the amino group by a hydroxylamino (—NH—OH) or methoxyamino (—NH—OCH_3) group. Two reactions can occur as shown in Figure IV-6. One pathway is addition to the 5,6 double bond (II) followed by replacement of the amino group (IV) to form the bis derivative and finally regeneration of the 5,6 double bond (III). The other mechanism is direct replacement of the amino group (I → III), which occurs most readily at lower reagent concentration and higher temperatures. At 40°C in 0.2 M pH 6 HA, direct replacement of the N^4 of C occurs three times faster than formation of the bis derivative,

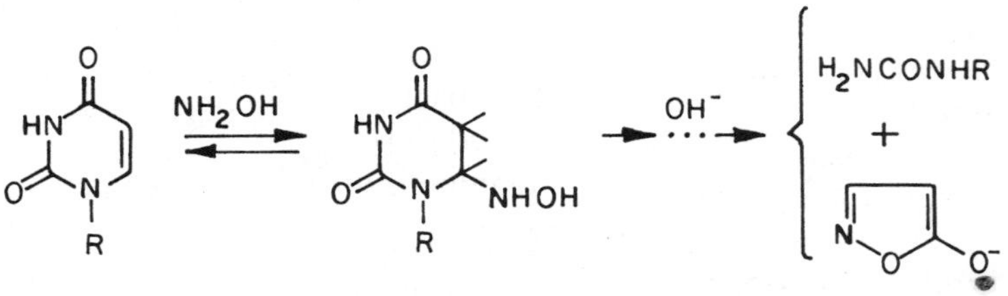

Figure IV-5. Reaction of uridine with hydroxylamine. At pH > 6, in addition to regeneration of the uracil base, there is cleavage of the N-1–C-6 and N3–C-4 linkages, with isoxazolone and ribosylurea being formed.

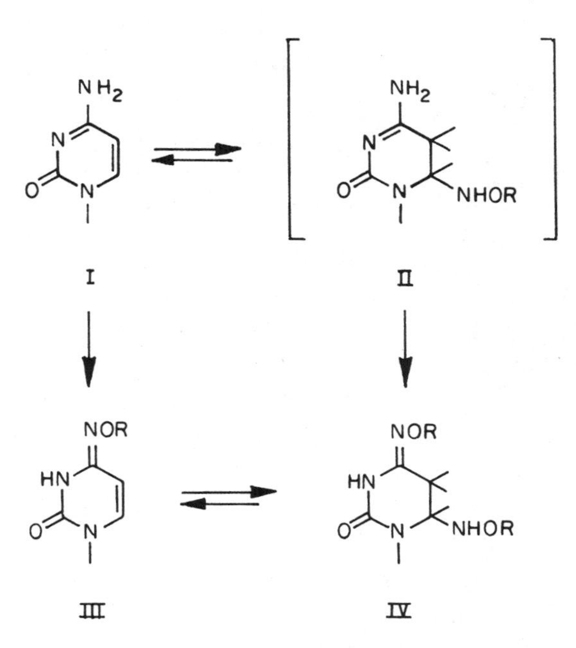

Figure IV-6. Reaction of hydroxylamine or *O*-methylhydroxylamine (methoxyamine) with cytidine. R = —NH—OH or —NH—OCH_3. Two reactions occur with either reagent. One is the primary addition of the reagent to the 5,6 double bond (II), which is followed by replacement of the amino group by the hydroxylamino or methoxyamino group (IV). The other reaction is direct replacement of the amino group (I → III). Derivative III is shown in the imino form since these derivatives exist predominately (N^4-hydroxy C) or totally (N^4-methoxy C) in the imino form.

primarily due to the reversibility of I → II. In the case of m^5C or hm^5C, only direct substitution of the exocyclic amino group is observed, proceeding at a rate about 1/10 that of reaction with C. MHA reaction is slower and the optimal pH is lower, but the same mechanism is found.

Reaction with adenosine is extremely slow, but both N^6-hydroxy Ado and N^6-methoxy Ado (Figure IV-7), as well as the di- and tri-phosphate have been characterized. It is obvious that direct replacement of the amino group is the only possible reaction mechanism.

As discussed in Chapter III, Section B, substitution of the N^4 of C or N^6 of A by a hydroxylamino or methoxyamino group shifts the tautomeric equilibrium toward the imino form. N^4-Methoxy C is entirely in the imino form and can substitute for U in transcription or hybridization experiments (see Chapter III). N^4-Hydroxy C exists as a mixture of amino and imino, with the imino predominating. Similarly, it can be shown that N^6-methoxy A shifts the tautomeric equilibrium so that about 10% of the time this derivative acts like G, but the effect of the hydroxy substituent is too small to be measured by the same techniques.

Reactivity of HA or MHA is greatly decreased in double-stranded nucleic

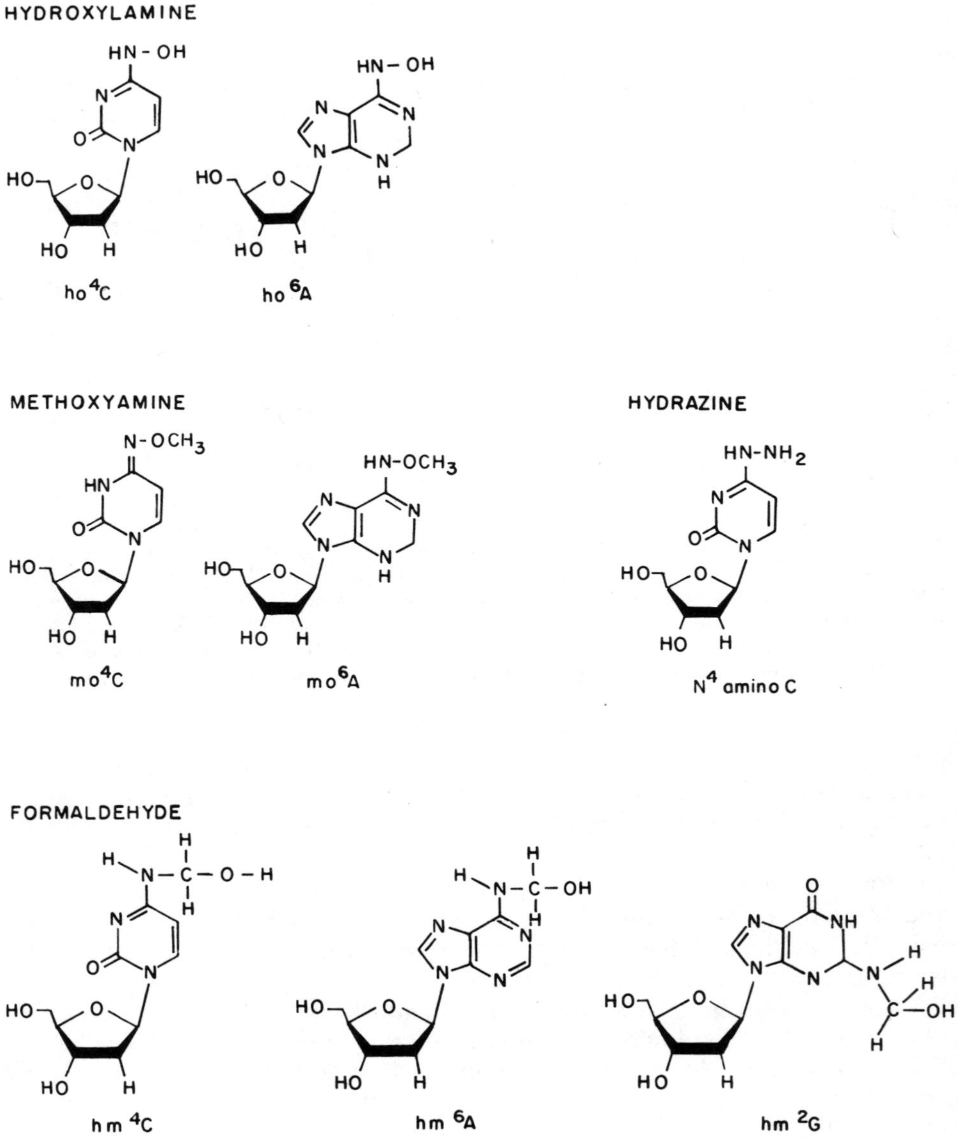

HYDROXYLAMINE

ho^4C ho^6A

METHOXYAMINE

mo^4C mo^6A

HYDRAZINE

N^4 amino C

FORMALDEHYDE

hm^4C hm^6A hm^2G

Figure IV-7. Reaction products of hydroxylamine, methoxyamine, hydrazine, and formaldehyde with amino groups of nucleosides.

acids and it is assumed that the mutagenic reaction in DNA is with single-stranded segments. Transforming DNA, phages T_4 and S_D, TMV RNA, herpes simplex virus DNA, and foot and mouth disease virus have all been highly mutated *in vitro*, but mutation is difficult to demonstrate when HA and MHA are directly tested under physiological conditions in bacteria or cells. HA, used at a concentration of 10^{-3}–10^{-4} M, readily undergoes oxidative conversions as well as forming free radicals. MHA is more stable.

4. *Hydrazine*

Hydrazine is a widely used industrial chemical that has been termed a weak mutagen *in vivo*, but is undoubtedly mutagenic when used to modify transforming DNA *in vitro*. It is also classified as a probable human carcinogen.

The major reaction at pH 6 is with cytidine, and results in a displacement at N^4 to generate N^4-amino C (Figure IV-7, middle right). This substitution is evidently a nucleophilic reaction that is sharply inhibited by protonation of the hydrazine. At pH 6, only ~0.4% of the reagent is nonprotonated, although the pK_a for hydrazine is 8.5. Thus, little reaction occurs in neutral aqueous solution. Data obtained with monomeric compounds indicate that N^4-amino C would seldom lead to changed basepairing since it has a $K_t \simeq 30$ in favor of the amino form, as contrasted to N^4-hydroxy C, which is predominately in the imino form.

Studies of hydrazine reactions with other pyrimidine bases indicate that the pyrimidine ring of U and T is rapidly degraded, so that in DNA, apyrimidinic sites might well be generated.

Recent experiments in which hydrazine was administered to rats and mice are illustrative of the difficulties of attributing carcinogenesis to an expected chemical reaction even when the chemistry has been rigorously studied. In two different laboratories a single exposure of 65–90 mg/kg body weight led to the formation of m^7G in DNA from rat and mouse liver. In rat liver O^6-methyl G was also quantitated. Since hydrazine has no alkyl group to donate and since the methylguanines are not normally present in DNA, it was concluded that hydrazine damaged the liver, which then led to an unknown series of events leading to *S*-adenosylmethionine being the methyl donor. It should also be borne in mind that hydrazine is highly reactive with molecular oxygen, forming peroxy radicals, which by additional reaction form peroxides.

5. *Formaldehyde*

This was first studied in the 1940s and found to be mutagenic in a variety of systems, but interest lagged when the chemistry was difficult to elucidate and fashions in mutation research changed. It is now clear that formaldehyde reacts reversibly with unbasepaired amino groups of nucleic acids, as has long been known for proteins.

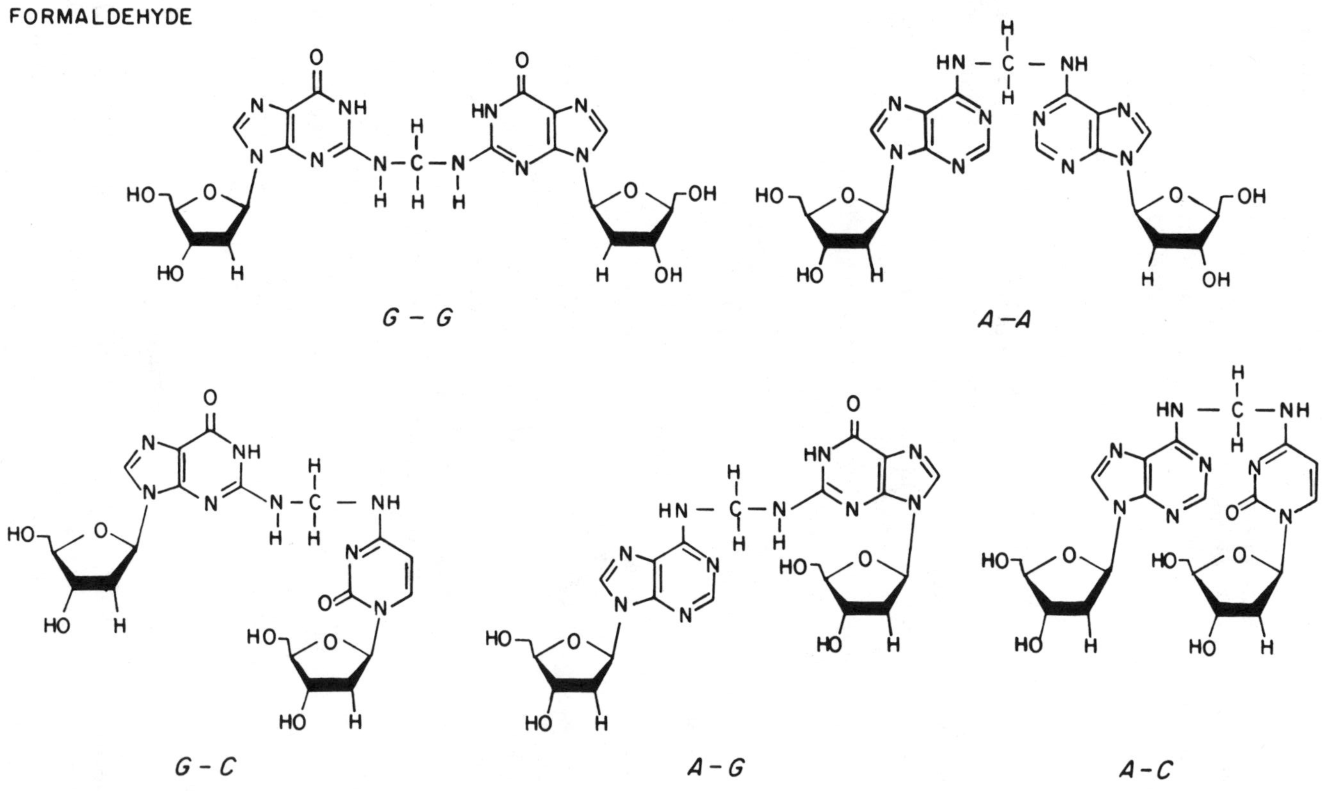

FORMALDEHYDE

G − G

A − A

G − C

A − G

A − C

Figure IV-8. Crosslinks between nucleic acid bases formed by formaldehyde. For simplification in drawing, some of the sugar residues are not stereochemically correct. All structures do contain D-deoxyribose.

With nucleic acids the reaction is strongly dependent on conformation, so that formaldehyde binding has been used as a probe of nucleic acid strandedness. In single-stranded nucleic acids, the hydroxymethyl derivatives of A, G, and C have been isolated (Figure IV-7, bottom). Although the two protons on the amino groups are equivalent, they rotate, but when substituted by a hydroxylamino group it has been shown that there is no rotation and the substituent is oriented *syn* to the basepairing side.

A second and slower reaction results in the formation of —CH_2= crosslinks, which were first shown to exist in formaldehyde-treated RNA. The unequivocal demonstration of crosslinks was provided in Feldman's and in Shapiro's laboratory. Until the isolation of five methylene crosslinks (G–G, A–A, A–G, A–C, G–C), it was not clear that cytidine could also react in that manner (Figure IV-8). Protein–DNA crosslinks can also occur with the same mechanism.

Although crosslinks are considered to have genetic effects, it is doubtful that the biological effects of formaldehyde are due to these reactions, but rather to those resulting from secondary effects on cellular components. There is both direct and indirect evidence that mutations may occur through the formation of organic peroxides or peroxidic radicals from formaldehyde, possibly aided by formaldehyde-induced catalase inhibition.

Dialdehydes, such as malonaldehyde, can react at both the N^2 of guanosine and the N-1, forming a fluorescent cyclic derivative, 1,N^2-(1-propenyl-3-ylidene)guanosine. However, the initial reaction is on the N^2 since 1-methylguanosine is unreactive.

Unsaturated monofunctional aldehydes (acrolein, crotonaldehyde) also form cyclic derivatives of guanine.

C. MONOFUNCTIONAL ALKYLATING AGENTS

1. *Introduction*

A large group of mutagens and carcinogens are simple mono- and bifunctional alkylating agents, although these differ greatly in their chemical reactions and biological effects. After the initial observation that mustard gas bound covalently with nucleic acids, and particularly to the N-7 of guanine, there has been a great effort to elucidate the chemical reactions of different classes of alkylating agents, with the specific aim of correlating a specific modification with mutagenesis or carcinogenesis.

In vitro, alkyl derivatives can be isolated and characterized, using classical organic chemistry, particularly from extensively reacted nucleosides. These methods are not generally applicable to the products obtained by hydrolysis of polynucleotides or nucleic acids alkylated under physiological conditions. Most particularly they cannot be used to identify the minor products that may be biologically significant. For this reason radioactively labeled compounds are generally employed. The more recently emerging techniques of fluorescence spectroscopy in appropriate cases and monoclonal antibodies against a particular derivative may be as sensitive, but focus only on selected alkyl nucleosides.

Although the use of radioactive alkylating agents has led to the detection of many alkyl derivatives in nucleic acids, the low extent of reaction *in vivo* ($\leqq$1 alkyl/10^5 DNA-P) makes identification of all derivatives formed possible only by means of using authentic unlabeled carrier derivatives. Unlabeled alkyl nucleosides or bases have, in the main part, been derived from modification of nucleosides, but some have also been chemically synthesized.

The definitive work on sites of modification of nucleosides has been primarily with, but not restricted to, monofunctional aliphatic methylating and ethylating agents. Bifunctional compounds react at the same sites but, in addition, can cyclize to form additional rings and can form crosslinks.

Appendix A contains UV spectral and pK_a data, as well as references to papers on synthesis, chromatographic properties, NMR and mass spectra, and physical constants of 560 derivatives. Representative spectra are shown in Appendix B.

2. *Alkylation of Nucleosides and Nucleotides*

Dimethylsulfate (Me_2SO_4) and methylmethanesulfonate (MeMS) react primarily in neutral, aqueous solution with the N-7 of G, N-1 and N-3 of A, and N-3 of C. The methylating agents are appreciably more reactive than the analogous ethylating agents. Nevertheless, diethylsulfate (Et_2SO_4) and ethylmethanesulfonate (EtMS) could be shown to have additional reactions at the O^6 of G, N^6 and N-7 of A, and O^2 and N^4 of C, although the absolute amount of such ethyl derivations was small. Neither uridine nor thymidine reacted significantly with alkylsulfates or alkylalkanesulfonates, but 3-alkyl Urd and dThd were detected.

The potent carcinogens methylnitrosourea (MeNU) and ethylnitrosourea (EtNU) have a great affinity for alkylating oxygens, and all four oxygens, the O^6 of G, the O^2 and O^4 of U or T, and the O^2 of C, can become alkylated. Methylation by MeNU also leads to substantial methylation of base oxygens. *N*-Methyl-*N*′-nitro-*N*-nitrosoguanidine (MNNG) and the analogous ethylating agent, *N*-ethyl-*N*′-nitro-*N*-nitrosoguanidine (ENNG) react similarly to nitrosoureas with the exception that virtually no reaction with guanosine is found. This is attributed to a requirement for intercalation, since guanosine in nucleic acids does react.

An example of the products of reaction with several alkylating agents is shown in Table IV-2. The two methylating agents Me_2SO_4 and MeNU under comparable conditions react with cytidine 85 and 20%, respectively, while ethylating agents react 1–7%. Reaction with both the exocyclic amino group and the O^2 is favored by ethylation, as is found with Et_2SO_4 and EtMS.

The only nitrogen that none of the alkylating agents discussed above apparently alkylates directly is the N^2 of G. However, *N*-nitroso-*N*-benzylurea will react to a low extent with guanosine, forming *N*^{2}-benzylguanosine, *O*^{6}-benzylguanosine, and 7-benzylguanosine (Table IV-3).

Table IV-2. Products of Reaction of Cytidine with Alkylating Agents[a]

Reagent	Percent reaction	Percent of total alkylation			Ratio N : O
		N-3	N^4	O^2	
Me_2SO_4	85	99	1	<0.05	
Et_2SO_4	1.2	63	11	26	2.9
EtMS	2.3	64	17	19	4.2
MeNU	29	69	4	24	3.0
EtNU	7	13	36	52	0.9

[a] Data from Singer (1976) *FEBS Lett.* **63,** 85. Reaction in aqueous solution at pH 6.1–6.3.

The secondary phosphates of nucleotides are readily esterified by alkylation and this is the major site of reaction with all nucleotides (not polynucleotides, which are tertiary phosphates). When UMP or dTMP is reacted in neutral aqueous solution with one of six alkylating agents, ethylation of phosphates was almost the only reaction observed, while methylating agents also reacted at the N-3 (Table IV-4). Ring oxygen alkylation was not found with nucleotides, probably because of competition with the extremely high reactivity of the phosphate.

Other classical alkylating agents are used for preparative purposes, but the reaction conditions do not make them useful for nucleic acid modification. Alkyl iodides, in the presence of dimethylsulfoxide and K_2CO_3, and diazoalkanes, under nonaqueous conditions, can modify nucleosides quantitatively. The rate and extent of ethylation are apparently the same as those of methylation, so that some useful comparisons may be made of relative affinities.

Methyl iodide and ethyl iodide react at the N-1, N-7, and O^6 of guanosine. Alkylation of both the O^6 and the N-7 is favored by ethyl iodide (Table IV-5). Two dialkyl derivatives are also formed, 1,7-dialkyl Guo and N^2,O^6-dialkyl Guo. The mechanism for formation of 1,7-dialkyl Guo is postulated to be the removal of a proton from the N-1 and, under these conditions, alkylation

Table IV-3. Products of Reaction of Guanosine with N-Nitroso-N-Benzylurea[a]

Reaction medium	Percent reaction	Percent of total alkylation				Ratio N : O
		N-1	N^2	N-7	O^6	
H_2O (1% DMF)[b]	0.6	nd	44	18	38	1.7
EtOH	0.7	13	7	50	30	2.4
DMF	1.9	13	nd	7	80	0.25

[a] Data from Moschel, Hudgins, and Dipple (1980) *J. Org. Chem.* **45,** 533.
[b] pH 6.8–7.4.

Table IV-4. Products of Reaction of Alkylating Agents with Pyrimidine Nucleotides[a]

		Percent of total alkylation			
Nucleotide	Reagent	3-Alkyl	Alkyl ester of 3-alkyl	Alkyl ester	Ratio N : O
UMP	Me_2SO_4	8	13	79	0.25
	MeMS	17	6	78	0.25
	MeNU	14	7	79	0.25
	Et_2SO_4	10	nd	90	0.1
	EtMS	<5	nd	>95	<0.05
	EtNU	<5	nd	>95	<0.05
dTMP	Me_2SO_4	2	13	85	0.17
	Et_2SO_4	1	4	95	0.09

[a] Data from Singer (1975) *Biochemistry* **14,** 4353. Reaction at pH 6.8 in aqueous solution. Under the same conditions of reaction, methylation is several times more efficient than ethylation.

occurs at the anion in preference to N-7. Further alkylation results in the disubstituted product. The mechanism for formation of N^2,O^6-dialkyl Guo is not clear since it was not possible to prepare this compound by alkylation of either N^2-ethyl Guo or O^6-ethyl Guo. In contrast, either 1-alkyl Guo or 7-alkyl Guo could be further alkylated to 1,7-dialkyl Guo.

Reaction of ethyl iodide with adenosine, in the presence of dimethylsulfoxide and K_2CO_3, yielded primarily 1-ethyl Ado and lesser amounts of 7-ethyl Ado, but no N^6-ethyl Ado, a product formed by Et_2SO_4 and EtMS. The same reaction conditions used to alkylate cytidine (50% reaction) led to the formation of N^4,N^4-diethyl Cyd, N^4-ethyl Cyd, and O^2-ethyl Cyd in addition to the major product, 3-ethyl Cyd, and the diethyl derivative, $3,N^4$-diethyl Cyd. Similar methylation of cytidine resulted in a quantitative yield of 3-methyl Cyd and no O^2 or N^4 derivative could be detected. Exhaustive reaction of 3-methyl Cyd with methyl iodide did result in a small amount of $3,N^4$-dimethyl Cyd but it is clear that the exocyclic group is much more readily ethylated than methylated.

Table IV-5. Products of Reaction of Guanosine with Alkyl Iodides[a]

	Percent of total alkylation					
Reagent	N-1	N-7	O^6	N-1,N-7[b]	N^2,O^6	Ratio N : O
Methyl iodide	51	nd	6	43	<1	16
Ethyl iodide	44	3	22	26	1	3.4

[a] Data from Singer (1972) *Biochemistry* **11,** 3939. Reaction in dimethylsulfoxide containing K_2CO_3.
[b] Includes ring-opened derivatives.

Substantial dialkylation occurred only with alkyl iodides, probably as a result of reaction conditions that were alkaline. This also led to imidazole ring-opened 7- and 1,7-alkyl guanosines.

Diazomethane, in ethereal solution, will react with methanol solutions of pyrimidines, as well as purines, to form *O*-methyl and *N*-methyl derivatives. Table IV-6 presents data for the reactivity of oxygens and nitrogens in bases and nucleosides toward diazomethane and diazoethane. The O^2 and O^4 of uridine and thymidine, the O^2 of cytosine, and the O^6 of deoxyguanosine are alkylated by diazoethane to a much higher extent than by diazomethane. The N : O ratio of ethylation is 0.4–1.2 while for methylation it is ~5, except for methylation of deoxyguanosine where there is a single oxygen to compete with two nitrogens. Although O^2-ethylcytosine and O^2-ethyldeoxycytidine could be prepared in good yield, neither O^2-methylcytidine nor O^2-methyldeoxycytidine were formed in appreciable amounts.

Ribose alkylated nucleosides can be prepared using diazoalkanes if stan-

Table IV-6. Products of Diazoalkane Reaction with Bases and Nucleosides in Methanol[a]

		Percent of total alkylation			
Pyrimidine	Diazoalkane	N-3	O^4	O^2	Ratio N : O
1-Methyluracil	Methyl	85	11	4	5.7
	Ethyl	50	29	21	1.0
Uridine[b]	Methyl	84	11	5	5.5
	Ethyl	48	23	26	1.0
1-Methylthymine	Ethyl	49	27	24	1.0
Thymidine	Methyl	84	10	6	5.3
	Ethyl	55	21	24	1.2
Cytosine	Methyl	*c*		≃5	
	Ethyl[d]	8		72	0.4
Deoxycytidine	Methyl[e]	trace		trace	
Purine		N-1	N-7	O^6	
Deoxyguanosine	Methyl[f]	38	27	35	1.9
	Ethyl[f]	16	15	69	0.45
	Butyl	16	16	68	0.47

[a] Data for pyrimidine reactions from Kuśmierek and Singer (1976) *Nucleic Acids Res.* **3**, 989, Singer (1976) *Nature (London)* **264**, 333 and unpublished. Data for purine reactions from Farmer, Foster, Jarman, and Tisdale (1973) *Biochem. J.* **135**, 203, and Singer, unpublished.
[b] Reaction with deoxyuridine gave similar results.
[c] 1,3-Dimethylcytosine was the major product but both 1-methylcytosine and 3-methylcytosine were also found.
[d] 20% 1-ethylcytosine formed.
[e] N^4-Ethyldeoxycytidine formed in trace amounts.
[f] N^2,O^6-Dialkyldeoxyguanosines and 1,7-dialkyldeoxyguanosines also identified as minor products.

nous chloride is present as a catalyst. The reported high yields of 2′(3′)-*O*-alkyl nucleosides suggest that other oxygen alkylation is minor.

Most *N*-methyl bases and nucleosides are commercially available. The *O*-methyl derivatives and all ethyl derivative generally must be prepared for use as UV markers when radiolabeling techniques are used to detect them in nucleic acids. The only derivatives that must be synthesized or isolated from alkylated nucleic acids are 3-alkyl adenines and guanines and N^2-alkylguanosine. The preferred methods for preparation of alkyl nucleosides (in quite wide-ranging yields) are as follows:

Diazoalkanes: *N*-1, *N*-7, O^6 methyl and ethyl Guo and dGuo; *N*-3, O^2, O^4 methyl and ethyl Urd and Thd.
Ethyl iodide: N^4-ethyl Cyd.
Methylnitrosourea and ethylnitrosourea: *N*-3, N^4, O^2 methyl and ethyl Cyd, dCyd.
Diethylsulfate or ethylmethanesulfonate: *N*-1, N^6, *N*-7 ethyl Ado.
Methylmethanesulfonate: *N*-7 methyl Ado.

Some unique chemical properties of alkyl nucleosides have been found to be useful in identification and necessary to know when isolating these derivatives. Thus, 1, 7, and O^6-alkylguanosines and 7-alkyladenosine are all fluorescent with characteristic spectra. 7- and 1,7-alkylpurine nucleosides are highly unstable in alkali and the imidazole ring opens. Ethylguanosines are more stable than the methyl derivatives and 7-alkylguanosines are 15–20 times as stable as 1,7-dialkylguanosines.

The stability of the glycosyl bonds of the 7-alkyl derivatives of guanosine, adenosine, and the deoxypurinenucleosides is sharply decreased compared to the unmodified nucleoside. There is little difference in the rate of glycosyl bond cleavage between methyl- and ethylguanosines, but, as would be predicted, 7-alkyldeoxyguanosine is very unstable compared to 7-alkylguanosine (see Chapter III, Section A). The glycosyl bond of 3-alkyl Ado, dAdo, Guo, and dGuo is extremely unstable (pH 1, 25°C, 3-MeAdo 17 min) and these derivatives have not been isolated after alkylation reactions as nucleosides, but only as bases. In nucleic acids the glycosyl bond is more stable (pH 7, 37°C, $t_{1/2}$ 3-MeAdo ~26 hr), enabling quantitation of these derivatives, although losses through depurination may result if the reaction time prior to nucleic acid isolation is long or if the reaction or subsequent treatment is at slightly acid pH.

The alkyl group on oxygens can be dealkylated in acid and/or alkali. The most stable *O*-alkyl derivative is O^6-alkyldeoxyguanosine in which the alkyl group is stable under conditions cleaving the glycosyl bond, but is dealkylated rapidly in 1 N HCl at 100°C. O^2-Alkyldeoxycytidine, heated at pH 7, 5 min at 100°C, is quantitatively converted to O^2-alkylcytosine, while the more stable glycosyl bond of O^2-alkylthymidine has a half-life of 2.7 hr (pH 7, 100°C). Although O^2-alkylation of pyrimidine deoxynucleosides labilizes the glycosyl bond compared to *N*-alkyl or unmodified deoxynucleosides, dealkylation of

O^2-alkyl Thd, but not O^2-alkyl dCyd, occurs rapidly in acid. Instead, the alkyl group of O^2-alkyl dCyd is very unstable at pH 12.5. In the case of O^4-alkyl Thd or dUrd, the glycosyl bond has the same stability as the parent deoxynucleoside and dealkylation is rapid (pH 1.5, 50°C, $t_{1/2}$ ~20 min). The alkyl groups of O^4-alkyl Urd and O^4-alkyl Ura are similarly labile.

The conventional methods of acid hydrolysis of nucleic acids will therefore dealkylate the *O*-alkyl pyrimidines. Thus, isolation of these derivatives, as well as ribose alkylated nucleosides, depends on the use of enzymes at neutral pH. Phosphate esters are also dealkylated, but only in strong acids such as 70% perchloric acid at 100°C for 1 hr.

A number of alkyl nucleosides and bases are quaternary bases and the resulting high pK_a is useful in separating and identifying these derivatives. 1-Alkyl Ado, 1-alkyl Ade, 7-alkyl Guo and Ado, 3-alkyl Cyd, 3-alkyl Cyt, O^2-alkyl Cyd, and the corresponding deoxynucleotides all have a $pK_a \geqq 7$. The low pK_a of O^2- and O^4-alkyl Urd, Ura, Thd, and Thy ($\leqq 1$) also aids in separation from other modified and unmodified constituents of nucleic acids. 3-Alkyl Ade and O^2-alkyl Cyt fall in an intermediate pK_a range of 5.5–6.5. Alkylation of ribose or phosphate does not markedly change the pK_a.

3. *Alkylation of Synthetic Polynucleotides*

There are at least two significant differences between alkylation of nucleosides and polynucleotides. One is the presence of phosphate in the phosphodiester bonds, which can form phosphotriesters, and the other is the effect of secondary structure on the availability of reactive sites. In general, the major sites of alkylation of single-stranded polymers are the same as for nucleosides, but the absolute extent of reaction is lower. Under conditions when the polymeric structure is preserved, total reaction seldom exceeds 1–2% for methylation and much less for ethylation. In addition, there are differences in extent of reactivity between ribo- and deoxyribopolynucleotides. For example, the reactivity of poly(dC) toward Me_2SO_4 or EtNU is much lower (~1/20) than that of poly(rC), which may be attributed to the highly organized secondary structure of poly(dC) under the conditions used for alkylation. Nevertheless, the reaction products with both methylating and ethylating agents were the same, with perhaps one exception (Table IV-7). Considerably

Table IV-7. Sites of Reaction of Ethylating Agents with Poly(rC)[a]

	Percent of total alkylation						
Reagent	N-3	N^4	N^4,N^4	O^2	PO_4	Ribose	Ratio N : O
Et_2SO_4	82	1	13	4			24
EtMS	78	7	11	4			24
EtNU	24	19	nd	39	13	≃5	0.8

[a] Data from Sun and Singer (1974) *Biochemistry* **13**, 1905.

more N^4,N^4-diethylcytidine than N^4-ethylcytidine was found when Et_2SO_4 or EtMS was used to modify poly(rC). No N^4,N^4-diethylcytidine was formed upon reaction of these reagents with cytidine (Table IV-2) or upon reaction of poly(rC) (Figure IV-9E) or poly(dC) with EtNU, although N^4-ethylcytidine was a major product.

The same order of specificity of alkylating agents for oxygens, as previously noted, was even more noticeable when poly(rU) was alkylated with radioactively labeled reagents (Table IV-8, Figure IV-9B,C). The two ring oxygens (O^2, O^4), phosphodiester, and 2′-O of ribose represented 97% of the sites of reaction when EtNU or ENNG was used. Even the reagent least reactive with oxygens, Me_2SO_4, formed 15% O-ethyl derivatives. Turning to reactions with poly(rA), in addition to forming phosphotriesters, EtNU alkylated the N-7 and N^6 (as well as N-1 and N-3), which were also products of EtNU reaction with adenosine. Figure IV-10 illustrates the use of HPLC to separate radioactively labeled alkyl bases. In this experiment, only the purines were analyzed. In other experiments it was shown that phosphodiesters are the main target. Thus, not only do EtNU and other N-nitroso compounds have a preference for oxygen reaction, but also for reacting at what might be termed "minor sites."

Because of the highly organized secondary structure of poly(rG) that makes this polymer double-stranded, poly(rG) is less reactive than other homopolymers. The ring positions found to be alkylated are the N-7, N-3, and O^6. All methylating reagents used (Me_2SO_4, MeNU, MNNG) reacted with the N-7 and, to a low extent, with the N-3. In addition, O^6-methylguanosine was a product of MeNU and MNNG reaction.

Double-stranded polynucleotides have been used as models for alkylation in DNA. It is in general agreement that sites involved in hydrogen-bonding are not significantly alkylated. This is not to say that no reaction occurs because even under physiological conditions there is thermal denaturation, and at a given moment up to 5% of the bases are unpaired, thus capable of modification

Table IV-8. Sites of Reaction of Alkylating Agents with Poly (U)[a]

Reagent	Percent of total alkylation: N-3	O^2	O^4	PO_4	Ribose	Ratio N : O
Me_2SO_4	85	nd	3	12	<1	5.7
MeNU	10	≃2	23	57	8	0.1
Et_2SO_4	28	≃2	13	56	≃1	0.4
EtMS	55	≃2	12	30	≃1	1.2
EtNU	3	18	27	36	16	0.03
ENNG	3	17	19	47	14	0.03

[a] Data from Kuśmierek and Singer (1976) *Biochim. Biophys. Acta* **442**, 420. Reactions were at pH 7 in aqueous solution. The total extent of reaction ranged from 0.4 to 1.4%.

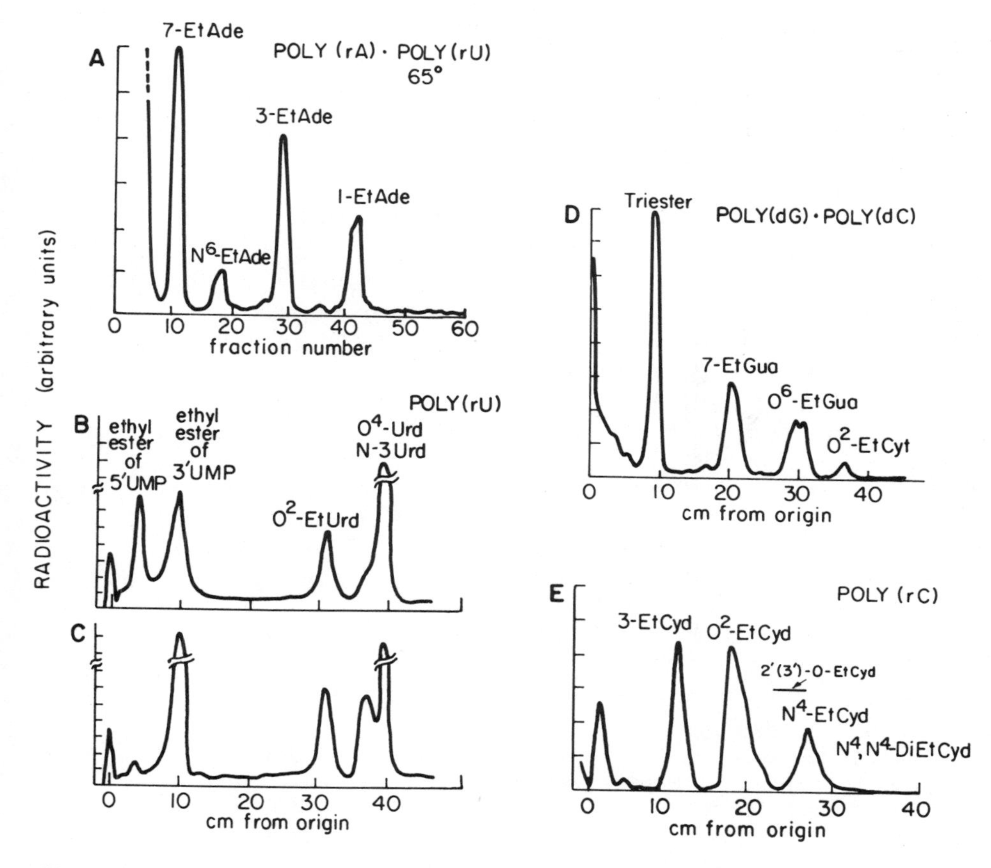

Figure IV-9. Separation of products from neutral, aqueous reaction of [^{14}C]ethylnitrosourea with polynucleotides. (A) HPLC fractionation of depurinated, thermally denatured poly(rA) · poly(rU). Bodell and Singer (1979) *Biochemistry* **13,** 2860. Note that the amount of 1-ethyladenine relative to the other modified adenines is lower than in the experiment using single-stranded poly(rA) (Figure IV-10), indicating that at 65°C there are still basepaired adenines. (B, C) Chromatographic separation of products of poly(rU) following hydrolysis with ribonuclease and phosphatases (B), and snake venom phosphodiesterase and phosphatases (C). Kuśmierek and Singer (1976) *Biochim. Biophys. Acta* **442,** 420. (D) Chromatographic separation of products from double-stranded poly(dG) · poly(dC). Depurination was followed by enzyme digestion. Singer *et al.* (1976) *Nature (London)* **276,** 85. (E) Chromatographic separation of enzyme digested poly(rC). Singer (1977) *J. Toxicol. Environ. Health* **2,** 1279.

at the N-1 of A or G, and N-3 of C, U, or T. The availability of the oxygens is unchanged since they have a free electron pair even when hydrogen-bonded. The double-stranded structure decreases the rate and extent of total modification, compared to single-stranded polymers. Identification and quantitation of minor products then depend on high specific radioactivity of the reagents used.

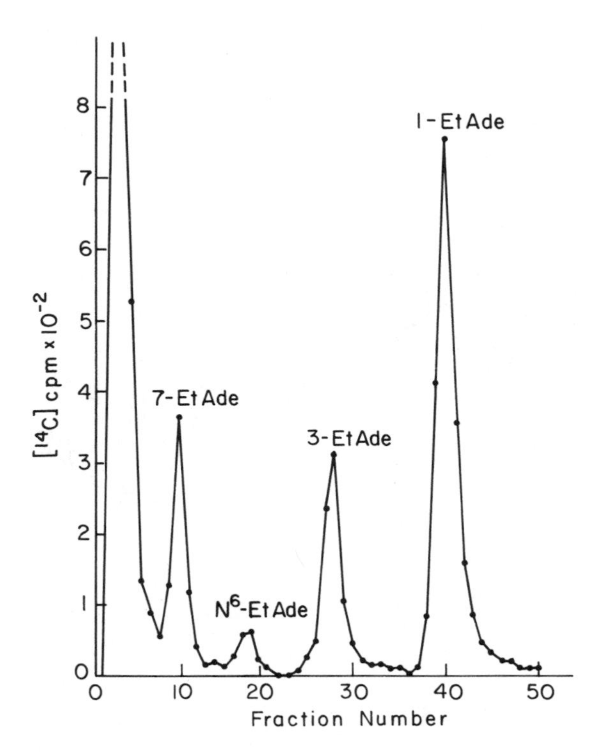

Figure IV-10. Radioactivity profile of ethylated adenines from [^{14}C]ethylnitrosourea-treated poly(rA) as analyzed by HPLC. The temperature of the neutral solution was 25°C. From Bodell and Singer (1979) *Biochemistry* **13,** 2860.

The alternating polymer, (dA-dT), or poly(rA) · poly(rU) and poly(dG) · poly(dC) have been reacted with alkyl sulfates, alkylalkane sulfonates, and *N*-nitrosoureas, but not in a completely systematic manner. The results are generally predictable on the basis of data obtained with single-stranded polymers, given the caveat regarding reactivity of hydrogen-bonded sites. Although the T_m of poly(rA) · poly(rU) is 56°C in 0.1 M NaCl, at 65°C there is still hydrogen-bonding as shown by the low amount of 1-EtA formed (Figure IV-9A), as compared to that in single-stranded poly(rA) (Figure IV-10).

Dimethylsulfate alkylated the N-3 and N-7 of A in poly(rA) · poly(rU) and the N-7 of G much more than N-3 of C in poly(rG) · poly(dC). Diethylsulfate and ethylmethanesulfonate, acting on the same polymers, also alkylated the N^6 of A and the N-3 and N^4 of C. N-1 ethylation of poly(rA) · poly(rU) was also found to occur in the EtNU-treated polymer, even at 4°C. When MeNU and EtNU were reacted with poly(dA-dT), the ring oxygen specificity of EtNU was very apparent. The data in Table IV-9 are illustrative of this point. While MeNU certainly reacted with phosphates as well as or better

Table IV-9. Sites of Reaction of Alkyl Nitrosoureas with Poly (dA-dT)[a]

	Percent of total alkylation							
	Adenine			Thymine				
Reagent	N-1	N-3	PO_4	N-3	O^2	O^4	PO_4	Ratio N : O
MeNU		22	43		3	1	39	0.3
EtNU	1	4	30	1	30	3	31	0.06

[a] Data from Jensen and Reed (1978) *Biochemistry* **17**, 5098, and Jensen (1978) *Biochemistry* **17**, 5108.

than EtNU, the proportion of O^2-Et Thd was 10-fold that of O^2-Me Thd. The experiment in Table IV-9 is one of the few that gives data for both bases in the double strand so that relative reactivities of sites in the two bases can be assessed. It is apparent that, excluding phosphotriesters, adenine reacts much more with MeNU while thymine reacts much more with EtNU. Another experiment, which particularly demonstrates the availability of oxygen in double-stranded polymers, is shown in Figure IV-9D. Poly(dG) · poly(dC), S_1 nuclease treated to remove any single-stranded tails, was reacted with EtNU at 37°C (the T_m is >100°C). Both O^6-EtG and O^2-EtC were formed. They were found, in proportion to 7-EtG, in amounts very similar to those determined in single-stranded nucleic acids.

4. *Alkylation of Nucleic Acids in Vitro*

The alkylating agents most extensively studied in terms of their reactions with nucleic acids are shown in Figure IV-11. Dialkylnitrosamines cannot be used *in vitro* since they must be metabolically activated and will be discussed in the section on *in vivo* reactions.

A variety of single-stranded nucleic acids have been methylated and ethylated and a compilation of the analyses is given in Table IV-10. In homopolynucleotides, reaction could be demonstrated at the N^6 of A, the N^4 of C, and, under specific conditions, the N^2 of G. Reaction at the N-1 of G was found only in nucleotides treated with alkyl iodides or RNA treated with an aqueous solution of diazomethane. In the latter case, it is likely that the solution became alkaline. Although looked for, none of these sites are sufficiently reactive in a heteropolymer to be detected. With the exception of EtNU (and ENNG) the N-7 of G is the most reactive site. The N-1 of A and N-3 of C are the next most reactive sites on the ring, except for EtNU, which reacts about 85% with oxygens. With ribonucleic acids, all four 2′-*O* ethylnucleosides have been characterized. It is difficult to demonstrate formation of O^6-MeG with Me_2SO_4 or MeMS.

Double-stranded nucleic acids have been studied to a greater extent than single-stranded. Data for alkylation of eukaryotic DNA are shown in Table

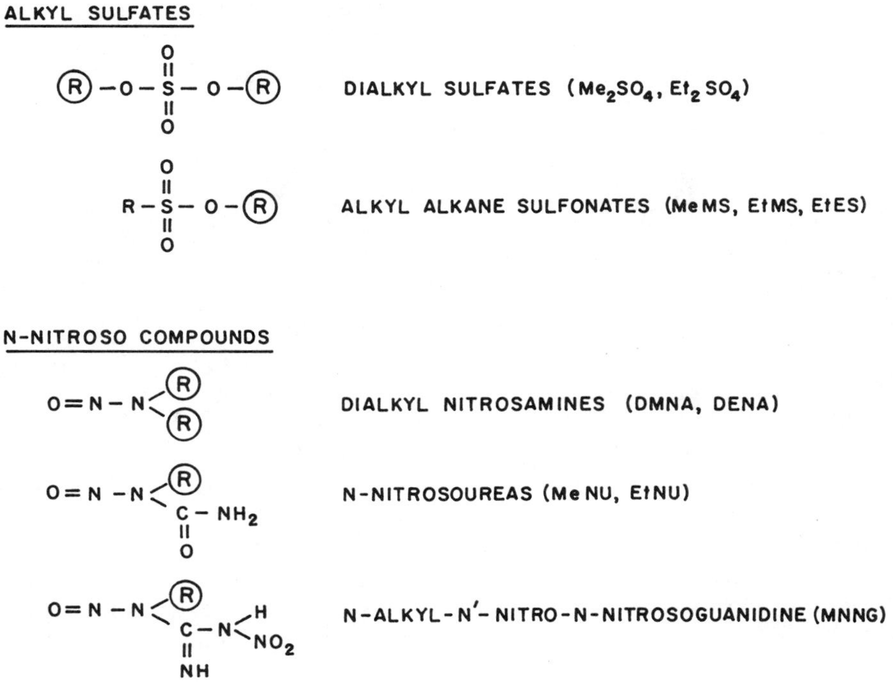

ALKYL SULFATES

(R)–O–S(=O)(=O)–O–(R) DIALKYL SULFATES (Me_2SO_4, Et_2SO_4)

R–S(=O)(=O)–O–(R) ALKYL ALKANE SULFONATES (MeMS, EtMS, EtES)

N-NITROSO COMPOUNDS

O=N–N<(R)(R) DIALKYL NITROSAMINES (DMNA, DENA)

O=N–N<(R)C(=O)–NH_2 N-NITROSOUREAS (MeNU, EtNU)

O=N–N<(R)C(=NH)–N<H NO_2 N-ALKYL-N′-NITRO-N-NITROSOGUANIDINE (MNNG)

Figure IV-11. Structural formulas of simple alkylating agents. All are direct acting except dialkylnitrosamines, which must be metabolically activated. See Figure IV-13.

IV-11. There are certain generalizations that may be made. The N-7 of G is, as in single-stranded nucleic acids, the most reactive site except for EtNU. The N-3 of A is alkylated in preference to the N-1, which is hydrogen-bonded. Nevertheless, both the N-1 of A and the N-3 of C can be methylated or ethylated detectably. This is a further indication of thermal fluctuation in double-stranded DNA. Oxygen alkylation of the ring is only significant for MeNU and EtNU. However, phosphotriesters are formed by all but Me_2SO_4.

There are several points in Tables IV-10 and 11 that should be particularly noted when discussing chemical reactivity. First, the tables deal with percent of total alkylation, or proportion, rather than absolute amounts. Methylation is about 20 times more efficient in reacting with nucleic acids than ethylation under the same reaction conditions. Thus, in Table IV-11, 6.3% O^6-MeG (MeNU) represents in moles about 15 times the amount of O^6-EtG (EtNU). Similarly 0.4% O^4-MeT (MeNU) is more in moles than 2.5% O^4-EtT (EtNU). Yet, as will be discussed in the next section, the biological effects of EtNU are greater than MeNU at similar dose levels. Regarding phosphotriesters, which are clearly the major product of EtNU reaction, here too, quantitatively there are as many or more methyltriesters than ethyltriesters, as is also the case when

Table IV-10. Alkylation of Single-Stranded Nucleic Acids in Vitro[a]

Alkylation site	Percent of total alkylation[b]					
	Me_2SO_4	MeMS	MeNU	Et_2SO_4	EtMS	EtNU
Adenine						
N-1	13.2	18	2.8	11	8	2
N-3	2.6	1.4	2.6	3	1	1.2
N-7	3.1	3.8	1.8	3	3	0.6
Guanine						
N-3	≃0.1	≃1	0.4	2	1	0.5
O^6	<0.2	nd	3	2	1	7
N-7	62	68	69	62	77	10
Uracil/thymine						
O^2				2		6
N-3				nd		
O^4				1		4
Cytosine						
O^2				2		5.5
N-3	9.5	10	2.3	11	5	1.7
Diester	<2	2	≃10	6	10	65
Ribose						12

[a] Analyses were from experiments using DNA from M13 phage and RNA from TMV, yeast, HeLa cells, animal ribosomes, and μ2 phage.
[b] The absolute amount of alkylation varied greatly but the proportion of derivatives was not noticeably affected. Methylation is more efficient than ethylation by 10- to 20-fold, using the same molarity of reagent. "nd" indicates that the derivative was not detected.

comparing triesters formed by MeMS and EtMS. All these comparisons, of course, are only true when comparing equimolar reactions under identical conditions of time, temperature, pH, etc.

To summarize: Ethylating agents, while less reactive than methylating agents, have a greater affinity for modifying oxygens. The reactive sites in single-stranded nucleic acids are generally the same as in double-stranded DNA, except that reaction at the hydrogen-bonded sites is low in single-stranded RNA or DNA. The specificities of eight reagents toward oxygen apparently correlate with the reported mutagenicity and carcinogenicity in various systems: EtNU, ENNG > MeNU, MNNG > EtMS > MeMS > Et_2SO_4 > Me_2SO_4.

Sites of substitution in double-stranded DNA by the poor carcinogen Me_2SO_4 are: N-7G > > N-3A > > N-1A, N-7A, N-3G > > O^6-G. Sites of substitution in double-stranded DNA by the potent carcinogen EtNU are: phosphate > > N-7G > O^2-T, O^6-G > N-3A > O^4-T, O^2-C > other N.

Possible sites of alkylation are shown in Figure IV-12. Thirteen reaction products have been found in polynucleotides or nucleic acids treated with EtNU in neutral aqueous solution. The other alkylating agents generally do not form as many derivatives and the sites and extent of modification are a function of each alkylating agent.

Table IV-11. Alkylation of Double-Stranded Nucleic Acids in Vitro[a]

Alkylation site	Percent of total alkylation[b]					
	Me_2SO_4	MeMS	MeNU	Et_2SO_4	EtMS	EtNU
Adenine						
N-1	1.9	3.8	1.3	2	1.7	0.2
N-3	18	10.4	9	10	4.9	4.0
N-7	1.9	(1.8)	1.7	1.5	1.1	0.3
Guanine						
N-3	1.1	(0.6)	0.8	0.9	0.9	0.6
O^6	0.2	(0.3)	6.3	0.2	2	7.8
N-7	74	83	67	67	65	11.5
Thymine						
O^2			0.11	nd	nd	7.4
N-3			0.3	nd	nd	0.8
O^4			0.4	nd	nd	2.5
Cytosine						
O^2	(nd)	(nd)	0.1	nd	nd	3.5
N-3	(<2)	(<1)	0.6	0.7	0.6	0.2
Diester		0.8	17	16	13	57

[a] Analyses are from experiments using DNA from salmon sperm, calf thymus, salmon testes, rat liver and brain, human fibroblasts, and HeLa and V79 cells.

[b] The absolute amount of alkylation varied greatly but the proportion of derivatives was not noticeably affected. Methylation is more efficient than ethylation by 10- to 20-fold, using the same molarity of reagent. "nd" indicates that the derivative was not detected. Parentheses indicate either a single value or the average of two very different values and thus less reliability than other data shown.

5. *Alkylation of Nucleic Acids in Vivo*

a. General Considerations

The dialkylsulfates and alkylakanesulfonates, which react primarily with ring nitrogens, are generally weak carcinogens, while the *N*-nitroso compounds, which react to a high extent with all oxygens, are potent carcinogens. For this reason, most of the analytical data have been obtained following administration of the methyl and ethyl *N*-nitrosoureas and *N*-nitrosamines. A major difference between the two groups is that the nitrosamines must be metabolically activated, as shown in Figure IV-13. The route of administration and the level of metabolic activity in organs determine to a great extent the cellular distribution of nitrosamine alkylation since the alkylating species is too reactive to be transported in significant amounts.

Dimethylnitrosamine (DMMA) or diethylnitrosamine (DENA), given by intravenous injection, is metabolized to an alkyldiazonium ion by the liver > kidney > lung and the extent of alkylation in each organ reflects this. Depending on the animal species and method of administration, the kidney or liver develops tumors.

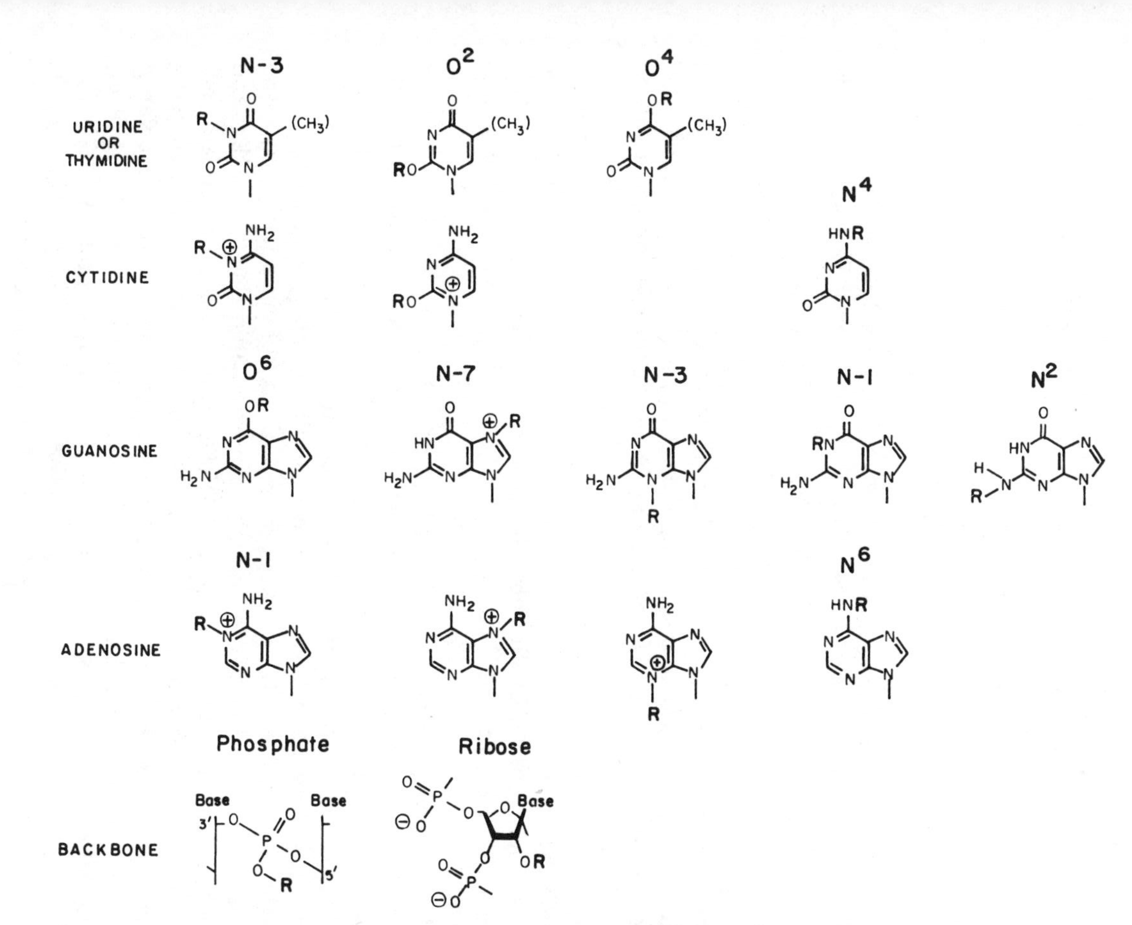

Figure IV-12. Sites of reaction of simple alkylating agents with nucleic acids or polynucleotides in neutral, aqueous solution. Reaction at the N^4 of C, N-1 of G, N^2 of G, and N^6 of A has only been demonstrated in polyribonucleotides.

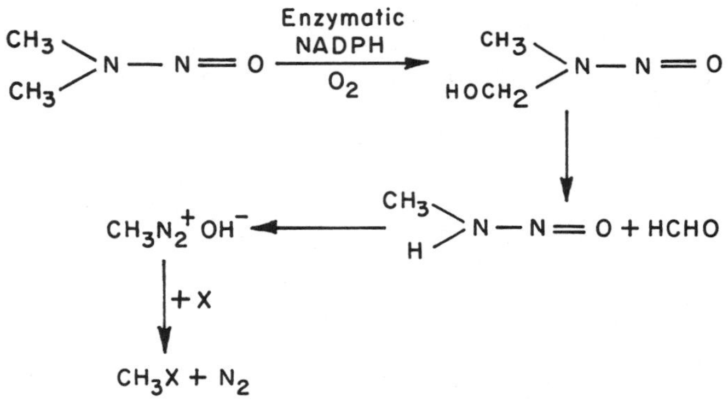

$$(CH_3)_2N{-}N{=}O \xrightarrow[O_2]{\text{Enzymatic NADPH}} (CH_3)(HOCH_2)N{-}N{=}O$$

$$(CH_3)(HOCH_2)N{-}N{=}O \rightarrow (CH_3)(H)N{-}N{=}O + HCHO \rightarrow CH_3N_2^+\,OH^-$$

$$CH_3N_2^+\,OH^- \xrightarrow{+X} CH_3X + N_2$$

Figure IV-13. Postulated mechanism for the metabolism of dimethylnitrosamine. From Pegg (1980) In *Molecular and Cellular Aspects of Carcinogen Screening Tests* (Montesano, Bartsch, and Tomatis, eds.), IARC Scientific Publication No. 27, pp. 3–22.

Nitrosamines are not only metabolized *in vivo* but are also biosynthesized in mammals from nitrite and secondary or tertiary amines. Biosynthesis and/or metabolism are enhanced by lowering the pH; ascorbate appears to inhibit biosynthesis.

1,2-Dimethylhydrazine (CH_3-NH-NH-CH_3, SDMH) is another type of methylating carcinogen that generates a methyldiazonium ion via the same pathway. Metabolism in the rat is highest in the liver, followed by colon, kidney, and ileum. SDMH induces a high incidence of colonic carcinoma in rats and mice.

The alkylnitrosoureas (MeNU, EtNU) directly form an alkyldiazonium ion (R-N_2^+) by a hydroxyl ion-catalyzed decomposition that occurs rapidly even at neutral pH. The half-life of EtNU *in vivo* is less than 8 min. All tissues are alkylated with virtually the same efficiency, but there is considerable organotropy which depends on age and species. In BD IX rats given a single transplacental dose of EtNU (75 mg/kg) on the 18th day of gestation, or by intravenous injection to 10-day-old animals, almost all animals will develop and die of neuroectodermal tumors (Figure IV-14). Adult rats are much less sensitive to the carcinogen and, as shown in Figure IV-15, a dose of 160 mg/kg was needed for a 50% yield of neurogenic malignancies. MeNU is similar in its induction of nervous system tumors in this strain and age of rat. However, in other species the target organ is quite different (see Table II-3) and can include liver, kidney, lung, lymph nodes, and the hemopoietic system. Gerbils given MeNU or EtNU develop melanomas but no other malignancy.

The related *N*-alkyl-*N'*-nitro-*N*-nitrosoguanidines (MNNG, ENNG) are nitrosoamidines that decompose to form relatively stable alkyldiazonium ions ($t_{1/2}$ 1–20 hr depending on the buffer). However, the presence of thiols greatly catalyzes decomposition and the $t_{1/2}$ can decrease to 1–2 min. In addition to

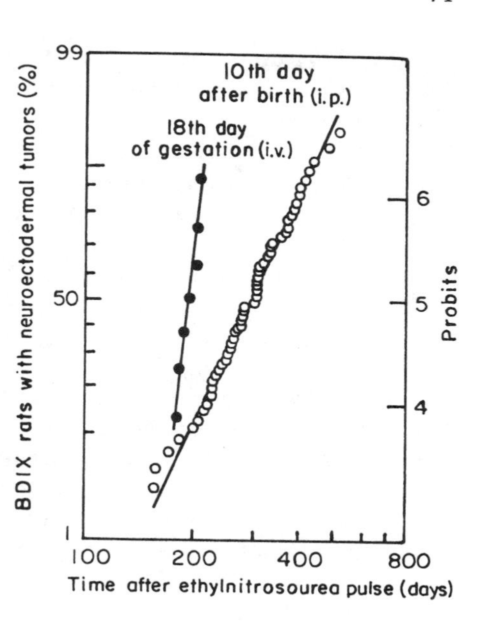

Figure IV-14. Mortality with neuroectodermal tumors in the offspring of BD IX rats exposed transplacentally to a single dose of 75 mg ethylnitrosourea/kg body wt, administered to pregnant females by i.v. injection on the 18th day of gestation (●) and in BD IX rats treated i.p. with the same dose of carcinogen at the age of 10 days (○). Each point represents one animal. Horizontal lines (probits) indicate one standard deviation of the T_{50} values. Adapted from Rajewsky *et al.* (1977) In *Origins of Human Cancer, Book B, Mechanisms of Carcinogenesis* (Hiatt, Watson, and Winsten, eds.), Cold Spring Harbor Laboratory, pp. 709–726.

the alkylating species, other products are formed that are likely to also react with cellular components. Since thiols are present in cells it is possible that some of the observed biological effects can be attributed to nonalkylating reactions. The lack of alkylation of certain tissues may reflect the absence of sulfhydryl reagents. MNNG and ENNG produced stomach tumors in rats or hamsters when given by stomach tube.

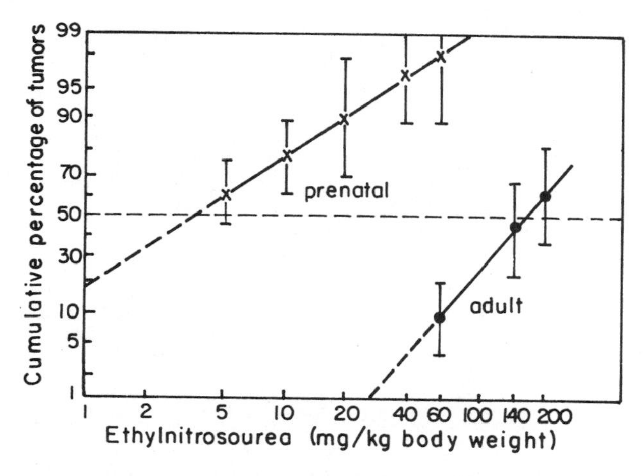

Figure IV-15. Dose–response relationships for the prenatal induction of malignant tumors of the nervous system compared with that in adult rats. Ethylnitrosourea was administered as a single dose. Adapted from Druckrey *et al.* (1969) *Ann. N.Y. Acad. Sci.* **163,** 676.

It is difficult to assess the relative mutagenicities of these *N*-nitroso compounds due to their varying requirements for activation to electrophiles which will elicit a mutation. In general, in mammalian cells, all are much more mutagenic than alkylsulfates or alkylalkanesulfonates (see Table II-4). The frequently used bacterial systems (e.g., the Ames test) differ from the eukaryotic test systems in that they require the addition of a mammalian liver fraction to contribute the enzymes necessary for metabolizing dialkylnitrosamines. Upon optimizing the conditions these compounds are definitely mutagenic. However, toxicity to the cells becomes an important factor in assessing mutation since data are generally presented in terms of survivors (Figure IV-16).

Toxicity itself elicits a biological response. This is illustrated by reports that various liver-damaging agents, including ethanol and hydrazine, lead to methylation of guanine in rat liver. The hypothesis that *S*-adenosylmethionine can transfer its methyl groups even to the O^6 and N-7 of G has been confirmed by *in vitro* experiments. On the other hand, enzymatic repair ability in rat liver is increased by levels of hepatotoxins that stimulate cell division.

In addition to the factors mentioned above, carcinogenic response is affected by hormones, age and species of animal, cell differentiation, and immunological and genetic determinants. Inhibition and promotion of damage are subjects of intense studies, discussion of which is beyond the scope of this book. Nevertheless, it must be kept in mind that the initiation of a

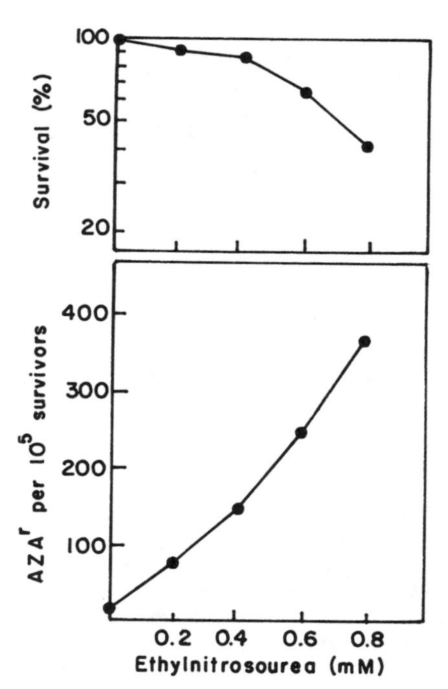

Figure IV-16. Toxicity and mutagenicity of ethylnitrosourea (EtNU) in V79 Chinese hamster cells. (Top) Survival as a function of concentration of EtNU. (Bottom) Mutation measured as resistance to 8-azaguanine. Adapted from Bartsch *et al.* (1980) In *Molecular and Cellular Aspects of Carcinogen Screening Tests* (Montesano, Bartsch, and Tomatis, eds.), IARC Scientific Publication No. 27, pp. 179–241.

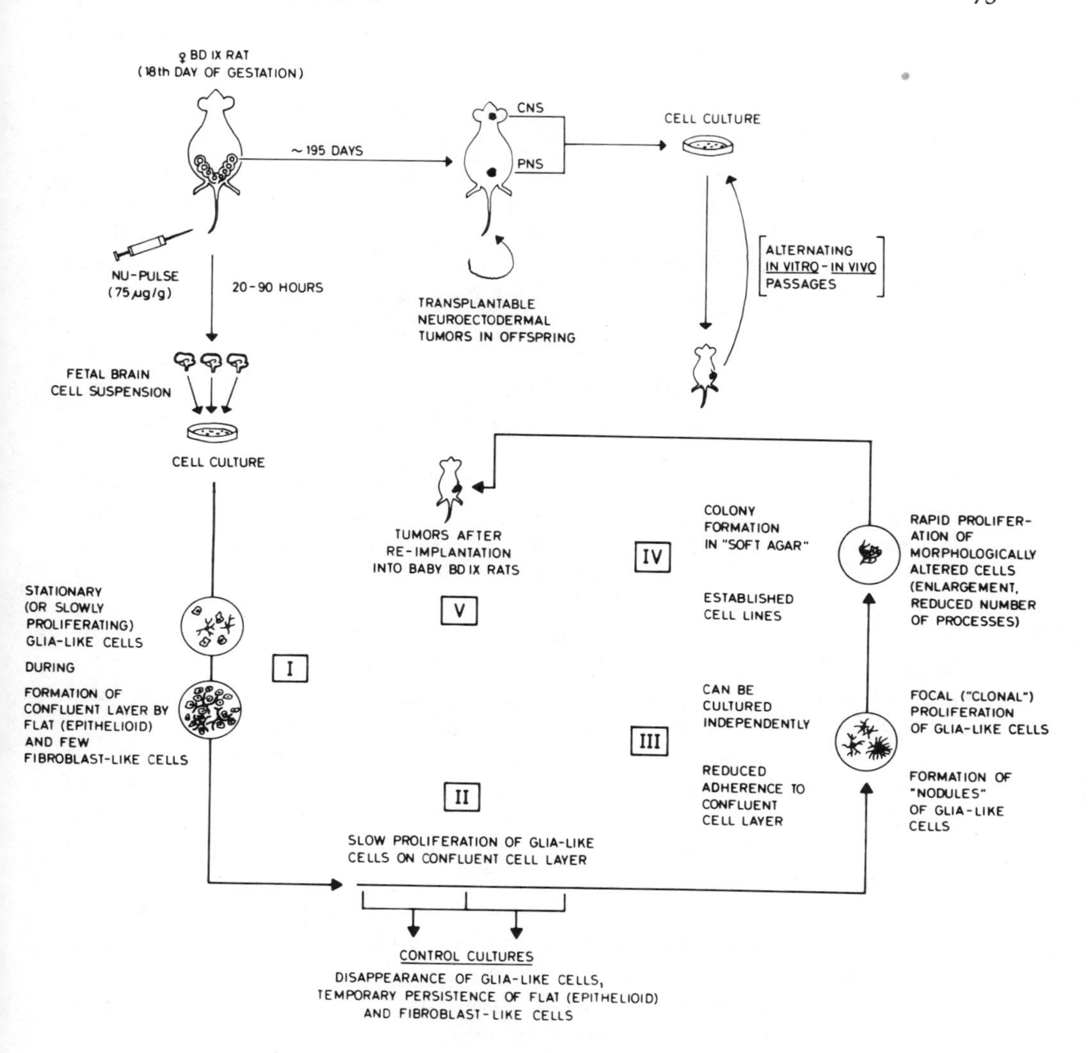

Figure IV-17. Diagrammatic representation of *in vivo–in vitro* system for neoplastic transformation of fetal BD IX rat brain cells in culture, after i.p. exposure to ethylnitrosourea *in vivo*. CNS, central nervous system; PNS, peripheral nervous system. Reprinted with permission from Laerum and Rajewsky (1975) *J. Natl. Cancer Inst.* **55,** 1177.

malignancy is likely to be due to a primary chemical modification of DNA. The elegant experiment shown in Figure IV-17 is strong confirmation of this since initiated cells will grow in tissue culture with the same morphological changes as *in vivo* and can be reimplanted to form a tumor. Thus, the chemical analysis of the products of carcinogens with nucleic acids in phages, cells, and whole animals is fundamental to understanding how carcinogens work.

b. CHEMICAL REACTIONS

Cells in culture or *in vivo* contain both RNA and DNA and both can be single- and double-stranded. Early work on comparisons of chemical modification between nucleic acids in replicating cells and *in vitro* was done primarily on RNA, assuming that it was in single-stranded form. Most analyses did not take into consideration the *O*-alkyl derivatives, but in spite of these shortcomings it was apparent that, within the limits of analytical methods available, alkylating agents reacted at the various sites in similar proportions *in vitro* and *in vivo* (Table IV-12). MeNU and DMNA formed 3–4% O^6-MeG while EtNU formed about 7% O^6-EtG, regardless of the RNA source. Similarly, MeMS, which forms little of the *O*-methyl products *in vitro*, reacted to a relatively high extent with the N-1 of A and N-3 of C in cells, phage, and rat liver.

Much more information is available on alkylation of DNA by *N*-nitroso compounds, but little on the less carcinogenic alkylsulfates and alkylalkanesulfonates. Unfortunately, even in the case of the much studied nitrosoureas and nitrosamines, attention has generally been focused on quantitation of three purine derivatives that, due to their labile glycosyl bonds, are readily depurinated from DNA. A simple column chromatographic method separates radiolabeled 3-alkyladenine, 7-alkylguanine, and O^6-alkylguanine (Figure IV-18). 7-Alkyladenine and 3-alkylguanine are also in the purine fraction and can be separated, but they are present in very low amounts.

Quantitation of other than labile purines requires complete hydrolysis of DNA. Strong acid dealkylates *O*-alkyl derivatives but is useful for analysis of 1-, 3-, and 7-alkyl A; 3- and 7-alkyl G and 3-alkyl C and T. Enzyme digestion releases all derivatives, but separation procedures are laborious regardless of whether columns, TLC, or HPLC are used.

Fluorescence spectrometry of O^6-methylguanine and 7-methylguanine yields data on their absolute amounts but not their proportion of total alkylation. When radioactivity is not or cannot be used, this method can show that alkylation of a nucleic acid occurs. The limit of detection of 7-MeG is 40 μM and of O^6-MeG, 2 μM per mole guanine.

A considerable number of monoclonal antibodies have been prepared (Table IV-13). As in the use of fluorescence spectrometry as an analytical tool, radioimmune assays do not require the administration of a radioactive alkylating agent. They are extremely sensitive, but at very low levels of modification cross-reactivity with unmodified mucleosides requires preliminary separation of the alkyl derivative. Using O^6-Et dGuo separated from ethylated DNA, as shown in Figure IV-18, 5×10^{-8} O^6-Et dGuo per Guo can be quantitated. Complete analysis of alkyl derivatives would require multiple antibodies.

The analytical data in Table IV-14 were obtained by using radioactively labeled alkylating agents. In the case of cultured mammalian cells or animals, the DNA was extracted after 1 hr of reaction, except for reactions with DMNA and DENA where the minimum time was generally 4–7 hr to allow for met-

Table IV-12. Alkylation of Single-Stranded Nucleic Acids[a]

Alkylating agent and state of RNA[b]		Percent of total alkylation[c]												
		Adenine			Guanine			Uracil/thymine			Cytosine			
		1	3	7	3	O^6	7	O^2	3	O^4	O^2	3	Diester	Ribose
Me_2SO_4	*in vitro*	13.2	2.6	3.1	≃0.1	<0.2	62					9.5	>2	
	cells	10.5	(1)	(1)	(<0.2)	<0.2	72					10		
	phage	12.5	<1	(1.5)	<0.3	<0.2	65		(nd)			11		
MeMS	*in vitro*	18	1.4	3.8	≃1	nd	68					10	2	
	cells	(10.3)	(2.1)		(nd)		(74)					(7)		
	phage	(11)	(1.5)		(nd)	(0.2)	(72)		(0.4)			(7)		
	rat liver	(9.6)	(2.6)	(0.8)		(nd)	(75)					(8.7)		
MeNU	*in vitro*	2.8	2.6	1.8	0.4	3	69					2.3	≃10	
	cells				(1)	(4)	(80)					(1.5)		
	phage	2.3	<1	(2.5)	0.8	3.8	80		(nd)			1.5		
DMNA	rat liver	2.1	1.0	(1.6)	(2)	3.7	79					2.7		
Et_2SO_4	*in vitro*	11	3	3	2	2	62	2	nd	1	2	11	6	
	cells													
EtMS	*in vitro*	8	1	3	1	1	77					5	10	
	phage	(4)	(0.2)			(1.4)	(82)					(5)		
EtNU[d]	*in vitro*	2	1.2	0.6	0.5	7	10	6	nd	4	5.5	1.7	65	12
	cells					(7.3)	(12)							
	phage	(1.3)	(0.9)		(1.2)	(7)	(28)				(2.3)	(1.5)		

[a] Data are primarily from Lawley, Shooter, Margison, O'Connor, Singer and their associates. Much of the data were compiled by Singer (1975) In *Prog. Nucleic Acid Res. Mol. Biol.* **15,** 219, and by Margison and O'Connor (1979) In *Chemical Carcinogenesis and DNA,* Vol. I (P.L. Grover, ed.), CRC Press. Other key references are Shooter *et al.* (1974) *Biochem. J.* **137,** 303; Shooter and House (1975) *Chem. Biol. Interact.* **11,** 563; Singer and Fraenkel-Conrat (1975) *Biochemistry* **14,** 772.

[b] Single-stranded nucleic acids used *in vitro* included M13 DNA and RNA from HeLa cells, TMV, yeast, animal ribosomes, and phage M2. Cells are rabbit reticulocytes and HeLa cells, and the phages are μ2 and R17. Abbreviations are the same as in the table on double-stranded nucleic acids.

[c] Total alkylation refers to addition and substitution reactions only and does not include metabolic labeling. Parentheses indicate either a single value or the average of two different values and thus less reliability than other data shown. There are considerably less data for RNA alkylation than for DNA. However, no repair of alkyl products has been demonstrated in RNA and for this reason cell, phage, or rat liver data are easier to obtain and interpret.

[d] Includes data for *N*-ethyl-*N'*-nitro-*N*-nitrosoguanidine (ENNG) reaction with TMV RNA.

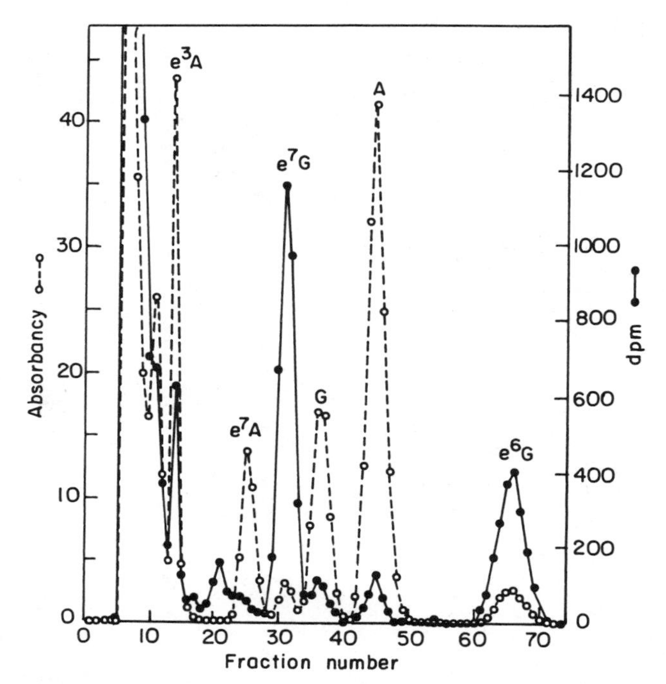

Figure IV-18. Separation of purine bases form BD IX rat brain DNA by radiochromatography on Sephadex G-10, 4 hr after pulse-ethylation by [^{14}C]ethylnitrosourea *in vivo*. Nonradioactive ethylated purine bases (e^{3}A, e^{7}G, and e^{6}G) were added as markers. The UV absorbency of A and G is from the DNA. Adapted from Goth and Rajewsky (1974) *Proc. Natl. Acad. Sci. USA* **71,** 639.

Table IV-13. Monoclonal Antibodies Directed Against Alkylated DNA Components[a]

Carcinogen-modified DNA component	Antibody affinity constant (1/mole)	Detection limit of RIA at 50% ITAB (pmole/100 μl RIA sample)	Cross-reactivity with the corresponding unmodified DNA component
O^6-Me dGuo	2.7×10^{10}	0.06	$>1.1 \times 10^7$
O^6-Et dGuo	2.0×10^{10}	0.04	$>2.5 \times 10^6$
O^6-Bu dGuo	9.1×10^9	0.06	2.5×10^6
O^6-Bu Guo	2.7×10^{10}	0.02	3.3×10^7
O^4-Et dThd	1.3×10^9	0.24	$>6.3 \times 10^5$
O^4-Bu dThd	8.8×10^8	0.3	2.2×10^6
O^2-Bu dThd	1.1×10^8	0.06	1.1×10^7

[a] Adapted from Müller and Rajewsky (1981) *J. Cancer Res. Clin. Oncol.* **102,** 99.

Table IV-14. Alkylation of Double-Stranded Nucleic Acids[a]

Alkylating agent[c] and state of DNA[a]		Percent of total alkylation[b]											
		Adenine			Guanine			Thymine			Cytosine		
		1[d]	3	7	3	O^6[e]	7	O^2	3[d]	O^4[e]	O^2[e]	3[d]	Diester
Me_2SO_4	*in vitro*	1.9	18	1.9	1.1	0.2	74				(nd)	(<2)	1.0
MeMS	*in vitro*	3.8	10.4	(1.8)	(0.6)	0.26	85	(nd)	(0.08)	(nd)	nd	<1	0.8
	cells		8.7	(2.4)	(1.3)	(0.3)	83					(<1)	
MeNU	*in vitro*	1.3	9	1.7	0.8	6.3	67	0.11	0.3	0.4	0.1	0.6	16
	cells		4.2	(3.1)	(0.5)	5.1	69					(0.4)	(9)
	rat liver		3.6			3.6[f]	70						
DMNA	rat liver	0.8	2.4	1.5	0.6	6.6[g]	69		(0.4)			(0.6)	(9)
Et_2SO_4	*in vitro*	2	10	1.5	0.9	0.2	67		nd			0.7	16
	phage T4	(0.3)	(4)	(0.5)	(0.2)	(1.6)	(71)					(0.5)	
EtMS	*in vitro*	1.7	4.9	1.1	0.9	2	65	nd	nd	nd	nd	0.6	13
	cells	(0.1)	2.1		0.7	2.1	69					(0.1)	15
	rat liver		3.3			1.5	70						
EtNU	*in vitro*	0.2	4.0	0.3	0.6	7.8	11.5	7.4	0.8	2.5	3.5	0.22	57
	cells	(0.1)	4.9	1.3	1.6	8.7	14	7.1	(nd)	2.8	1.8		54
	rat liver		4.1	0.6	1.4	7.2	13.5	7.4		2.3	1.3		60
DENA	rat liver		(3.7)			(5.6)[g]	15	6.0		(0.7)			

[a] Data are primarily from Lawley, Brookes, Roberts, Margison, O'Connor, Craddock, Swenson, Scherer, Pegg, Singer and their associates. Some of the data were compiled by Singer (1975) In *Prog. Nucleic Acid Res. Mol. Biol.* **15,** 219. Other key references are: Swenson *et al.* (1980) *Carcinogenesis* **1,** 931; Beranek *et al.* (1980) *Carcinogenesis* **1,** 595; Shackelton *et al.* (1979) *Eur. J. Biochem.* **77,** 425; Newbold *et al.* (1980) *Nature (London)* **283,** 596; Sun and Singer (1975) *Biochemistry* **14,** 1795; Scherer *et al.* (1980) *Cancer Lett.* **10,** 1; Singer *et al.* (1978) *Nature (London)* **276,** 85; Singer (1976) *Nature (London)* **264,** 333; *J. Natl. Cancer Inst.* **62,** 1329 (1979); Singer *et al.* (1981) *Carcinogenesis* **2,** 1069; Lawley *et al.* (1975) *Biochem. J.* **145,** 73; Pegg *et al.* (1978) *Biochim. Biophys. Acta* **520,** 671; Goth and Rajewsky (1974) *Z. Krebsforsch.* **82,** 37; Margison *et al.* (1976) *Biochem. J.* **157,** 627; Craddock (1975) *Chem. Biol. Interact.* **10,** 323. *In vitro* experiments were done using DNA from salmon sperm, calf thymus, salmon testes, rat liver and brain, human fibroblasts, and HeLa cells. Cells refer to mammalian cells only, in culture, and include V79 cells, human fibroblasts, and L cells. Rat liver data are used as being representative of *in vivo* alkylation and also in order to compare metabolically activated nitrosamines, which primarily alkylate liver, with directly acting alkylating agents, using a single administration in all cases. In general, there is little or no tissue or organ difference in sites of alkylation or proportion of derivative. The data are, however, limited for *in vivo* alkylation of most organs other than liver. It is recognized that the liver has been demonstrated to remove or repair most alkyl derivatives and the 1-hr reaction with directly acting agents or up to 7 hr with DMNA and DENA may not be maximal.

[b] Total alkylation refers to addition and substitution reactions only and does not include metabolic labeling. Different investigators present data in different fashions and some values have thus been recalculated. Underlining indicates that an arbitrary value has been assigned to 7-alkyl G (based on data for cells or *in vitro*) and the other derivatives are calculated on that basis. Values in parentheses are either a single determination or an average of two very different values and thus of less reliability than others. All other data are averages of 2–9 separate experiments, generally from different laboratories.

[c] Abbreviations of alkylating agents are Me_2SO_4 and Et_2SO_4, dimethylsulfate and diethylsulfate; MeMS and EtMS, methylmethanesulfonate and ethylmethanesulfonate; MeNU and EtNU, methylnitrosourea and ethylnitrosourea; DMNA and DENA, dimethylnitrosamine and diethylnitrosamine.

[d] These positions are hydrogen-bonded in double-stranded nucleic acids. Although thermal denaturation at 37°C can be demonstrated, the reactivity of the N-1 of A is much greater than would be expected on such grounds.

[e] These positions, while hydrogen-bonded in double-stranded DNA, have an unbonded electron pair that is free to react and the proportion of these *O*-alkyl derivatives is similar in both single- and double-stranded nucleic acids.

[f] Rat and mouse liver repair O^6-MeG rapidly (single-dose conditions), while hamster liver has lower repair capability. The time at which the liver is removed is extremely important in quantitating O^6-MeG and this relatively low value is probably due to such variations.

[g] The amount of O^6-alkyl G found is highly dose-dependent, as well as time-dependent. Pegg and Balog [*Cancer Res.* **39,** 5003 (1979)] show that repair is most efficient at 0.5 mg DENA/kg or 0.02 mg DMNA/kg. The data used in the table are for relatively high amounts of carcinogen.

abolic activation. The extent of alkylation *in vitro* was generally higher than in cell culture or in animals where total alkylation was as low as 1 alkyl group/ 10^6 DNA-P. Nevertheless, where there are sufficient data for comparison, the alkylation patterns for each alkylating agent are remarkably similar. Generally, lower levels of O^6-alkyl G and 3-alkyl A are found in rat liver DNA as compared to *in vitro* data. This is attributed to repair enzymes (discussed in Chapter VIII).

Reaction of EtNU with DNA has been studied in a variety of systems and individual experiments are shown in Table IV-15. The major derivatives formed in DNA under all conditions are 3-EtA, O^6-EtG, 7-EtG, O^2-EtT, O^2-EtC, and ethylphosphotriesters. In addition, there is clear evidence for reaction at the O^4 of T, N-1 and N-7 of A, N-3 of G, and N-3 of T and C. Thus, as *in vitro*, all nitrogens and oxygens can be ethylated in mammalian cells or organs.

To summarize: All simple, direct-acting alkylating agents react with nucleic acid *in vivo* on the same sites as *in vitro*. For each alkylating agent the amount and distribution of alkyl groups are characteristic. The proportion of reaction at each site is, within experimental accuracy, similar, except for cells in which rapid repair of O^6-alkyl G or 3-alkyl A takes place.

D. Bifunctional Alkylating Agents

Many of the cyclic alkylating agents to be discussed in Section E are also bifunctional, but only on ring opening. There exist, however, a group of therapeutic nitrosoureas that are bifunctional in carrying a halogen and a nitroso group. In addition to those shown in Figure IV-19 [BCNU, bis(chloroethyl)nitrosourea; CCNU, *N*-(2-chloroethyl) -*N'*-cyclohexyl-*N*-nitrosourea; BFNU, bis(fluorethyl)nitrosourea], several related compounds have been reacted with nucleosides and DNA. All are mutagenic when tested for 6-thioguanine resistance in V79 Chinese hamster lung cells. In contrast to alkyl nitrosoureas, which are mutagenic at low toxicity, halonitrosoureas are mutagenic only at high toxicities.

While metabolic activation is not necessary for reaction of these compounds, the spectrum of identified derivatives indicates that aqueous decomposition of the halonitrosoureas occurs simultaneously in different ways and forms such intermediates as cyclic oxazolidines and oxadiazolines. Figure IV-20 shows some possible pathways that could account for the multiple derivatives of nucleosides.

As shown in Figure IV-21, BCNU reacts *in vitro* with DNA at the N-7 of G to form chloroethyl, hydroxyethyl, and aminoethyl compounds. Reaction to form mono-hydroxyethyl adducts also occurs at the O^6 of G and N-3 of C. Two ethano compounds are also formed that have a 3,N^4 ethano ring on C or a 1,N^6 ethano ring on A. Model experiments indicate that the mechanism

Table IV-15. Products of Ethylnitrosourea Reaction with DNA

State of DNA	Percent of total ethylation											
	Adenine			Guanine			Thymine			Cytosine		
	1	3	7	3	O^6	7	O^2	3	O^4	O^2	3	Diester
In vitro[a]	0.2	4.0	0.3	0.6	7.8	11.5	7.4	0.8	2.5	3.5	0.22	57
Cell culture												
Human fibroblasts[b]		4.5			9.2	13.6	7.1		2.1	4.5		51
Fetal rat brain[b]		4.7			7.6	12.3	7.0		4.3	3.4		49
Chinese hamster V79[c]		5.0	1.3	1.7	9.1	12.3	8.5		1.4	1.8		66
HeLa[d]	0.1	4.6	1.3	1.5	7.5	17.0		0.4			0.3	56
In vivo												
BD IX rats												
Liver (10 days)[e]		2.7			5.7	14.0	7.2		2.4	1.4		61
Brain (10 days)[e]		4.6			5.9	12.3	7.8		2.0	0.7		59
Liver (adult)[f]		4.1			5.7	13.0[h]						
C57 BL mice (8–10 weeks)												
Liver[g]		3.6	0.5	1.3	6.5	13.0[h]			2.2			56
Brain[g]		3.8	0.5	1.0	7.4	13.0[h]			1.7			48

[a] Singer, (1979) *J. Natl. Cancer Inst.* **62,** 1329, Beranek, Weis, and Swenson (1980) *Carcinogenesis* **1,** 595.
[b] Singer, Bodell, Cleaver, Thomas, Rajewsky, and Thon (1978) *Nature (London)* **276,** 85.
[c] Swenson, Harbach, Trzos (1980) *Carcinogenesis* **1,** 931.
[d] Sun and Singer (1975) *Biochemistry* **14,** 1795.
[e] Singer, Spengler, and Bodell (1981) *Carcinogenesis* **2,** 1069.
[f] Goth and Rajewsky (1974) *Z. Krebsforsch.* **82,** 37.
[g] Frei, Swenson, Warren, and Lawley (1978) *Biochem. J.* **174,** 1031.
[h] Arbitrary value assigned to 7-ethyl G (based on data obtained with DNA from 10-day-old BD IX rats), and the other derivatives are calculated on that basis.

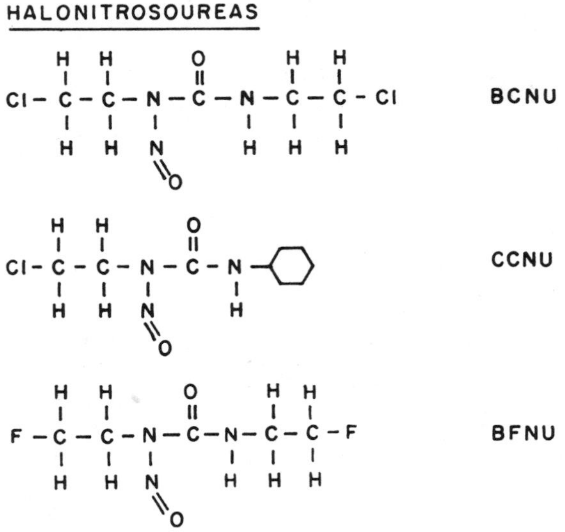

HALONITROSOUREAS

Cl–CH2–CH2–N(N=O)–C(=O)–NH–CH2–CH2–Cl   BCNU

Cl–CH2–CH2–N(N=O)–C(=O)–NH–(cyclohexyl)   CCNU

F–CH2–CH2–N(N=O)–C(=O)–NH–CH2–CH2–F   BFNU

Figure IV-19. Structural formulas of representative halonitrosoureas.

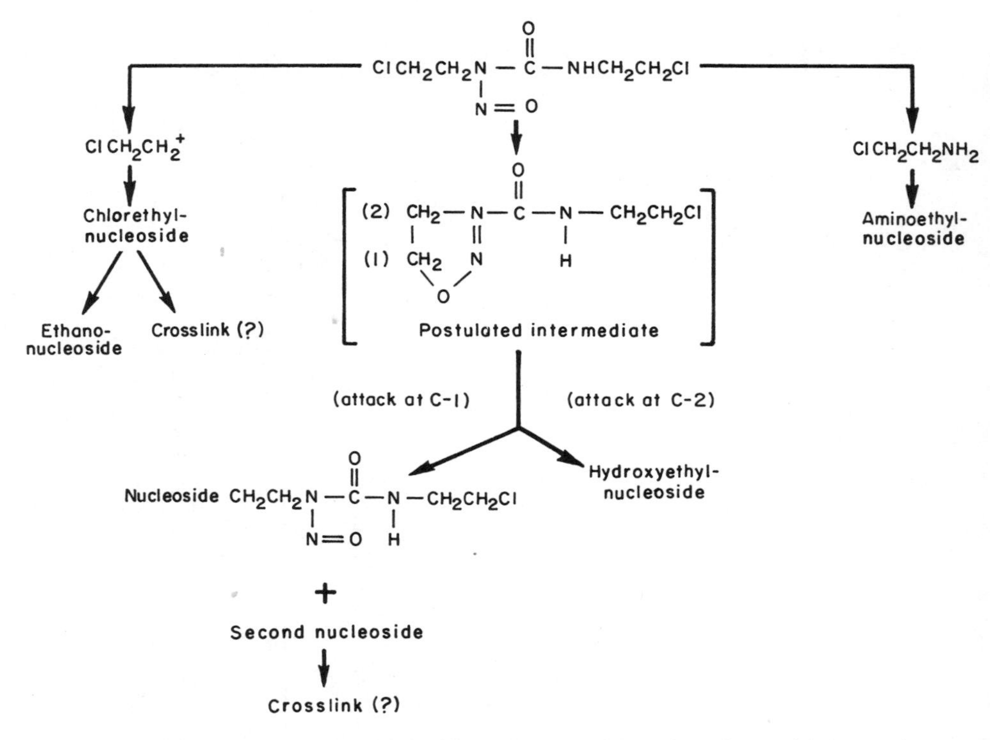

Figure IV-20. Scheme proposed indicating how BCNU could produce the modifications shown in Figure IV-21. From Ludlum and Tong (1981) In *Nitrosoureas: Current Status and New Developments* (Prestayko, Crooke, Baker, Carter, and Schein, eds.), Academic Press, pp. 85–94.

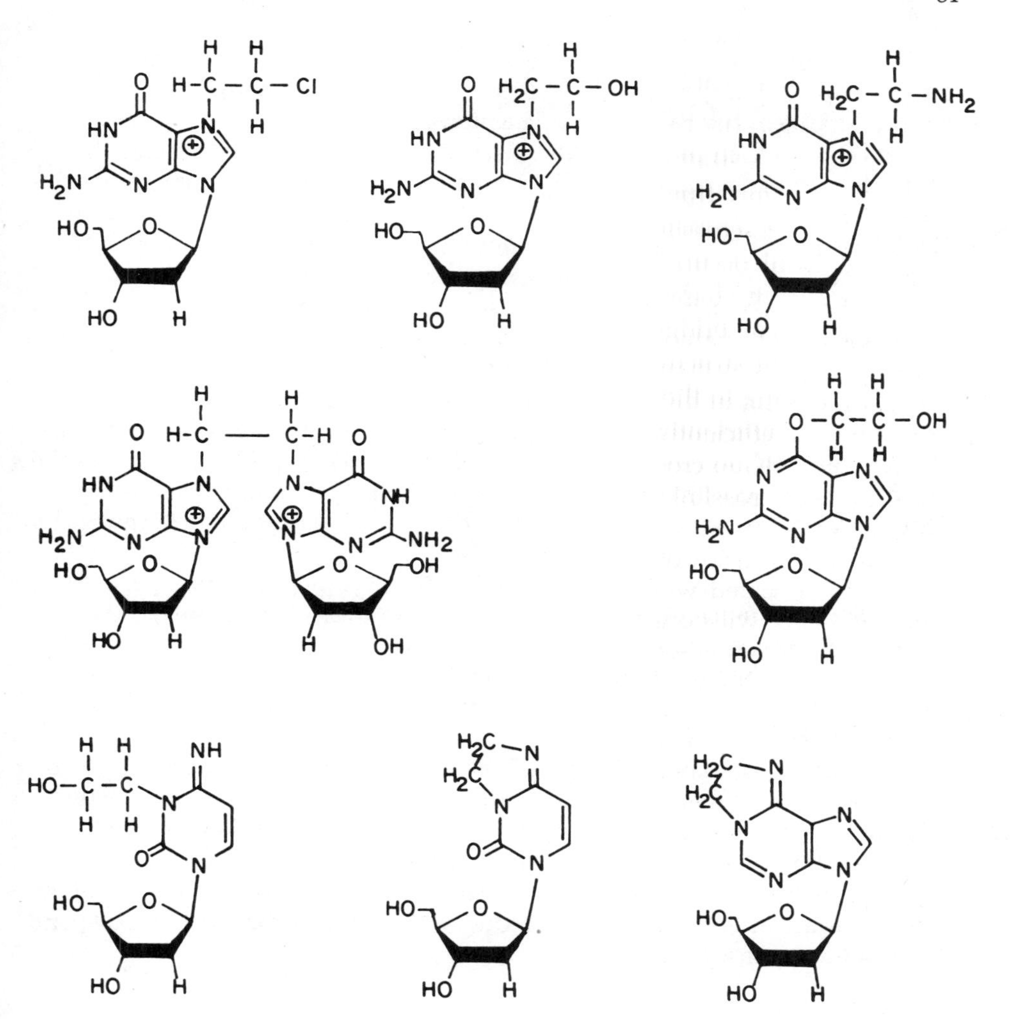

Figure IV-21. Products identified from reaction of BCNU with nucleic acids. The top row shows chloroethyl, hydroxyethyl, and aminoethyl modification of the N-7 of G. The second row shows the N-7–N-7 crosslinked diguanyl derivative and O^6-hydroxyethyl G. In the third row are 3-hydroxyethyl C and the two cyclic ethano derivatives, 3,N^4-ethano C and 1,N^6-ethano A. The crosslink between the N-3 of C and the N-1 of G is illustrated in Figure IV-22 (bottom). See Figure IV-20 for a possible mechanism for the formation of these derivatives. BCNU = 1,3-bis(2-chloroethyl)-1-nitrosourea.

of formation of ethano derivatives is substitution of a haloethyl carbonium ion on the N-3 of C or N-1 of A, followed by rapid cyclization.

Like other bifunctional agents, crosslinks can be formed between reactive groups. The crosslink between guanines, diguanyl ethane, is through the N-7. Since a 2-carbon bridge is not sufficiently long to span the distance between two guanines of the double helix, an interstrand crosslink probably results.

The more recently described crosslink between the N-3 of C and the N-1 of G, (1-[*N*-3-deoxycytidyl],2-[*N*-1-deoxyguanosinyl]-ethan), could be intrastrand, linking a basepaired C–G. The mechanism by which the C–G crosslink is formed is not clear but probably arises through the transfer of a chloroethyl group to one of the bases, followed by a second reaction of this group with the other DNA strand (Figure IV-22). A calculation of 1 C–G crosslink/10^4 DNA bases comes from DNA reacted with a high concentration of BCNU for 18 hr. The G–G crosslink is appreciably more abundant.

A crosslink occurring in halonitrosourea-treated mammalian cells has not been isolated or characterized chemically but is suggested to be a crosslink with an ethylene bridge between the O^6 of G and the N^4 of C. The evidence for this type of structure comes primarily from the decrease in cytotoxicity and crosslinking in those cells that are termed mer^+ because they repair O^6-alkylguanine efficiently. As shown in Figure IV-22, a possible route to the dCyd-$(CH_2)_2$-dGuo crosslink involves initial chloroethylation of the O^6 of G. Thus, *in vivo* crosslinks may be the same or related to those determined *in vitro*.

While only a few of the derivatives produced by BCNU have been isolated from DNA treated with the other halonitrosoureas, it is probable that the same reactions will occur, but not necessarily to the same extents. Thus, FCNU [*N*-(2-fluoroethyl)-*N*′-cyclohexyl-*N*-nitrosourea] does not form detectable diguanyl crosslinks, while CNU (chloroethyl nitrosourea) produces relatively more 7-hydroxyethylguanine.

The chemotherapeutic effect of this class of compound appears to correlate with the cytotoxicity and is attributed to their crosslinking activity, whereas the other reactions are likely to lead to depurination and chain breakage, and thus would probably be mutagenic and carcinogenic. It is for this reason that highly cytotoxic levels are used in chemotherapy where halonitrosoureas are effective in treating Hodgkin's disease, other lymphomas, and some solid tumors.

E. Cyclic Alkylating Agents

The cyclic alkylating agents (Figure IV-23), which comprise a number of unrelated mutagens and carcinogens, owe their alkylating ability to reactive unstable ring structures. Several of them have a second potentially reactive group, and can therefore crosslink, both intra- and interstrand, as well as form cyclic derivatives.

1. *Mustards*

Earliest studied were the nitrogen and sulfur mustards, the former of which (HN-2) is also used therapeutically while the later (mustard gas) is classified as a human carcinogen. Both are inefficient mutagens in bacterial

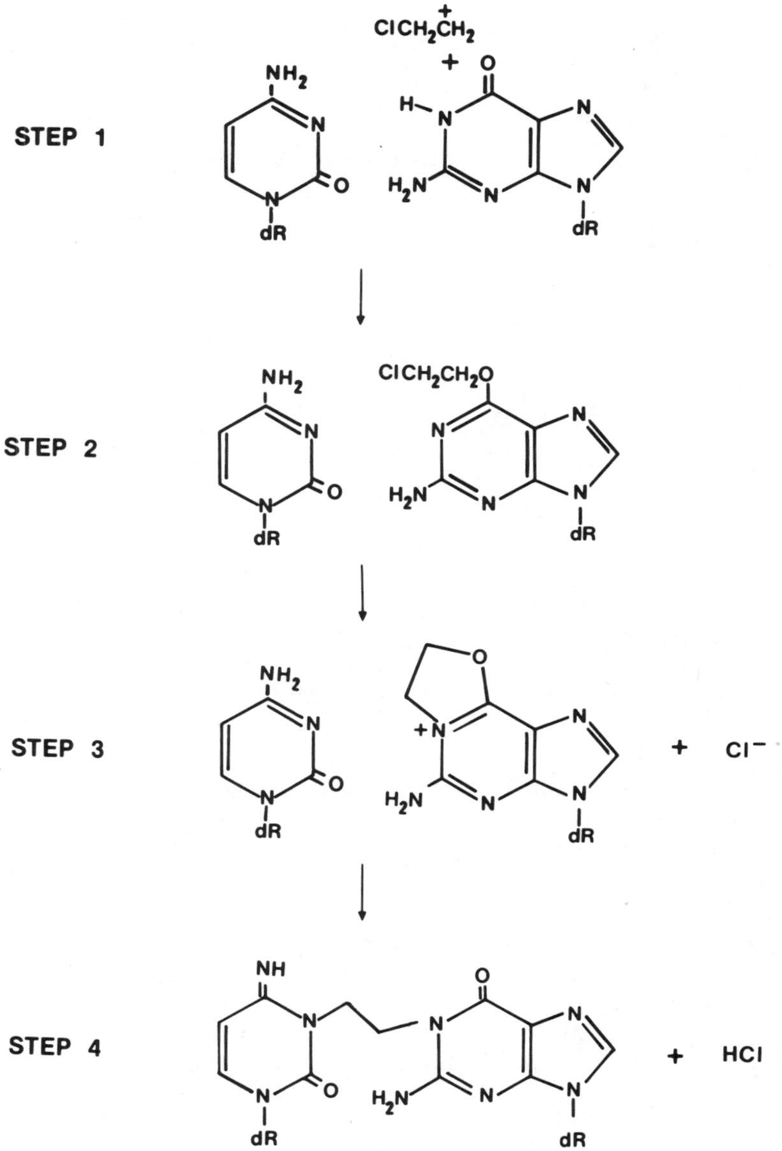

Figure IV-22. Possible route to the formation of the dCyd-CH_2CH_2-dGuo crosslink. In step 1, a chloroethyl carbonium ion attacks the O^6 position of deoxyguanosine to form the O^6-chloroethyl-substituted deoxyguanosine shown in step 2. This derivative rearranges as shown in step 3 and subsequently crosslinks with the opposite nucleoside, deoxycytidine, as shown in step 4. Reprinted with permission from Tong *et al.* (1982) *Cancer Res.* **42,** 3102.

CYCLIC COMPOUNDS

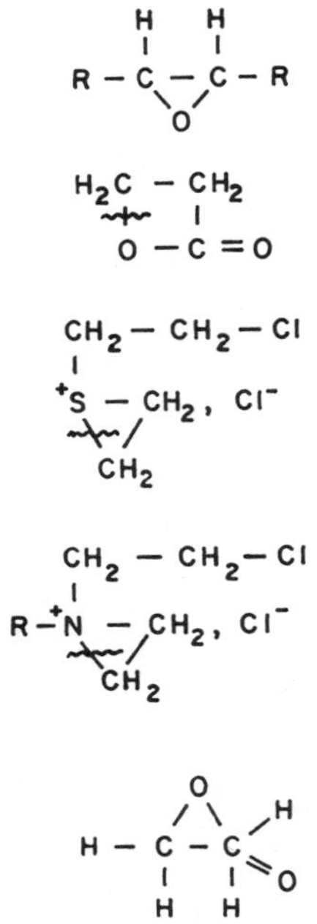

EPOXIDES (CHLOROETHYLENE OXIDE, ETHYLENE OXIDE)

LACTONES (β-PROPIOLACTONE)

S-MUSTARDS (MUSTARD GAS)

N-MUSTARDS (HN-2)

GLYCIDALDEHYDE

Figure IV-23. Structural formulas of representative cyclic alkylating agents.

test systems. Brookes and Lawley, in a pioneering series of studies, found that both mustards reacted with the N-7 of G and a monoadduct and a crosslinked adduct were formed (Figure IV-24). In phage T7, one out of every three or four diguanyl derivatives per molecule crosslinks the DNA strands and there is a reasonable approximation between lethality and crosslinks.

Later, derivatives resulting from reaction at the N-1 and N-3 of A and N-3 of C were identified, so it now appears that sulfur mustard resembles alkylsulfates in reactions except, being bifunctional, crosslinks are also formed.

In mammalian cells ^{35}S-labeled mustard gas formed the same products as those obtained in *in vitro*-treated DNA. There is some evidence that ^{35}S-labeled mustard gas at a dose of 0.5 μg/g mouse ascites cells was tumor-inhibiting.

Phosphoramide mustard, which is a mutagen formed by metabolism of the chemotherapeutic agent (as well as carcinogenic) cyclophosphamide, has been reacted with guanosine to form a highly unstable N-7 derivative ($t_{1/2}$ 2.3 hr at 37°C, pH 7.4). Cyclophosphamide is a nitrogen mustard (Figure IV-23, $R = -P(=O)(-NH_2)(-OH)$) that is effective against a variety of neoplasms and it is

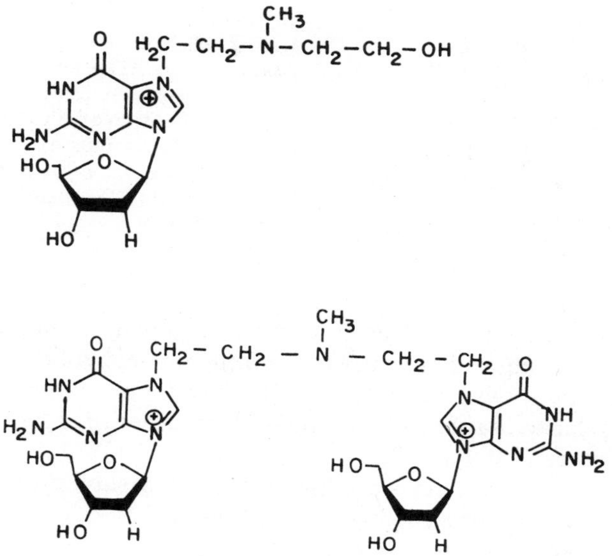

Figure IV-24. Reaction products of nitrogen mustard with the N-7 of G. Both the monoadduct (above) and the crosslink (below) have been identified *in vivo*. Mustard gas forms analogous derivatives.

believed that the antitumor activity is due to phosphoramide mustard. However, the lability of the one known N-7 derivative makes it unlikely that its biological effect is due to crosslinking.

A sulfur mustard with four methyl substituents α, α' to the sulfur completely fails to produce crosslinks due to the steric inhibition of the substituents.

2. *Lactones*

The first reported carcinogenic β-lactone was β-propiolactone (βPL), which has been the most studied in this class. Nevertheless, this compound is a relatively poor carcinogen although it is both an initiator and a promoter of tumors on mouse skin. The rapid decomposition of βPL in H_2O may be a factor in the high dose necessary to produce tumors. Mutagenicity has been shown in a variety of test systems.

βPL generally behaves as a typical alkylating agent and reacts with polynucleotides and DNA at the N-7 of G and N-1 of A to form carboxyethyl derivatives. Reaction with thymidine-5′-phosphate results in diester formation. The fact that modification of the N-3 of A was not detected may be due to the lability of the glycosyl bond of such compounds.

It has been recently proposed that an additional cyclizing reaction with

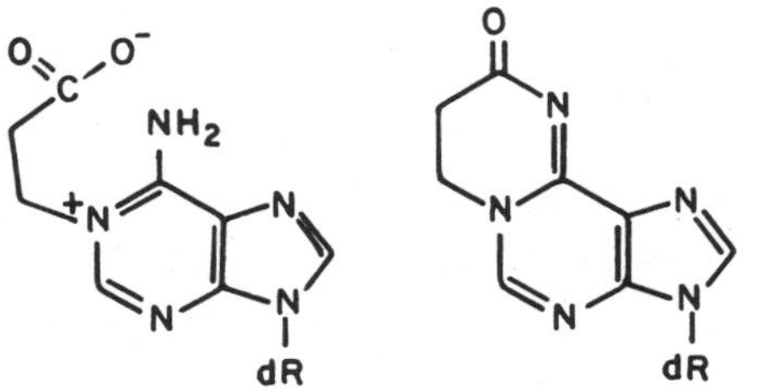

Figure IV-25. Reaction of deoxyadenosine with β-propiolactone. The left structure is 1-(2-carboxyethyl)deoxyadenosine and on the right is a fluorescent derivative with a suggested structure of 3-(β-D-deoxyribosyl)-7,8-dihydropyrimido[2,1-*i*]purine-9-one. Maté *et al.* (1977) *Chem. Biol. Interact.* **18,** 327; Chen *et al.* (1981) *Carcinogenesis* **2,** 73.

adenosine occurs, yielding a lactam on the loss of water. This is a fluorescent derivative with a suggested structure of 3-(β-D-2-deoxyribosyl)-7,8-dihydropyrimido[2,1-*i*]purine-9-one (Figure IV-25). The involvement of the N-1 and N^6 of A is similar to that found with halonitrosoureas or chloroacetaldehyde, both of which are bifunctional.

There is also indirect evidence that βPL can crosslink protein and nucleic acid. Reaction with cysteine to form *S*-2-carboxyethyl-L-cysteine in mouse skin protein has been known for some years.

β-Butyrolactone has not been studied in chemical terms but is somewhat less carcinogenic than the propyl analog. γ-Lactones are not reactive alkylating agents. All the three compounds tested of this class were not carcinogenic, even under lifetime skin application to mice.

3. *Epoxides*

Epoxides, being three-membered rings, are highly reactive alkylating agents. Such rings are the metabolic intermediates of many carcinogens. This section will deal only with aliphatic and low-molecular-weight epoxides, such as ethylene oxide and glycidaldehyde (Figure IV-23), while those derived from polyaromatic hydrocarbons will be discussed in Chapter VI. There are also several simple carcinogenic chemicals that form epoxides *in vivo*. These include vinyl chloride and ethyl carbamate, which are in the following section.

The simplest epoxides, ethylene oxide and propylene oxide, are used in industry and are weakly mutagenic. Ethylene oxide is widely used as a sterilant although there is some evidence that increased rates of leukemia result. The epoxides resemble typical alkylating agents and react with DNA and RNA at the N-7 of G and the N-1 and N-3 of A, to form hydroxyethyl or hydroxypropyl derivatives. No O^6-alkyl derivative of G was detected using propylene oxide.

Glycidaldehyde is bifunctional in that it has a reactive aldehyde group in addition to the epoxide group. Although carcinogenic in mice and rats under specific conditions, it is probably less carcinogenic than βPL. Glycidaldehyde forms a highly fluorescent Δ-imidazoline ring derivative of guan-

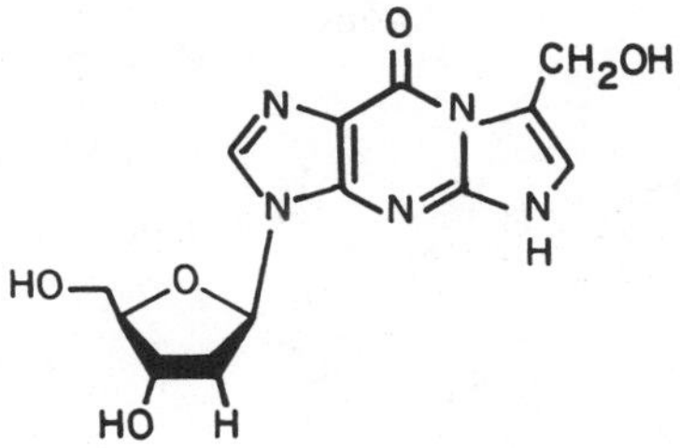

Figure IV-26. Product of glycidaldehyde reaction with deoxyguanosine in calf thymus DNA. Van Duuren and Lowengart (1977) *J. Biol. Chem.* **252,** 5370.

osine in DNA *in vitro* (Figure IV-26). No other derivatives are known, but similar ring structures with adenosine and cytidine would be expected.

Diepoxides, such as diepoxybutane $(\overbrace{CH_2—CH}^{O}—\underbrace{CH—CH_2}_{O})$, exist as stereoisomers that react at the N-7 of G. Crosslinking between the N-7 of two guanines on two strands of a double helix occurs readily with both D and L isomers but is strongly inhibited by the *meso* compound. The inhibition results from steric hindrance of the optically inactive compound. L-Isomer crosslinks are formed with preference, probably due to a subsequent conformational change.

A very promising chemotherapeutic diepoxide, 1,2 : 5,6-dianhydrogalactitol (DAG), reacts preferentially with Yoshida sarcoma cell DNA in rats bearing this tumor. Three products have been identified *in vivo,* all on the N-7 of G (Figure IV-27). The ratio of mono- to diguanyl derivatives is approximately 2, regardless of the extent of reaction or time after administration (1–24 hr). Although similar in some respects to the mustards, i.e., *in vivo* alkylation of Lettré–Ehrlich tumor cells yields about the same extent of alkylation and proportion of crosslinks, there may be differences in excision of crosslinks. Also, no adenine derivative was detected with DAG, while the N-1 and N-3 of A react with mustards and simple epoxides.

F. VINYL HALIDES

Vinyl chloride is carcinogenic in man and experimental animals. The usual vinyl chloride tumor is an angiosarcoma of the liver, but other sites of tumors are the brain, lung, and hematolymphopoietic system in humans. Industrial use of vinyl chloride is estimated (1982) to be about 4×10^9 kg/year in the United States. Exposure to humans is generally by inhalation. The analog, vinyl bromide, which is a less used industrial chemical, also appears to be hepatotoxic and carcinogenic. The biological effects of these compounds depend on metabolic activation by the microsomal P-450 monooxygenase sys-

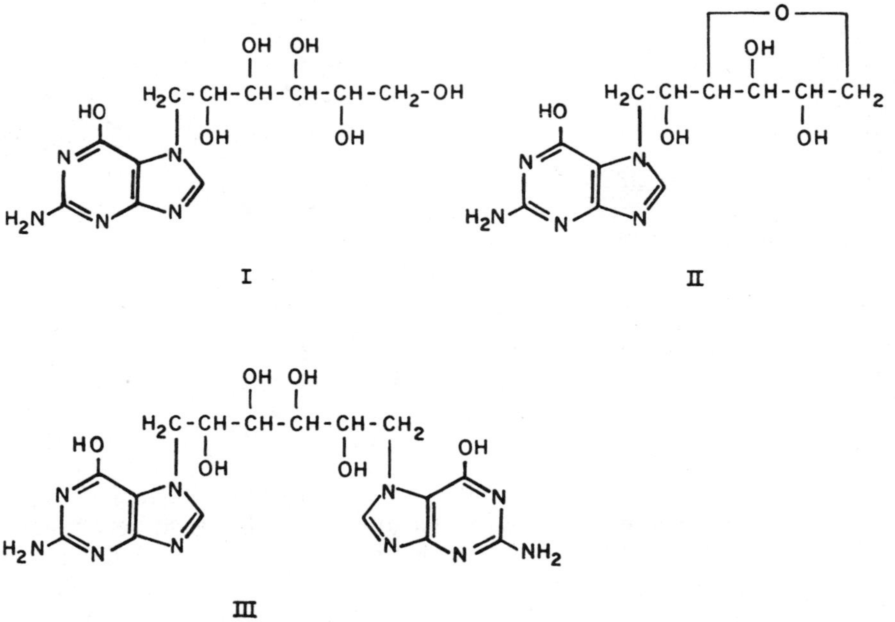

Figure IV-27. Products of reaction of 1,2 : 5,6-dianhydrogalactitol (DAG) with DNA *in vivo*. I, 7-(1-deoxygalactit-1-yl)guanine; II, 7-(1-deoxyanhydrogalactit-1-yl)guanine; III, 1,6-di(guanin-7-yl)-1,6-dideoxygalactitol. Institoris and Tamas (1980) *Biochem. J.* **185,** 659; Institoris (1981) *Chem. Biol. Interact.* **35,** 207.

tem. Vinyl chloride, the better studied of this class, forms a highly reactive metabolite, chloroethylene oxide (CEO), again an epoxide, which rapidly rearranges, nonenzymatically, to chloroacetaldehyde (CAA) (Figure IV-28). The $t_{1/2}$ in 0.1 M pH 7.4 buffer at 37°C is 0.87 min. There is indirect evidence that vinyl bromide is similarly activated but the comparative rates are not clear. Both CEO and CAA are mutagenic and carcinogenic, but the former is more potent.

1. *Reaction of Metabolites with Monomers*

The stable metabolite, chloroacetaldehyde, has long been used to prepare fluorescent derivatives of nucleosides and nucleotides (Figure IV-29). 1,N^6-Ethenoadenosine (εA), 3,N^4-ethenocytidine (εC), and 1,N^2-ethenoguanosine can be prepared in good yield in aqueous solution at mild pH and temperatures. However, the N^2,3-etheno derivative of guanosine could only be made by reacting an O^6-alkylguanosine followed by dealkylation. Guanine reacts directly with CAA to form N^2,3-ethenoguanine. In the case of εA, X-ray analysis of model derivatives was interpreted as showing that the N-1 rather than the N-6 reacts with the α carbon of the aldehyde. In Figure IV-30 the

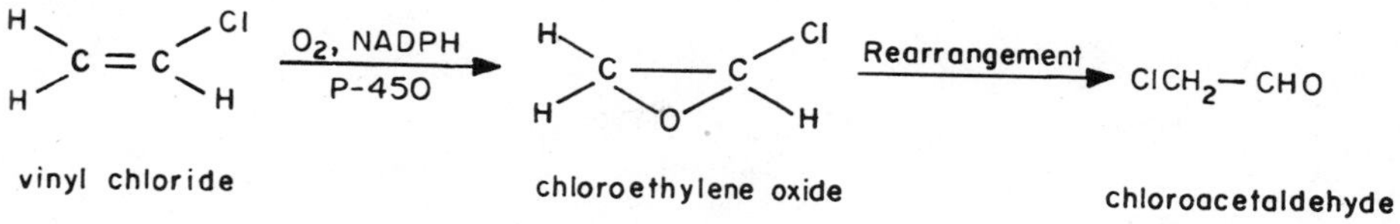

$$\mathrm{H_2C{=}CHCl} \xrightarrow[\text{P-450}]{\mathrm{O_2,\ NADPH}} \text{chloroethylene oxide} \xrightarrow{\text{Rearrangement}} \mathrm{ClCH_2{-}CHO}$$

vinyl chloride — chloroethylene oxide — chloroacetaldehyde

Figure IV-28. Postulated mechanism for the metabolism of vinyl chloride. Adapted from Zajdela *et al.* (1980) *Cancer Res.* **40,** 352.

alternative primary site of reaction is shown as being the imino group. This is supported by the experimentally determined reversibility of step 2 → 1, which could not occur if alkylation of the endocyclic nitrogen were indeed the first step. In step 3 cyclization yields a nonfluorescent hydrated compound.

Dehydration of hydrated etheno Ado or Cyd to fluorescent etheno derivatives occurs rapidly in strong acid, but is greatly decreased at neutrality. There is relatively little dependence of the rate over the range of pH 3–7. The rate constant is more sensitive to changes in pH in the case of the cytidine intermediate than in that of adenosine, but it is increased for both as the hydrogen ion concentration increases. Under physiological conditions

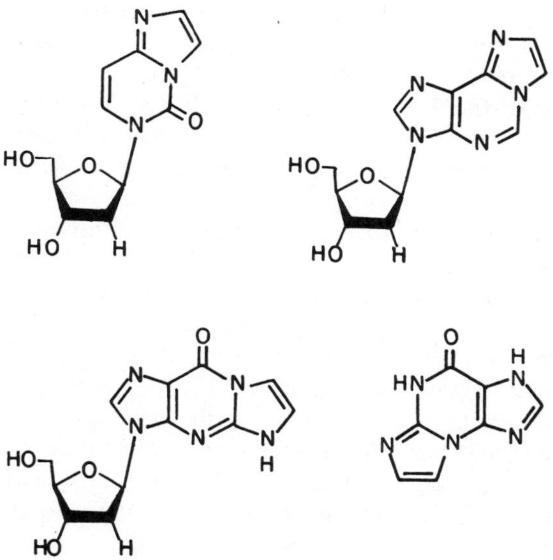

Figure IV-29. Etheno derivatives formed by chloroacetaldehyde reaction. On the top line are 3,N^4-ethenocytidine and 1,N^6-ethenoadenosine, both found in DNA or RNA from vinyl chloride-treated rats. On the bottom line are 1,N^2-ethenoguanosine and N^2,3-ethenoguanine. While the 1,N^2 derivative has not been found to be a product of reaction in polynucleotides, the base, N^2,3-ethenoguanine, is present in low amounts in chloroacetaldehyde-treated DNA.

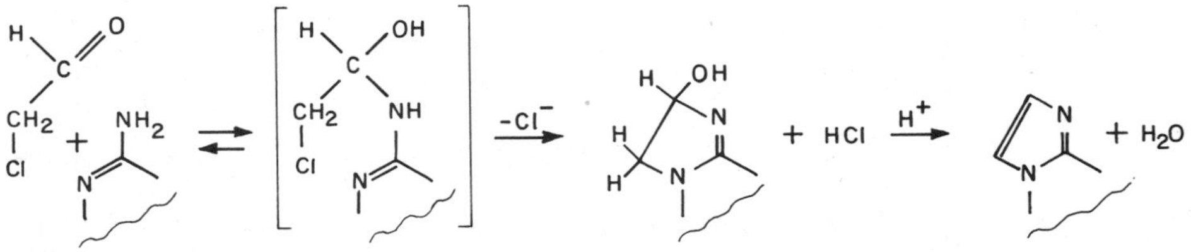

Figure IV-30. Reaction of chloroacetaldehyde with adenosine or cytidine. Only the reactive portion of the base is shown. The rate of conversion of the stable intermediate **3** to the dehydrated etheno derivative **4** is dependent on pH.

(pH 7.25) the half-life of the εA hydrate is about 1.5 hr at 37°C while that of the εC hydrate is about 5 hr at 50°C. Over the entire pH range studied, the adenosine intermediate was dehydrated 6–14 times faster than the cytidine intermediate.

Bromoacetaldehyde reactions have been studied less, but both adenosine-5′-phosphate and cytidine-5′-phosphate form etheno derivatives, including the hydrated intermediate. The reaction rates are about 10-fold faster than with chloroacetaldehyde.

The etheno nucleotides of A and C have been used to mimic the structural features of natural coenzymes. 1,N^6-Etheno ATP showed considerable substrate activity as a replacement of ATP with adenylate kinase while the diphosphate was an excellent substitute for ADP in the pyruvate kinase system. The introduction of the second ring on cytidine gives the molecule a similar spatial outline and similar potential binding areas to those of adenosine. In the enzymatic phosphorylation of 3-phosphoglyceric acid, εCTP is essentially equivalent to ATP.

Chloroethylene oxide differs from chloroacetaldehyde in its site(s) of reaction with nucleosides. As previously noted, CAA is of low reactivity toward guanosine. In contrast, guanosine (in glacial acetic acid) reacts almost quantitatively in 1–2 min to form 7-(2-oxoethyl)guanosine (Figure IV-31). It is suggested, on the basis of proton NMR spectra, that this derivative could exist in equilibrium with a cyclic hemi-acetal, O^6,7-(1′-hydroxyethano)guanosine.

Although no other reactions of CEO with nucleosides are known, it might be predicted that the other sites modified by typical alkylating agents would also be reactive. The major difficulty in studying monomer reactions is the rapid hydrolysis of CEO. The rate constant for reaction of CEO with H_2O (37°C) is 2×10^{-6}, which is five orders of magnitude greater than that of CAA.

2. *Reaction of Chloroacetaldehyde with Polymers*

CAA modifies adenine and cytosine residues in single-stranded ribo- and deoxyribopolynucleosides, but C residues are more reactive than A residues,

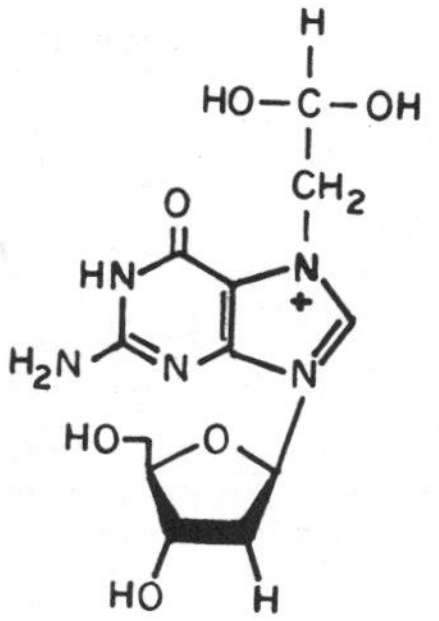

Figure IV-31. Reaction product of chloroethylene oxide with deoxyguanosine. Scherer *et al.* (1981) *Carcinogenesis* **2,** 671. The base derived from this derivative, 7-(2-hydroxyethyl)guanine, has been identified in DNA and RNA from animals given vinyl chloride. Ostermann-Golkar *et al.* (1977) *Biochem. Biophys. Res. Commun.* **76,** 259; Laib *et al.* (1981) *Chem. Biol. Interact.* **37,** 219.

as they are for monomers, over the entire pH range 4–8. Below pH 6, reactivity is decreased by about 10-fold. In view of the fact that the reactivity of monomeric C is almost the same over the pH range 5–7, the great decrease in reactivity of poly(rC) can be attributed to the formation of the protonated secondary structure of poly(rC). Since poly(rA) precipitates at pH 5 and below, no similar conclusions can be drawn although it is likely that protonation plays a role.

The reactivity of poly(rC) is similar to that of poly(dC), whereas poly(rA) is much more reactive than poly(dA). Guanine residues in polymers were not detectably reactive at pH 5–7, although 1,N^2-ethenoguanosine is the major product of monomer reaction and N^2,3-ethenoguanine has been isolated from CAA-treated DNA (Figure IV-29). Uracil reaction has not been reported.

The reaction mechanism for εA and εC is that shown in Figure IV-30. This is substantiated by evidence that the hydrated intermediates (**3**) can be isolated from enzyme hydrolysates and are the major products (>90%) from CAA-reacted poly(rA), poly(dA), poly(rC), and poly(dC). After separation of the hydrated etheno compounds, they can be dehydrated to εC or εA by heating. This is illustrated in Figures IV-32 and IV-33, which also show the characteristic UV spectra of products **3** and **4** (Figure IV-30). Similarly, hydrated etheno compounds can be converted, by heating or lowered pH, to etheno derivatives in polynucleotides and $tRNA^{Phe}$. The dehydration rates in polymers are similar to those of hydrates in monomers.

The stability of the glycosyl bond of εA or εC in polymers is the same as that of the unmodified A or C. N^2,3-Ethenodeoxyguanosine in DNA was depurinated under the same conditions as dG and did not exhibit the instability of glycosyl bonds found with N-7 or N-3 derivatives.

Only single-stranded polymers react with CAA. However, thermal instability leads to "breathing" or momentary single-strandedness of normally double-stranded polymers. These non-hydrogen-bonded regions permit some reaction. The higher the T_m, the less likely that reaction will occur. There is no detectable modification at 37°C of C in poly(rC) · poly(rG) ($T_m > 100$°C) whereas in poly(rA) · poly(rU) (T_m 56°C), adenine reacted about 1/10 as much as in poly(rA). Only a few A or C residues in tRNA react.

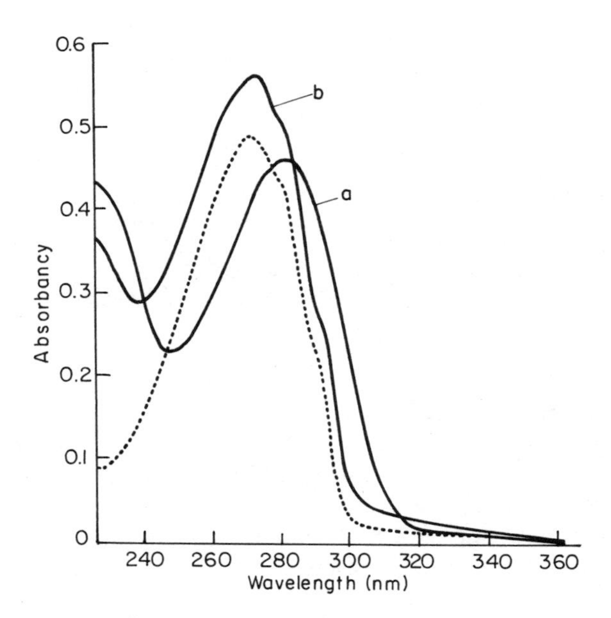

Figure IV-32. UV spectra at pH 7 of products of chloroacetaldehyde-treated poly(rC), after enzyme digestion to nucleosides. HPLC fractionation yields two products: 3,N^4-ethenocytidine (εC) (----) and the hydrated intermediate (a). Upon heating, the spectrum of (a) is converted to (b), which is identical to that of εC. The relative extinction coefficients are shown only by the solid lines. From Kuśmierek and Singer (1982) *Biochemistry* **21,** 5717.

CAA, like formaldehyde and bisulfite (Section B), has been suggested as a tool for probing secondary structures.

3. *Reaction of Vinyl Halides in Vivo*

Both vinyl chloride and vinyl bromide react with RNA and DNA in rats and mice given the ^{14}C-labeled carcinogen by inhalation for short times. The earliest products identified were 1,N^6-ethenoadenine and 3,N^4-ethenocytosine, which were also produced when unlabeled vinyl chloride was administered to rats for 2 years in drinking water. These etheno derivatives were also formed on incubation of polynucleotides with rat liver microsomes, NADPH, and vinyl chloride or vinyl bromide. As discussed earlier, the stable metabolic products of vinyl halides are haloaldehydes that produce etheno compounds *in vitro.* It had generally been considered that the labile haloepoxides did not play a role *in vivo.* However, there was evidence, which has now been substantiated, that 7-(2-hydroxyethyl)guanine is an *in vivo* product in DNA, and this derivative is apparently only formed by CEO reaction. Since the N-7 of guanine is not involved in hydrogen-bonding, this site can be reactive in double-stranded nucleic acids, while the etheno compounds can

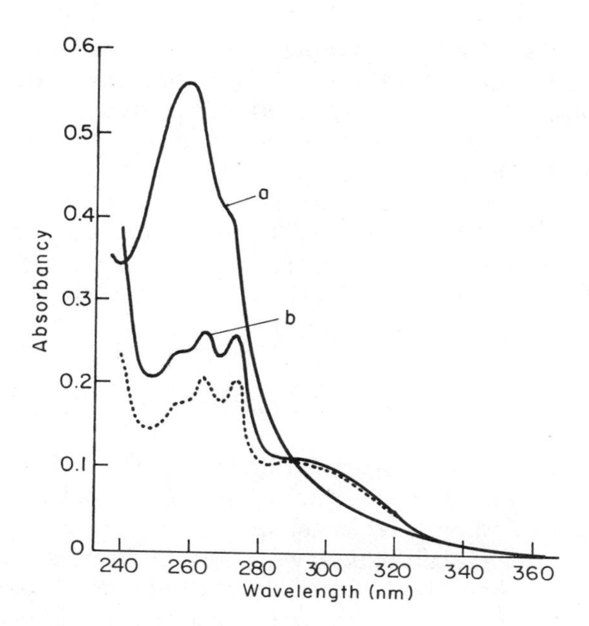

Figure IV-33. UV spectra at pH 7 of products of chloroacetaldehyde-treated poly(rA), after enzyme digestion to nucleosides. HPLC fractionation yields two products: 1,N^6-ethenoadenosine (εA) (----) and the hydrated intermediate (a). Upon heating, the spectrum of (a) is converted to (b), which is identical to that of εA. The relative extinction coefficients are shown only by the solid lines. From Kuśmierek and Singer (1982) *Biochemistry* **21,** 5717.

only be formed in single-stranded regions. It now appears that 7-(2-hydroxyethyl)guanine (Figure IV-31) is the major product of reaction of vinyl halides with DNA *in vivo* and *in vitro* after activation of vinyl chloride with rat liver microsomes and an NADPH-regenerating system.

It is clear that the two major metabolites of vinyl halides each react with nucleic acids at different sites. Haloaldehydes form etheno derivatives in single-stranded regions of DNA or RNA by the same mechanism as shown in Figure IV-30. Hydrated etheno compounds must be expected to be present but there are no reports of attempted identification. On the other hand, the N-7 derivative resulting from reaction with the haloepoxide indicates that this metabolite is a typical alkylating agent and is likely to react at all sites previously found reactive with simple epoxides (Section E3).

There appears to be a discrepancy between the short-term inhalation experiments and the long-term oral administration experiment, in that the latter data report εA and εC in liver DNA while the former experiments state that no etheno compounds are found in DNA, but only in RNA. This difference can, in part, be attributed to the extremely slow reaction of CAA with double-stranded DNA. Over the 2-year experimental period, if not repaired, etheno compounds could be accumulated and thus detectable by mass spectrometry. In contrast, 7-alkyldeoxyguanosines have a labile glycosyl bond ($t_{1/2}$ 7-Me dGuo 155 hr) and would be continually formed and depurinated.

In most experiments, reaction of DNA with CAA will lead to some etheno compounds. In the one *in vitro* experiment using DNA where only 7-(2-hydroxyethyl)guanine was produced with metabolites of vinyl chloride, a special technique was devised (transmembrane alkylation) to prevent nuclease-mediated alteration of the double strand. It is more reasonable to assume that *in vivo* DNA contains single-stranded regions capable of being modified on normally basepaired positions, including the N^2 of G.

G. ALKYL CARBAMATES

Ethyl carbamate, formerly termed urethan (CH_3—CH_2—O—CO—NH_2), is carcinogenic in several species and it has been speculated that it may be metabolically activated by dehydrogenation to vinyl carbamate (CH_2=CH—O—CO—NH_2), which is subsequently epoxidated. Vinyl carbamate is a more potent carcinogen than ethyl carbamate and induces a wide spectrum of tumors. Although vinyl carbamate was not detected as a metabolite in the mouse *in vivo*, it has been deduced that this pathway exists. The evidence is primarily on the basis that injection of a single dose of [*ethyl*-1-^{14}C]ethyl carbamate into rats or mice resulted in the formation of labeled 1,N^6-ethenoadenosine and 3,N^4-ethenocytidine adducts in liver RNA. Presumably, vinyl carbamate epoxide or the stable aldehyde can react with adenosine and cytidine to form etheno derivatives in the same way as they are formed with CAA.

An early report that ethylphosphotriesters were the major product of ethyl carbamate reaction with DNA has not been supported by more recent work. This does not preclude that some reaction occurs with the phosphodiesters.

SELECTED REFERENCES

Auerbach, C., Moutschen-Dahmen, M., and Moutschen, J. (1977) Genetic and cytogenetical effects of formaldehyde and related compounds. *Mutat. Res.* **39,** 317–362.

Barrows, L. R., and Magee, P. N. (1982) Nonenzymatic methylation of DNA by *S*-adenosylmethionine *in vitro. Carcinogenesis* **3,** 349–351.

Bartsch, H., Malaveille, C., Barbin, A., and Planche, G. (1971) Mutagenic and alkylating metabolites of halo-ethylenes, chlorobutadienes and dichlorobutenes produced by rodent or human liver tissues. *Arch. Toxicol.* **41,** 249–277.

Brown, D. M. (1974) Chemical reactions of polynucleotides and nucleic acids. In *Basic Principles in Nucleic Acid Chemistry,* Vol. II (P.O.P. Ts'o, ed.), Academic Press, New York, pp. 1–90.

Budowsky, E. I. (1976) The mechanism of the mutagenic action of hydroxylamines. *Prog. Nucleic Acid Res. Mol. Biol.* **16,** 125–188.

Chung, F.-L., and Hecht, S. S. (1983) Formation of cyclic 1, N^2-adducts by reaction of deoxyguanosine with α-acetoxy-*N*-nitrosopyrrolidine, 4-(carboxy-nitrosamino)butanal, or crotonaldehyde. *Cancer Res.* **43,** 1230–1235.

Druckrey, H., Kruse, H., Preussmann, R., Ivankovic, S. and Landschütz, C. (1970) Cancerogene alkylierende Substanzen. III. Alkyl-halogenide, -sulfate, -sulfonate und ringgespannte Heterocyclen. *Z. Krebsforsch.* **74,** 241–270.

Druckrey, H., Ivankovic, S., and Gimmy, J. (1973) Cancerogene Wirkung von Methyl- und Athylnitrosoharnstoff (MNH und ANH) nach einmaliger intracerebraler bzw. intracarotidaler Injektion bei neugeborenen und jungen BD-Ratten. *Z. Krebsforsch.* **79,** 282 –297.

Fraenkel-Conrat, H. (1981) Chemical modification of viruses. In *Comprehensive Virology,* Vol. 17 (H. Fraenkel-Conrat and R. R. Wagner, eds.), Plenum Press, New York, pp. 245–283.

Guengerich, F. P. (1982) Metabolism of vinyl halides: *In vitro* studies on roles of potential activated metabolites. In *Biological Reactive Intermediates—II,* Part A (Snyder *et al.*, eds.), Plenum Press, New York, pp. 685–692.

Ivankovic, S., and Druckrey, H. (1968) Transplacentare Erzeugung maligner Tumoren des Nervensystems. *Z. Krebsforsch.* **71,** 320–360.

Kimball, R. F. (1977) The mutagenicity of hydrazine and some of its derivatives. *Mutat. Res.* **39,** 111–126.

Kircher, M., Fleer, R., Ruhland, A., and Brendel, M. (1979) Biological and chemical effects of mustard gas in yeast. *Mutat. Res.* **63,** 273–289.

Kleihues, P., Lantos, P. L., and Magee, P. N. (1976) Chemical carcinogenesis in the nervous system. In *International Review of Experimental Pathology,* Vol. 15, Academic Press, New York, pp. 153–232.

Kohn, K. W. (1981) Molecular mechanisms of cross-linking by alkylating agents and platinum complexes. In *Molecular Actions and Targets for Cancer Chemotherapeutic Agents* (A. C. Sartorelli, J. S. Lazo, and J. R. Bertino, eds.), Academic Press, New York, pp. 3–16.

Kuśmierek, J. T., and Singer, B. (1982) Chloroacetaldehyde-treated ribo- and deoxyribopolynucleotides. 1. Reaction products. *Biochemistry* **21,** 5717–5722.

Laib, R. J., Gwinner, L. M., and Bolt, H. M. (1981) DNA alkylation by vinyl chloride metabolites: Etheno derivatives or 7-alkylation of guanine? *Chem. Biol. Interact.* **37,** 219–231.

Lindahl, T., Rydberg, B., Hjelmgren, T., Olsson, M., and Jacobson, A. (1982) Cellular defense mechanisms against alkylation of DNA. In *Molecular and Cellular Mechanisms of Mutagenesis* (J. F. Lemontt and W. M. Generoso, eds.), Plenum Press, New York, pp. 89–102.

Lown, J. W., McLaughlin, L. W., and Chang, Y.-M. (1978) Mechanism of action of 2-haloethylnitrosoureas on DNA and its relation to their antileukemic properties. *Bioorg. Chem.* **7,** 97–110.

Ludlum, D. B., and Tong, W. P. (1981) Modification of DNA and RNA bases. In *Nitrosoureas: Current Status and New Developments* (A. W. Prestayko, S. T. Crooke, L. H. Baker, S. K. Carter, and P. S. Schein, eds.), Academic Press, New York, pp. 85–94.

Montesano, R., and Bartsch, H. (1976) Mutagenic and carcinogenic *N*-nitroso compounds: Possible environmental hazards. *Mutat. Res.* **32,** 179–228.

Müller, R., and Rajewsky, M. F. (1981) Guest editorial: Antibodies specific for DNA components structurally modified by chemical carcinogens. *J. Cancer Res. Clin. Oncol.* **102,** 99.

Neale, S. (1976) Mutagenicity of nitrosamides and nitrosamidines in micro-organisms and plants. *Mutat. Res.* **32,** 229–266.

Pegg, A. E. (1977) Formation and metabolism of alkylated nucleosides: Possible role in carcinogenesis by nitroso compounds and alkylating agents. *Adv. Cancer Res.* **5,** 195–269.

Pegg, A. E. (1980) Metabolism of *N*-nitrosodimethylamine. In *Molecular and Cellular Aspects of Carcinogen Screening Tests* (R. Montesano, H. Bartsch, and L. Tomatis, eds.), IARC Scientific Publication No. 27, International Agency for Research on Cancer, Lyon, France, pp. 3–22.

Pullman, B. (1979) The macromolecular electrostatic effect in biochemical reactivity of the nucleic acids. In *Catalysis in Chemistry and Biochemistry: Theory and Experiment* (B Pullman, ed.), Reidel, Dordrecht, pp. 1–10.

Ribovich, M. L., Miller, J. A., Miller, E. C., and Timmins, L. G. (1982) Labeled 1,N^6-ethenoadenosine and 3,N^4-ethenocytidine in hepatic RNA of mice given [ethyl-1,2-^{3}H or ethyl-1-^{14}C] ethyl carbamate (urethan). *Carcinogenesis* **3,** 539–546.

Schoental, R., and Bensted, J. P. M. (1964) Gastro-intestinal tumors in rats and mice following various routes of administration of *N*-methyl-*N*-nitroso-*N'*-nitroguanidine and *N*-ethyl-*N*-nitroso-*N'*-nitroguanidine. *Br. J. Cancer* **23,** 757–764.

Shapiro, R. (1968) Chemistry of guanine and its biologically significant derivatives. *Prog. Nucleic Acid Res. Mol. Biol.* **8,** 73–112.

Shapiro, R. (1982) Genetic effects of bisulfite: Implications for environmental protection. In *Induced Mutagenesis: Molecular Mechanisms and Their Implications for Environmental Protection* (C. W. Lawrence, L. Prakash, and F. Sherman, eds.), Plenum Press, New York, pp. 35–60.

Singer, B. (1975) The chemical effects of nucleic acid alkylation and their relation to mutagenesis and carcinogenesis. *Prog. Nucleic Acid Res. Mol. Biol.* **15,** 219–284, 330–332.

Singer, B. (1976) All oxygens in nucleic acids react with carcinogenic ethylating agents. *Nature (London)* **264,** 333–339.

Singer, B. (1979) Guest editorial: *N*-Nitroso alkylating agents: Formation and persistence of alkyl derivatives in mammalian nucleic acids as contributing factors in carcinogenesis. *J. Natl. Cancer Inst.* **62,** 1329–1339.

Singer, B. (1982) Mutagenesis from a chemical perspective: Nucleic acid reactions, repair, translation and transcription. In *Molecular and Cellular Mechanisms of Mutagenesis* (J. F. Lemontt and R. Generoso, eds.), Plenum Press, New York, pp. 1–42.

Singer, B., and Kuśmierek, J. T. (1982) Chemical mutagenesis. *Annu. Rev. Biochem.* **51,** 655–693.

Tong, W. P., Kohn, K. W., and Ludlum, D. B. (1982) Modifications of DNA by different haloethylnitrosoureas. *Cancer Res.* **42,** 4460–4464.

Van Duuren, B. L. (1969) Carcinogenic epoxides, lactones and halo-ethers and their mode of action. In *Biological Effects of Alkylating Agents, Ann. N.Y. Acad. Sci.* **163,** 633–651.

V

Metabolic Activation of Carcinogens and Mutagens

In this chapter we will first describe enzymes that are involved in the metabolic activation of polycyclic aromatic hydrocarbons and *N*-substituted aromatic compounds. Further, we will discuss some of the metabolic pathways and metabolites that are formed from these indirectly acting carcinogens.

Metabolism of vinyl chloride and of the aliphatic, but not cyclic, nitrosamines was discussed in Chapter IV. Since the metabolic activation of some naturally occurring carcinogens, such as aflatoxin, and that of the products of protein pyrolysis are not that well studied and understood, it seemed to us more appropriate to describe them in Chapter VI in connection with their formation and interaction with DNA. On the other hand, metabolism of some other carcinogens is not included here, but the interested reader may find reference to these compounds at the end of the chapter.

A. Enzymatic Activation of Polycyclic Aromatic Hydrocarbons

1. *Introduction*

Polycyclic aromatic hydrocarbons (PAHs) are widespread environmental pollutants of the atmosphere, water, and food chain. They are formed predominantly through sufficient combustion and pyrolysis of fossil fuels and other organic materials. Epidemiological studies suggest that PAHs in the environment could be a significant determinant in the incidence of human cancer. Many PAHs in cigarette smoke, for example, are powerful carcinogens in experimental animals and thus apparently play an important role in the etiology of lung and skin cancer in humans.

One of the widely distributed and studied PAHs is benzo[*a*]pyrene. It was identified as a carcinogenic constituent of coal tar by Kenneway in 1938. Structures of some PAHs are presented in Figure V-1.

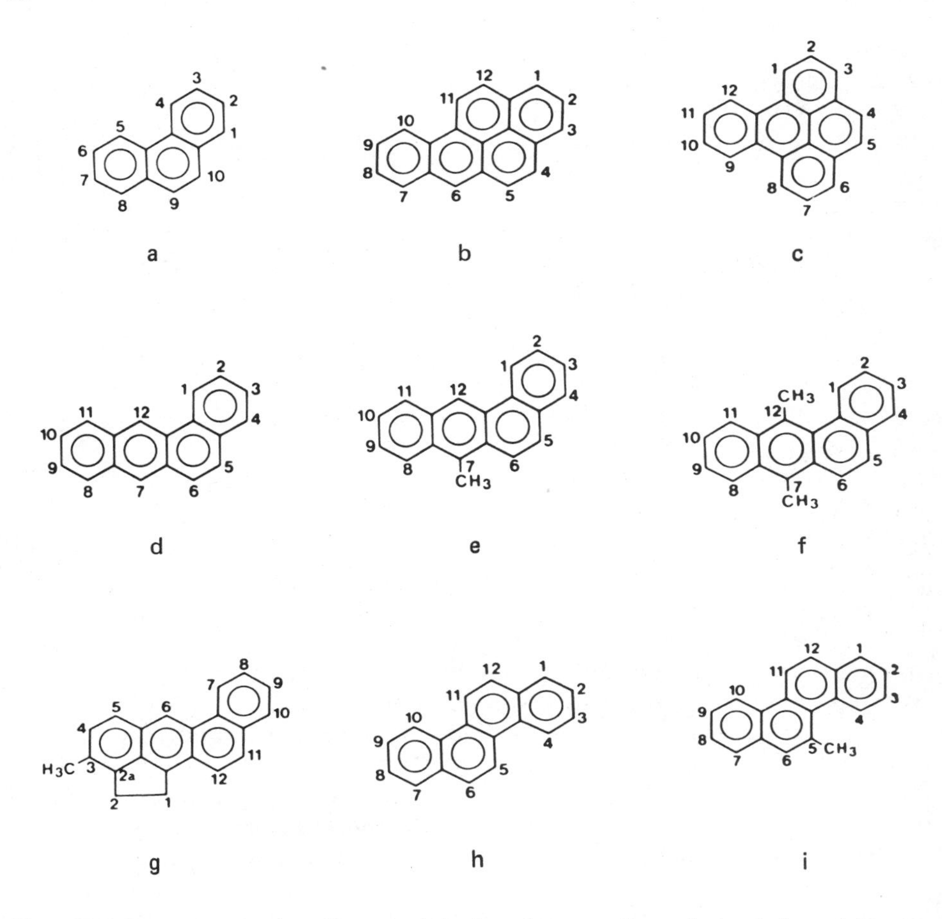

Figure V-1. Structures of polycyclic aromatic hydrocarbons. a, phenanthrene; b, benzo[*a*]pyrene; c, benzo[*e*]pyrene; d, benz[*a*]anthracene; e, 7-methylbenz[*a*]anthracene; f, 7,12-dimethylbenz[*a*]anthracene; g, 3-methylcholanthrene; h, chrysene; i, 5-methylchrysene.

Most mutagenic or carcinogenic PAHs are usually hydrophobic chemicals that are metabolized by enzyme systems localized mainly in the liver. The enzyme systems involved in the metabolism of PAHs can be divided into two groups: phase I and phase II. During phase I metabolism, one or more polar groups (such as hydroxyl) are introduced into the hydrophobic molecule. By this reaction the parent compound becomes a suitable substrate for phase II, conjugating enzymes, such as UDP-glucuronyl transferase. The conjugated products are then sufficiently polar and are readily excreted from the cell and from the body.

Although this metabolic pathway represents a detoxification mechanism, in some cases it leads to formation of carcinogenic products. It was first recognized by E. C. and J. A. Miller that most chemical carcinogens must be metabolically activated to derivatives termed "proximate" and "ultimate" carcinogens, which are the actual initiators of carcinogenesis. The balance between enzymes that potentiate and those that detoxify the highly reactive intermediates is one of the critical factors in the carcinogenic potencies of chemicals. The ultimate carcinogens are strong electrophilic reactants containing electron-deficient atoms capable of reacting covalently with electron pairs in electron-rich or nucleophilic atoms in cellular components, particularly in nucleic acids and proteins.

2. *Monooxygenases or Arylhydrocarbon Hydrolases*

a. DISTRIBUTION

One of the most important of the phase I enzyme systems is a group of enzymes known as the cytochrome P-450-mediated mixed-function oxidases (MFO) or monooxygenases (MO) located in the microsomes of the endoplasmic reticulum of the cell. In the case of PAHs the reactions catalyzed by these enzymes are hydroxylation reactions and therefore they are called arylhydrocarbon hydroxylases (AHHs).

AHH is an almost ubiquitous enzyme system. It has been found in 90% of tissues examined including liver, lung, gastrointestinal tract, and kidney in rats, mice, hamsters, and monkeys. The enzyme was also detected in human liver, lung, placenta, lymphocytes, monocytes, and alveolar macrophages. Although this system is primarily localized in microsomes, it is also present in nuclear and mitochondrial fractions of the cells. The membrane-bound multicomponent system involves at least two proteins: cytochrome P-450 (actually a family of different cytochromes) and a protein called NADPH-cytochrome P-450 reductase, as well as molecular oxygen and NADPH.

The other component necessary for the activity in the reconstituted enzyme system is a phospholipid, phosphatidylcholine. It is required to facilitate interaction between the cytochrome P-450 and the NADPH-cytochrome P-450 reductase which is essential for AHH activity.

b. ASSAY OF ARYLHYDROCARBON HYDROLASES

Most techniques used for assaying AHHs utilize benzo[*a*]pyrene (BP) as substrate. The assay is simple and very sensitive. It measures the fluorescence of the 3-hydroxy-BP and other phenols formed upon incubation of BP with microsomes and cofactors. The products of the reaction are extractable from an organic phase by alkali. The other metabolites formed, such as diols, quinones, and tetrols, are not extracted into the alkaline phase, or when partially extracted, they contribute very little fluorescence compared to the phenols. Thus, although this method is not an absolute measure of BP metabolism it essentially measures phenol formation. By this method 1 pmole of phenol product can be detected. Another assay utilizes a radioactive substrate. The reaction is stopped with 0.5 M NaOH in 80% ethanol and then the mixture is extracted with hexane. After this extraction approximately 99.5% of the remaining substrate is in the upper phase while about 94% of the products formed remain in the lower phase and can be easily measured by scintillation counting. It is surprising that the radioactive assay is less sensitive than the fluorescence method. It generally requires between 40 and 100 pmole phenol products for analysis.

3. *Cytochrome P-450*

The critical first step in the metabolism of PAHs is an oxidative reaction involving an electron transport system in which cytochrome P-450 serves as the oxidation–reduction component operational in bringing together substrate, oxygen, and reducing equivalents.

In the monooxygenation reactions, one atom of molecular oxygen is reduced to water while the other is incorporated into the substrate. Since the electrons involved in the reduction of cytochrome P-450 are derived from NADPH, the overall reaction can be written in the following way (RH represents the substrate):

$$RH + O_2 + NADPH + H^+ \rightarrow ROH + H_2O + NADP^+$$

Cytochrome P-450 is a hemoprotein best characterized by the presence of a relatively intense absorbance at about 450 nm for the CO derivative of the reduced hemoprotein (Figure V-2).

a. MECHANISM OF INDUCTION

After exposure of mammalian tissues or cells in culture to certain chemicals, the enzyme activity of the AHH system increases and different forms of cytochrome P-450 are synthesized. This induction process may provide an important mechanism for detoxification of the chemicals on one hand, or in some cases it provides a pathway for formation of carcinogenic intermediates.

There are two major classes of compounds that induce different cytochromes P-450. Phenobarbital (PB) represents one class and PAHs such as 3-methylcholanthrene (3MC) the second class of compounds. Liver microsomes isolated from PB-treated animals differ with respect to substrate specificity and the absorption maxima of their reduced CO complex from microsomes prepared from 3MC-treated species. 3MC treatment induced synthesis of a new hemoprotein designated "cytochrome P-448" or "cytochrome P_1-450." When PB and 3MC are administered simultaneously to the animal the induction of both types of cytochromes is additive. Therefore, it is suggested that the synthesis of cytochrome P-450 and P-448 are under separate genetic control.

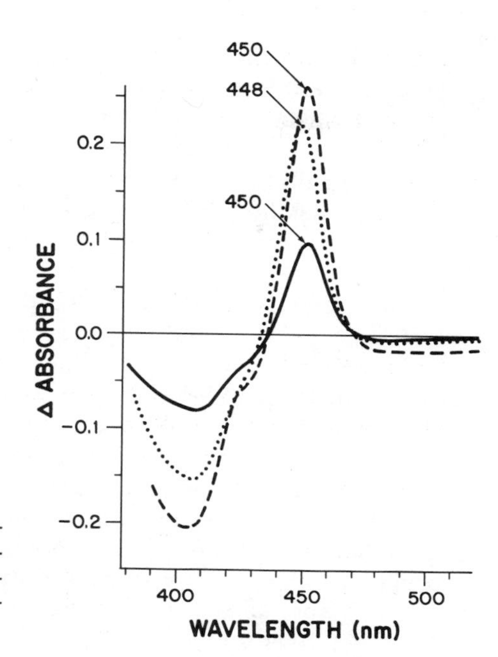

Figure V-2. Absorption spectra of the CO complex of reduced cytochrome P-450 from microsomes of untreated (——), 3-methylcholanthrene-treated (· · · · ·), and phenobarbital-treated rats (- - - -).

The existence of multiple forms of cytochrome P-450 that exhibit different, but often overlapping, substrate specificities provides a reasonable explanation for the remarkably broad substrate specificity of the liver monooxygenase system. Variations in the levels of the individual forms of the cytochromes can readily account for the differences in catalytic activity observed in microsomal preparations from animals of different age, sex, strain, and type of chemicals used for induction.

Similarly to the preferential induction of specific forms of cytochrome P-450 by various inducers, different forms of cytochromes are differentially inhibited. The 3MC-induced form of rat liver cytochrome P-450 is inhibited by 7,8-benzoflavone, but the phenobarbital (PB) induced or noninduced form is only weakly affected. Thus, induction or inhibition of the cytochrome system may alter the balance of the metabolic pathways of chemicals.

By the use of puromycin and actinomycin D, inhibitors of protein and RNA synthesis, it was established that induction of cytochrome P-450 requires continuous protein synthesis but RNA synthesis is required only during the initial 30–60 min of induction. Therefore, it is suggested that induction requires an initial phase of RNA but not protein synthesis and a later phase of protein but not RNA synthesis.

Cytochrome P-450 has been purified from a variety of sources and has been well characterized. By means of different chromatographic techniques, especially that of polyacrylamide gel electrophoresis in the presence of SDS, or by spectral analyses, peptide mapping, and amino acid sequence determinations, it is now clear that multiple forms of microsomal cytochrome P-

450 exist. Unfortunately, there is no unified, generally accepted nomenclature for the different forms of P-450. The terms P-450 and P-448 (or P_1-450), which refer to the absorption maximum of the reduced CO complex of the cytochrome and which have been used to specify cytochromes present in liver of untreated or PB-treated vs. 3MC-treated animals, are not sufficient since the absorption maxima of the induced forms are not all 450 or 448 nm and very often different forms of cytochrome P-450 have an identical absorption maximum.

Since there is no generally accepted nomenclature for different forms of cytochrome P-450, a session on P-450 nomenclature was held at the Third International Congress on the Biochemistry, Biophysics, and Regulation of Cytochrome P-450 in Saltsjöbaden, Sweden, June 1980. It was agreed that the term "cytochrome" should be dropped and that these monooxygenases appear to behave like isozymes with overlapping substrate specificities. The best terminology agreed upon was "polysubstrate monooxygenase," trivial name P-450. After a form of the enzyme has been well characterized (genetically, electrophoretically, and/or immunologically) in a given species, tissue, and subcellular fraction, a hyphen and number should be added. For example, forms 2, 4, and 3b of liver microsomal P-450, viz. LM_2, LM_4, LM_{3b}, should be called rabbit liver microsomal polysubstrate monooxygenases 2, 4, and 3b. P-450b and P-450c should be called rat liver microsomal polysubstrate monooxygenases 2 and 4, respectively, as they correspond approximately to the rabbit liver forms. Thus far this nomenclature has not been adopted by investigators in this field. Therefore, we will use in this chapter for the designation of multiple forms of cytochrome P-450 the terms used by the original investigators.

b. Rabbit Microsomal Cytochromes P-450

From rabbit liver and lung at least six different forms of cytochrome P-450 have been isolated. *Form 1* was highly purified from liver microsomes of 3MC 2,3,7,8-tetrachlorodibenzo *p*-dioxin (TCDD)-treated rabbits. A similar preparation termed LM_1 was obtained from liver microsomes of untreated rabbits. Antibodies produced against forms 2, 4, and 6 do not react with form 1. *Form 2* is the major species of cytochrome P-450 of PB-treated rabbits. It has been characterized clinically and immunologically. It is inactive in metabolizing PAHs and is termed LM_2. An immunologically indistinguishable form was also isolated from the lung microsomes of untreated rabbits. *Form 3* was obtained mainly from untreated rabbits and it is very active in catalyzing *N*-demethylation of aminopyrine but it is rather inactive in metabolism of BP. This cytochrome, termed LM_{3b}, has a subunit molecular weight of 52,000 and the absorption maximum of its reduced CO complex is 451 nm. *Form 4* is the major liver microsomal form induced by TCDD and β-naphthoflavone. It has been well characterized and has been called LM_4, LM_{4b}, *P-448*, and *P-448*$_1$. The absorption maximum of its reduced CO difference spectrum is 447 to 448 nm and the subunit molecular weight is between 51,000 and 55,000. Form 4 from TCDD- or β-naphthoflavone-treated rabbits, in contrast to cytochrome P-450 isolated from the microsomes of 3MC-treated rabbits, is rather inactive in metabolism of PAHs. *Form 6* is induced in rabbit liver by TCDD. It has a subunit molecular weight of 57,500 and absorption maximum of CO complex at 448 nm. It is about 100-fold more active in metabolizing BP than forms 2, 3, and 4. A distinct form of cytochrome P-450 isolated from untreated rabbits, called LM_7, is related to form 6. Several other cytochrome P-450 fractions have been isolated from rabbit liver microsomes but their relationship to the preceding five forms has not been determined.

c. Rat Microsomal Cytochromes P-450

From rat liver microsomes five to seven forms of cytochrome P-450 have been isolated and well characterized. The major PB-inducible form is also induced by Aroclor 1254. Subunit molecular weights range from 49,000 to 53,000 and the absorption maximum of its reduced CO spectrum is 450 to 451 nm. This form catalyzes the metabolism of such substrates as benzphetamine and it corresponds to P-450_b species. It is immunologically distinct from the major PB-inducible form in rabbit liver microsomes, form 2.

Another major form induced by 3MC and polychlorinated biphenyls such as Aroclor 1254, termed P-450_c or P-448, has a subunit molecular weight of 53,500–56,000 and an absorption maximum at 447 to 448 nm. It is one of the most efficient of the cytochromes P-450 in catalyzing the metabolism of a variety of PAHs. This form is immunologically distinct from the major 3MC-inducible form in rabbit liver microsomes, form 4.

A different species of cytochrome P-450 has been purified from PB-, 3MC-, and Aroclor-treated rats with a molecular weight of 48,000 and absorption maximum at 452 nm. It is termed P-450_a and it catalyzes the hydroxylation of testosterone. The partial amino acid sequence appears to be different from that of either the P-450_b or the P-450_c form. Isosafrole treatment of rats induced a species called P-450_d, which has been isolated as an isosafrole metabolic-cytochrome complex. This cytochrome P-450 shows partial identity with P-450_c but does not cross-react with antibodies prepared against cytochrome P-450_a or P-450_b. Peptide maps of proteolytic digests of cytochrome P-450_d indicate that its primary structure differs from that of P-450_a, P-450_b, and P-450_c.

Besides these well-characterized cytochromes, other P-450 fractions have been isolated from rat liver microsomes. However, it is not known which of these preparations correspond to any of the fractions described.

d. Mouse Microsomal Cytochromes P-450

There is now immunological evidence in mice for at least six different forms of P-450. Four forms of liver microsomal cytochrome P-450 from mouse have been purified. Two forms have been highly purified from PB-treated mice. The molecular weight of the A_2 form is 50,000 and of the C_2 form, 56,000. The absorption maximum of their reduced CO complexes is at 451 nm for A_2 and at 450 nm for C_2.

Two forms of cytochrome P-450 referred to as P-448 and P_1-450 have been purified by Nebert and co-workers from 3MC-treated C57BL/6N (B6) or other responsive mice strains. Although the subunit molecular weights of both forms are very similar, the absorption maximum of the CO complexes is at 449 nm for P_1-450 and at 448 nm for P-448. However, they differ in their peptide maps, immunological properties, and abilities to metabolize BP and acetanilide.

On the other hand, induction of P_1-450 is absent in liver and other tissues in the DBA/2N (D2) and other nonresponsive inbred mouse strains. This responsiveness to PAHs is associated with the *[Ah]* complex, which is viewed by Nebert as a combination of regulatory, structural, and probably temporal genes that may or may not be linked. thus, the *[Ah]* complex of the mouse regulates the PAH-dependent induction of numerous drug-metabolizing enzyme activities in the liver as well as other tissues. The induction of AHH activity and P_1-450 that occurs in 3MC-treated B6 and is absent or always much lower in 3MC-treated D2 is associated with the occurrence of a cytosolic receptor. This receptor, a product of the *[Ah]* regulatory gene, is capable of binding specifically to inducers such as 3MC, PB, and TCDD. The B6 mouse appears to have an at least 50-fold higher affinity toward inducers of P_1-450 than the D2 mouse. The formed inducer–receptor complex is further translocated into the nucleus. The mechanism of the expression of the complex in the nucleus is not well understood but it is supposed to be similar in many ways to the complexes between steroids and their cytosolic receptors. Thus, the response in the nucleus is transcription of specific mRNAs, translation of the mRNAs into specific enzymes such as P_1-450, and incorporation of P_1-450 into cellular membranes.

With the use of an antibody to P_1-450, at least two mRNA species associated with AHH induction have been shown to be greatly increased in 3MC-treated B6 inbred mice, but absent or present in low concentration in 3MC-treated *[Ah]* nonresponsive D2 and control untreated B6 and D2 mice. The mechanism of P_1-450 induction was directly determined by the use of a DNA clone with inserted sequence of the structural gene encoding 3MC-induced P_1-450. Similarly a complementary DNA of PB-inducible P-450 was cloned in *E. coli* pBR322. From the nucleotide sequence the entire amino acid sequence composed of 491 amino acids was determined.

The good correlation between the induction of mRNA and P_1-450 strongly infers that the induction process is under transcriptional control. It is, however, not clear whether the process reflects an increased rate of transcription or a decreased rate of mRNA degradation. In any case, the level of mRNA appears to be closely related to the level of induced AHH activity.

e. HUMAN LIVER MICROSOMAL CYTOCHROMES P-450

Several laboratories have recently reported on the purification of cytochrome P-450 from human liver. The catalytic activity of these preparations are generally low. Column chromatography has allowed separation of several P-450 fractions, which suggests that human liver also contains multiple forms of cytochrome P-450. However, fractions isolated from different individuals have significantly different subunit molecular weights. Human and rat cytochrome P-450 are different proteins as assayed immunologically but human cytochromes P-450 are immunologically more similar to the major PB-inducible form of rat cytochrome P-450 than to the major 3MC-inducible form. Thus, it is clear that human cytochrome P-450 forms a family of isoenzymes that are distinct from the cytochromes P-450 isolated from experimental animals.

4. *NADPH-Cytochrome P-450 Reductase*

Electrons from NADPH involved in the reduction of cytochrome P-450 are transferred to cytochrome P-450 by a flavoprotein enzyme known as NADPH-cytochrome P-450 reductase. The use of various detergents to disrupt the membrane fractions of microsomes coupled with high-specific-affinity chromatographic methods has permitted the isolation of this enzyme from rat, rabbit, steer, and human hepatic microsomal fractions. The molecular weight of rabbit and rat enzymes is around 75,000. In contrast to cytochrome P-450, only one form of the reductase exists in rabbit liver microsomes independently of the inducers used. Rabbit and rat reductases have a similar amino acid composition. Both enzymes form high-molecular-weight aggregates in detergent-free aqueous solutions. This is a consequence of the amphiphilic nature of the protein, which contains a hydrophobic and a hydrophilic domain. The hydrophobic region, which is responsible for binding to the membrane and for the interaction with cytochrome P-450, is situated at the amino-terminal of the protein. NADPH-cytochrome P-450 reductase contains one molecule of flavin mononucleotide (FMN) and one molecule of flavin-adenine dinucleotide (FAD) per enzyme polypeptide. The best evidence that this reductase is an essential component of the monooxygenase system comes from experiments showing that antibodies to reductase inhibited the enzyme complex activity.

5. *Epoxide Hydrolase*

a. DISTRIBUTION

Many chemical carcinogens, including PAHs, are metabolized by MFO to very reactive epoxides that can then react chemically with nucleophilic moieties of cellular macromolecules, or can be further metabolized by epoxide hydrolase. Epoxide hydrolase (EH) is the name recommended most recently by the Nomenclature Committee of the International Union of Biochemistry. It has also been called epoxide hydrase and epoxide hydratase.

EH is present in all organs and tissues of mammals. The levels of activity are generally low in lower vertebrates, with increased activity seen in rodents and higher mammals. The organ distribution of EH is broad. The organ distribution in different species shows some quantitative differences. The highest activity was detected in liver, followed by testis, kidney, lung, and other organs.

The highest specific activity of EH is found in the endoplasmic reticulum (microsomes) fraction where the enzyme is localized to function in coordination with the P-450 monooxygenase system. The enzyme is also found in nuclear membrane. Kinetic and immunological criteria have shown that nuclear EH is probably identical to microsomal EH. A cytoplasmic hydrolase also exists but its substrate specificity is different from the microsomal enzyme. The physical properties of the cytoplasmic and membrane-bound enzymes, including molecular weights and behavior during gel electrophoresis, suggest that they are distinct proteins.

The difficulties encountered in the isolation of purified membrane-bound EH are due to its hydrophobic character. By using detergents for solubilizing EH from rat and human liver microsomes and by further chromatographic purifications, an enzyme preparation was obtained that was homogeneous by gel electrophoretical, analytical, and immunological criteria. The pure enzyme has a molecular weight of 44,000 and contains a high percentage (56%) of hydrophobic amino acid residues. The homogeneous enzyme has a broad specificity for substrate epoxides. Despite the similarities the human and rat enzymes represent different proteins as judged by their immunochemical properties as well as their selective catalytic activities toward certain substrates.

b. Assay of Epoxide Hydrolase

The EH assay is based on the use of [^{3}H]styrene oxide, [^{3}H]benzo[*a*]pyrene-4,5-oxide, or 3-methylcholanthrene-11,12-oxide as substrate. The products of the reactions are extracted into ethyl acetate and measured by scintillation counting It seems that all these substrates are metabolized by the same enzyme. Antibodies raised in rabbits against pure rat liver EH precipitated the enzyme for both substrates, styrene oxide and BP-4,5-oxide.

c. Regulation of Epoxide Hydrolase Induction

The regulation of EH levels and that of certain P-450s, although similar in many aspects, does not correspond exactly. Rat liver microsomal EH is very low during the prenatal period and then increases, reaching a maximum at about 50 days of age. PB, which induces some forms of P-450, also induces rat hepatic EH two- to threefold. On the other hand, induction of EH by 3MC is only very slight. It thus appears that EH and cytochrome P-450 are not under common control and the induction of EH is not associated with *[Ah]* locus.

The attempt to find compounds that selectively increase EH activity with little or no increase in the activities of other microsomal enzymes led to the discovery of *trans*-stilbene oxide. This compound is an effective inducer of EH, inducing the enzyme three- to fourfold with little or no effect on monooxygeneses. This selective induction is seen only in the liver of rats and hamsters but not in mice. Kidney EH is induced in the rat by *trans*-stilbene oxide but not in the mouse or hamster. Treatment of rats with this inducer results not only in higher enzyme activity but also in greater amounts of enzyme protein. Inhibitors of RNA and protein synthesis such as actinomycin D and cycloheximide prevent the induction of EH activity.

The existence of multiple forms of cytochrome P-450 would imply that various forms of EH would also exist. Although three separate purified EH fractions were recently isolated from rat liver, the physical and biochemical differences among these distinct forms appear to be less significant than those among the multiple forms of cytochrome P-450.

d. MECHANISM OF EPOXIDE HYDROLASE ACTION

EH plays an important role in the formation of vicinal diol epoxides, which are the ultimate carcinogenic forms of PAHs. Figure V-3 shows the formation of arene oxides by monooxygenation on a benzo ring and its further hydration by EH. The product of this reaction is almost exclusively *trans*-dihydrodiol. Thus, the enzyme activity is highly stereoselective, resulting in high optical purity of products. The dihydrodiol is then the substrate for a second monooxygenation on the same ring. This reaction yields two isomers, the *anti*- and *syn*-dihydrodiol epoxides. The resultant vicinal diol epoxide is either rapidly detoxified to a tetrol or can bind covalently to nucleophilic sites on the cellular macromolecules. Thus, EH plays a dual role for compounds possessing an angular polycyclic structure: either inactivating epoxides or producing precursor molecules for the formation of the dihydrodiol epoxides. This can be illustrated with BP and it will be discussed in more detail in the section on BP metabolism. With other aromatic or olefinic compounds that do not contain such angular polycyclic structure, EH plays only a simple detoxifying role.

Simple arene oxides are good substrates for EH but the diol epoxides appear to be poor substrates. Nonenzymatic hydration of diol epoxides to tetrols proceeds rapidly in aqueous media in the absence of EH. Since the presence of high levels of pure EH has little influence on the mutagenicity of diol epoxides, this indirectly indicates that EH has little activity toward diol

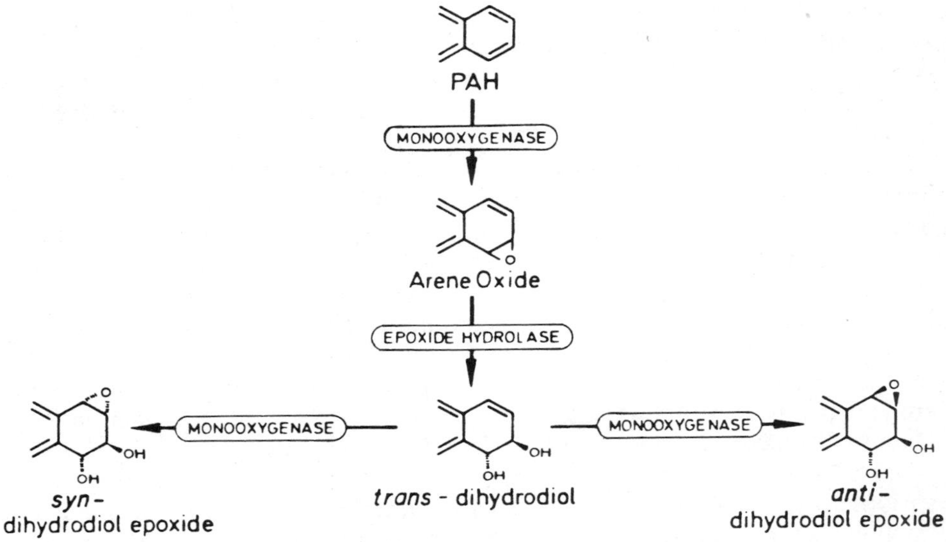

Figure V-3. Scheme of oxidation of polycyclic aromatic hydrocarbons (PAH) to arene oxide and its hydration by epoxide hydrolase to *trans*-dihydrodiol.

epoxides. It could be concluded that the specificity of EH is such that epoxides on saturated benzo rings carrying hydroxyl groups vicinal to the epoxide ring are no longer good substrates for this enzyme.

6. Conjugation Enzymes

Compounds containing functional groups such as hydroxyl, amino, carboxyl, epoxide, or halogen, as well as metabolites of phase I reactions, can undergo reactions referred to as conjugations, or phase II reactions. In these conjugation reactions the metabolite is coupled with a readily available endogenous agent derived from a carbohydrate or protein. The products of the conjugating enzymes are usually more polar and less lipid-soluble and are readily excreted from the organism. Thus, the important role of these enzymes is in metabolic detoxification of foreign compounds.

a. Glutathione-S-Transferase

Detoxification of reactive intermediates of PAH oxides proceeds in two ways. One involves hydration by microsomal EH as described previously. The other is through conjugation of compounds possessing electrophilic centers with the reduced form of glutathione (GSH) catalyzed by glutathione-S-transferase. These enzymes are multifunctional proteins, and they catalyze conjugation reactions between GSH and a very wide range of second substrates. These second substrates are usually hydrophobic electrophiles. Their conjugation with GSH increases their solubility in water. Furthermore, by this reaction the chemical reactivity of the hydrophobic substrate is decreased. Another function of the GSH-S-transferase proteins in detoxification is related to their ability to act as trapping agents for strong electrophiles such as polycyclic hydrocarbon derivatives. One example is the removal of the *trans*-7,8-diol of BP by conjugation to water-soluble products, thereby avoiding its further oxidation to active 7,8-diol-9,10-epoxide-BP. GSH-S-transferase is also involved in conversion of *anti*-7,8-diol-9,10-oxide-BP and of *anti*-8,9-diol-10,11-oxide of benz[*a*]anthracene into a GSH conjugate in the presence of rat liver supernatant fraction and GSH. This enzyme may also exert some indirect control over the extent of binding of PAH metabolites to DNA.

The GSH moiety from conjugates with PAHs has not been detected in urine because the conjugate undergoes mercapturic acid biosynthesis. The mercapturic acid conjugate is formed after digestion of the glutamyl moiety of the GSH conjugate with γ-glutamyltranspeptidase (Figure V-4). This is followed by removal of glycine with cysteinylglycinase and acetylation with *N*-acetylase. Most of the GSH conjugates are preferentially excreted through the bile, whereas the mercapturic acid is formed in urine. The GSH-S-transferases are distributed quite widely in nature. They have been detected in almost all species examined, ranging from bacteria to human tissues. The

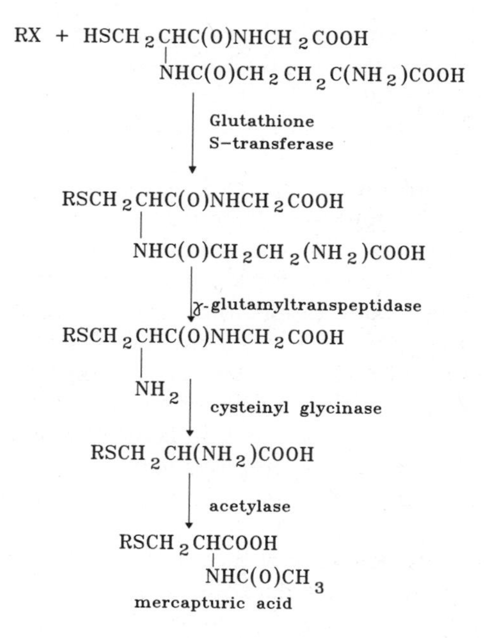

Figure V-4. Metabolic degradation of glutathione conjugates to mercapturic acid. R = polycyclic aromatic hydrocarbon. Based on Boyland *et al.* (1961) *Biochem. J.* **78**, 376.

highest concentration of the enzymes is found in the soluble supernatant fraction of the liver. The GSH-S-transferase activity detected in the endoplasmic reticulum is distinct from that of the cytoplasm. The enzyme activities in nuclei and mitochondria are low.

All cytosolic GSH-S-transferases have similar overall structures with molecular weights of 40,000–50,000. Each enzyme is composed of two subunits that are classified according to their apparent size in SDS/polyacrylamide gel electrophoresis. The subunits are called Ya, Yb, and Yc (or a, b, c) and have molecular weights of approximately 23,000, 23,500, and 25,000, respectively. Chromatography of rat liver cytosol on a DEAE-cellulose column, followed by a CM-cellulose column, separated six peaks upon elution with a gradient of increasing salt concentration. The peaks are designated as AA, A, B, C, D, and E. The enzyme B is composed of subunits a + a and a + c. Transferase AA has a subunit composition c + c, whereas transferase A, C, D, and E are all composed of b subunits. The difference between these enzymes are presumably due to differences in primary structures. A close relationship appears to exist between B pair and AA transferases. There is immunological cross-reactivity between A and C enzymes, but the tryptic map for A is quite different from that of the B enzyme.

GSH-S-transferase activity was also detected in rat liver microsomes. The activity of this enzyme, in contrast to the cytosolic forms of the enzyme, can be stimulated severalfold by treatment with sulfhydryl reagents *in vitro*. The enzyme is a component of the endoplasmic reticulum.

The microsomal enzyme solubilized with detergents was purified by chromatography on hydroxyapatite and CM-Sepharose. The purified protein has an apparent molecular weight of 14,000 by SDS gel electrophoresis. Antibodies against soluble GSH-S-transferases A, B, and C do not react with the purified microsomal enzyme.

The microsomal enzyme might have a special role in detoxification, since the cytochrome P-450 system, which produces many reactive mutagenic and carcinogenic intermediates, is localized in the same membrane.

The activity of GSH-S-transferase has been measured using K-region epoxides such as BP-4,5-oxide as substrate. The difficulty in this assay system, based on the measuring of the disappearance of GSH is in the autooxidation of GSH. Small changes in GSH can be determined by reaction with iodate. Radiometric analysis using labeled GSH or substrates are more sensitive. The products are separated by paper or thin-layer chromatography. The other method using labeled substrates employs extraction of unreacted substrates by organic solvents and measurement of the radioactivity remaining in the aqueous phase.

When specific activities of GSH-S-transferase were determined for epoxides of phenanthrene, benz[*a*]anthracene, mono- and 7,12-dimethylbenz[*a*]anthracene, and 3MC, it was concluded that the specific activities with the K-region were high, whereas non-K-region epoxides were poor substrates.

The binding of BP-7,8-diol to DNA is inhibited to a small extent by GSH alone and to a much greater extent by GSH and purified cytosolic GSH-transferases B and E. GSH-transferases A and C are less active. Inhibition of binding of BP-7,8-diol metabolites to DNA catalyzed by GSH-transferases is associated with the formation of GSH conjugates. Since the extent of inhibition of binding is similar with nucleic acid alone, nuclei, and rat liver microsomes, or with calf thymus DNA and rat liver microsomes, this indicates that reactive metabolites of BP-7,8-diol formed either by nuclei or by microsomes are readily accessible to soluble GSH-transferases. The binding of reactive metabolites from 9-hydroxy BP is also inhibited by GSH and cytosol. All of the isolated enzymes are effective with the K-region epoxide of BP. Human liver enzyme also catalyzes reaction between BP-4,5-oxide and GSH. Thus, in the hepatocyte, GSH and GSH-transferases may be important in protecting DNA from electrophilic attack by reactive BP metabolites.

The level of GSH-S-transferase activity is influenced by drug treatment. There is, however, diversity of responses of the enzyme activity to drug application among animal species. For example, rat liver GSH-S-transferase activity was increased by 50% of the control after application of PB, BP, and 3MC. On the other hand, 3MC treatment was without any effect on mouse hepatic enzyme activity. Single injection of PB increased GSH-transferase B mRNA levels after 16–24 hr. Interestingly, the synthesis of the two subunits of GSH-transferase B is directed by different mRNAs. Thus, the mRNA specific for the low-molecular-weight subunit was elevated markedly by PB administration whereas the mRNA specific for the high-molecular-weight subunit was only slightly increased.

The most efficient inducers of the enzyme are antioxidants such as butylated hydroxyanisol (BHA) or butylated hydroxytoluene (BHT), which increase the basal enzyme level in mouse liver up to 11-fold.

b. SULFOTRANSFERASE

This enzyme plays an important role in detoxification. It is also localized in the cytosol fraction. The transferase esterifies a wide variety of compounds such as steroid, carbohydrates, and alkyl and aryl compounds to form conjugates with sulfate. Sulfate ion is activated by cytosol enzymes to form 3′-phosphoadenosine-5′-phosphosulfate (PAPS) in the presence of ATP and then transferred to phenolic substrates according to the following scheme:

$$SO_4^{2-} + \text{ATP} \xrightarrow[\text{ATP sulfurylase}]{Mg^{2+}} \text{adenosine-5}'\text{-phosphosulfate} + \text{PP}$$

$$\text{Adenosine-5}'\text{-phosphosulfate} + \text{ATP} \xrightarrow[\text{adenosine-5}'\text{-phosphosulfate kinase}]{Mg^{2+}} \text{PAPS} + \text{ADP}$$

$$\text{PAPS} + \text{phenol} \rightarrow \text{phenyl sulfate} + \text{ADP}$$

The activity of the sulfotransferase is determined mainly by adding $Na_2{}^{35}SO_4$ to the reaction mixture as a sulfate donor. Because of the high reactivity of the intermediate PAPS the characterization of the sulfotransferase is more difficult. Changes in enzymatic activity after administration of chemicals have not been detected. In PAH metabolism, sulfate conjugates of oxygenated BP metabolites were bound as major polar metabolites in several lines of cells in tissue culture. In rat hepatic cytosol, sulfate is transferred to all phenol and quinone metabolites of BP. The enzyme has lower activity with dihydrodiols. Although there are a number of sulfotransferases, none has been obtained in a pure state.

c. UDP-Glucuronyl Transferase

Conjugation of substrates with glucuronic acid requires two initial reactions by cytosol enzymes. The first is the formation of an activated intermediate, UDG-glucose (UDPG) from glucose-1-phosphate, followed by formation of UDP-glucuronic acid (UDPGA) according to the scheme:

$$\text{Glucose-1-phosphate} + \text{UTP} \xrightarrow[\text{pyrophosphorylase}]{\text{UDP-glucose}} \text{UDPG} + \text{PP}$$

$$\text{UDPG} + 2\text{NAD} \xrightarrow[\text{UDPG dehydrogenase}]{H_2O} \text{UDPGA} + 2\text{NADH}$$

The cosubstrates for the conjugation catalyzed by UDP-glucuronyl transferase (GT) are mainly hydroxylated compounds, but in some cases amino- and SH-compounds are the functional groups.

GT probably consists of a family of closely related but functionally different forms. These different forms (GT_1, GT_2, etc.) represent the functional heterogeneity of the enzyme for various groups of inducers upon induction, during enzyme purification, or during development. The enzyme forms do not distinguish the various functional groups (-OH, -COOH, $-NH_2$, etc.), but rather the different molecular configurations of the substrates. The enzyme, in contrast to other conjugation enzymes, is located in endoplasmic reticular membranes and is found in latent or activated states. It can be activated either by UDP-*N*-acetylglucosamine or by various alterations of the microsomal membrane structure.

GT is distributed in various tissues such as liver, lung, kidney, stomach, and skin, but has not been observed in brain and spleen. The enzymatic activity of GT can be monitored using UV spectroscopy to measure the decrease in the absorption or by fluorescence of a substrate such as 3-hydroxy-BP. If either UDPGA or the cosubstrate is radioactively labeled, the product can be isolated by chromatography and the radioactivity of the product determined. The different enzyme forms, GT_1 and GT_2, are characterized by differential induction of groups of conjugation reactions by 3MC and PB, respectively. The separation of these forms was achieved by DEAE-cellulose chromatography to apparent homogeneity. The 3MC-inducible GT_1 has high activity toward aromatic compounds, such as BP and its metabolites, e.g., 3-hydroxy-BP and BP-3,6-quinone. It shows relatively low activity toward BP-7,8-dihydrodiol.

In C57BL/6 mice, a "responsive strain" for the induction of AHH activity, GT was also induced by 3MC. By contrast, in DBA/2 mice neither AHH nor the transferase activity was influenced by 3MC. This suggests that induction of AHH and the UDP-GT is genetically linked. The coordinate induction of AHH and GT may greatly facilitate the inactivation of aromatic carcinogens.

d. RELATIONSHIP BETWEEN CONJUGATION REACTIONS AND CARCINOGENICITY

Evaluation of the relationship of detoxifying enzymes and carcinogenicity is difficult. The reason is that treatment of animals with chemicals such as PB or 3MC increases the enzymatic activity for later activation and detoxification. Investigations by Wattenberg and his co-workers showed that antioxidants such as 2-*tert*-butyl-4-hydroxyanisole (BHA) inhibited BP-induced neoplasia of the mouse forestomach. The same compounds increased the GSH-S-transferase activity and the acid-soluble sulfhydryl levels in the forestomach by 78–183%. These relationships would indicate that enhancement of this enzyme activity is associated with a reduced carcinogenic response of the forestomach to BP. Antioxidants also elevate the level of UDP-GT. It is possible that antioxidants inhibit carcinogenesis by enhancing the activity of conjugating enzymes that catalyze the binding of a wide variety of ultimate electrophilic forms of carcinogens to sulfhydryl groups of GSH or UDP-glucuronyl residues and in this manner may prevent carcinogens from reaching, or reacting with, critical target sites in DNA.

A number of naturally occurring organic compounds having the capacity to inhibit the neoplastic effects of chemical carcinogens have also been identified. Organic isothiocyanates are constituents of cruciferous vegetables such as brussels sprouts, cabbage, cauliflower, and broccoli and as such they are likely to inhibit PAH carcinogen-induced neoplasia. Although the mechanism of inhibition by this class of compounds is not established, it is suggested that they increase the activity of detoxifying conjugating enzymes. On the other hand, the possibility remains that the conjugated metabolites are transported to target tissues, cleaved by digestive or other enzymes and then become available for further metabolic activation, thus actually enhancing the carcinogenicity of these compounds. Another possibility is that the antioxidants will potentiate the activity of microsomal monooxygenases and by this mechanism increase the neoplastic response.

The diversity of enzyme inductions is similarly reported for other compounds such as TCDD and β-naphthoflavone. TCDD increased the activity of major detoxification enzymes including UDP-GT, GSH-S-transferase and EH. In the process of induction, TCDD binds to a cytoplasmic receptor. The

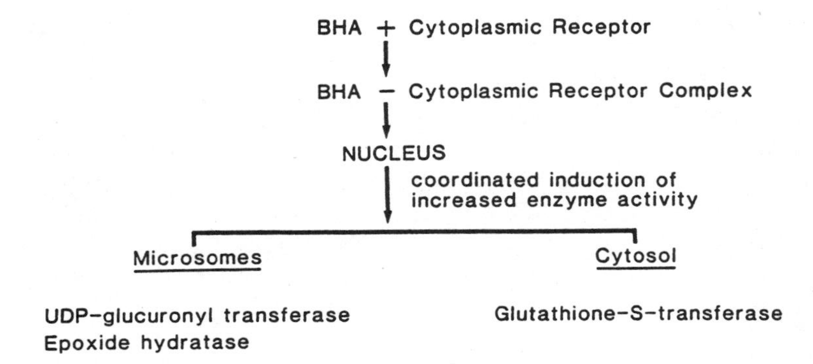

Figure V-5. Scheme of antioxidant action. Based on Watenberg (1980) *J. Environ. Pathol. Toxicol.* **3**, 35.

resulting complex enters the nucleus and causes the induction of enzymes by a mechanism similar to that of steroids. Although no cytoplasmic receptors have thus far been detected for antioxidants, it is possible that they act by a similar series of events as TCDD. A possible scheme of these reactions is shown in Figure V-5.

Considerations of the optimal role that antioxidants and other naturally occurring inhibitors might play are especially complicated by the multifunctional nature of these compounds. Thus, inhibitors that alter the microsomal monooxygenase system and/or the conjugating enzymes may increase detoxification of some carcinogens but also enhance the neoplastic response under other conditions. Therefore, more research is required before any recommendation can be made concerning the use of some of these compounds as inhibitors of carcinogenesis.

B. Specific Metabolites of Polycyclic Aromatic Hydrocarbons

1. *Benzo[a]pyrene*

The application of HPLC for the separation of BP metabolites was a major breakthrough in the determination of BP metabolism. When BP is incubated with cells or tissue preparations, various metabolites are formed. The overall metabolism of BP presented in Figure V-6 represents HPLC separation of BP derivatives.

The primary metabolites of BP are three epoxides: the 4,5-epoxide, the 7,8-epoxide, and the 9,10-epoxide. In addition, five phenols have been isolated as metabolites: 1-OH, 3-OH, 6-OH, 7-OH, and 9-OH. The phenols are converted to quinones; those isolated are the 1,6-, 3,6-, and 6,12-quinones. The primary epoxides can be conjugated to GSH-S conjugates catalyzed by GSH-S-transferase. The phenols and quinones are excreted principally as their more

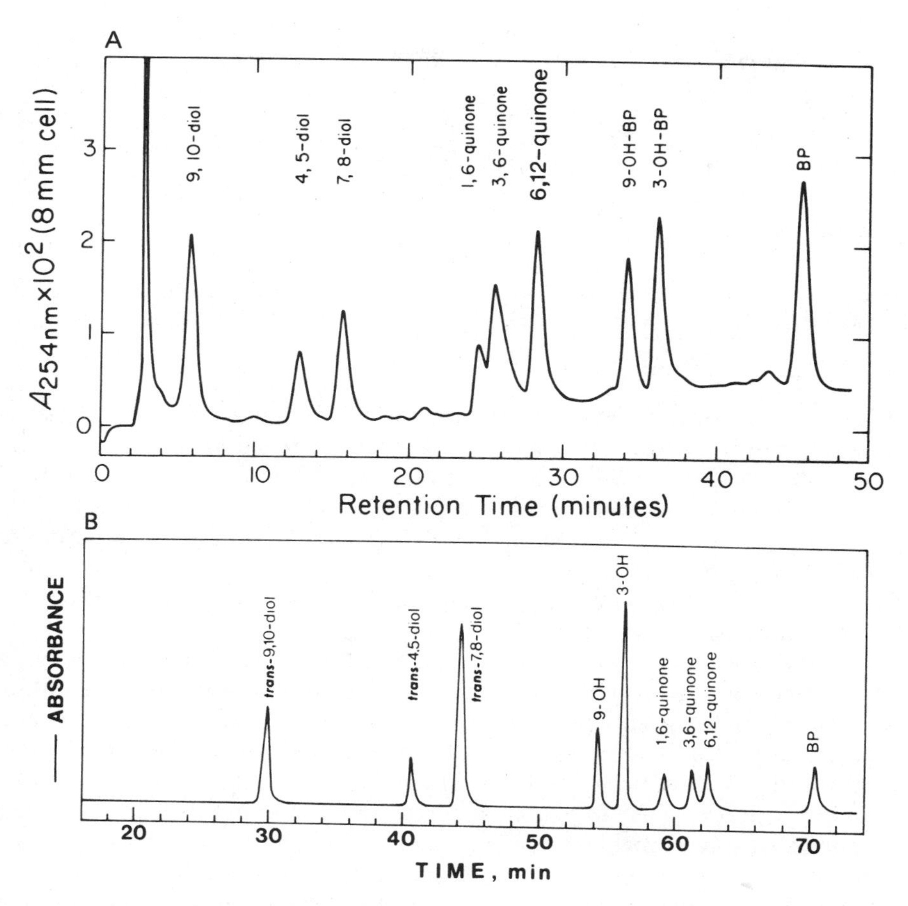

Figure V-6. HPLC separation of BP derivatives. (A) With a 2.1-mm by 1-m DuPont Permaphase-ODS column; (B) with a 6.2-mm by 25-cm DuPont Zorbax-ODS column, eluted with a 20-min linear gradient from 60% methanol to 100% methanol at a solvent flow rate of 0.8 ml/min at ambient temperature. From Gelboin (1980) *Physiol. Rev.* **60,** 1107.

water-soluble glucuronic and/or sulfate esters catalyzed by UDP-GT and/or sulfotransferase, respectively. The three epoxides are converted by the action of EH into 4,5-diol, 7,8-diol, and 9,10-diol.

The evidence for the mechanism of dihydrodiol formation by EH from the diol epoxides is that the EH inhibitor 1,2-epoxy-3,3,3-trichloropropane (TCPO) completely eliminates dihydrodiol formation. The readdition of EH results in the formation of the diols.

Although all three BP dihydrodiols can exist in *cis* and *trans* configurational isomers, the 7,8-diols formed enzymatically are *trans* and are (−) enantiomers of high optical activity. Thus, the epoxidation of BP by AHH followed by EH-catalyzed hydration is highly stereoselective, resulting in pure

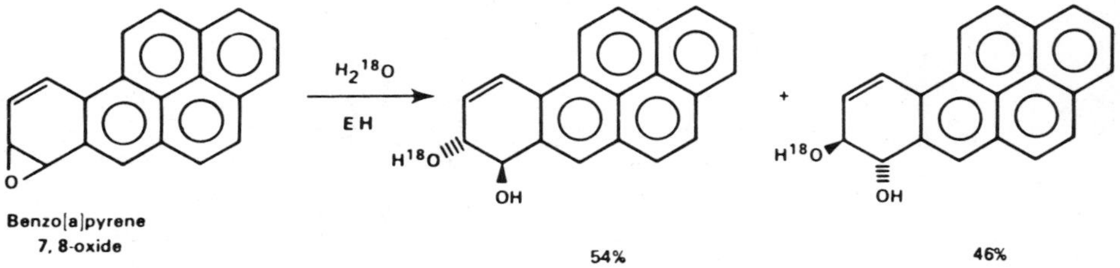

Figure V-7. Formation of ^{18}O-labeled BP-*trans*-7,8-diol from racemic BP-7,8-oxide by epoxide hydrolase (EH). Based on Thakker *et al.* (1977) *J. Biol. Chem.* **252,** 6328.

optical isomers of the *trans* configuration. The stereoselectivity of EH-catalyzed hydration was proven by the use of H_2O in reaction with racemic BP-7,8-epoxide. It was found by mass spectral analysis of the hydration products that the ^{18}O was exclusively at the C-8 position in both the (+)- and the (−)-*trans*-7,8-diol (Figure V-7). Thus, both 7,8-epoxide enantiomers are cleaved by EH at the C-8–0 bond to form the *trans*-7,8-diol, indicating that each of the 7,8-epoxide enantiomers is stereospecifically hydrated to a specific *trans*-7,8-diol enantiomer (Figure V-8).

After the observation that the 7,8-diol metabolite binds covalently to DNA in the presence of microsomes to a 10-fold greater extent than BP, attention was shifted to the characterization of the formed metabolites. Chromatographic and fluorescence spectral evidence indicated that the reactive metabolite bound to DNA was 7,8-diol-9,10-epoxide. The high tumorigenic and mutagenic activity of the epoxide suggested the possibility that diol epoxides are the ultimate carcinogenic metabolites of PAHs. Since 7,8-diol formed from BP by liver microsomes is the *trans* isomer, the diol epoxides derived from this diol exist as a pair of diastereomers in which the benzylic hydroxyl group

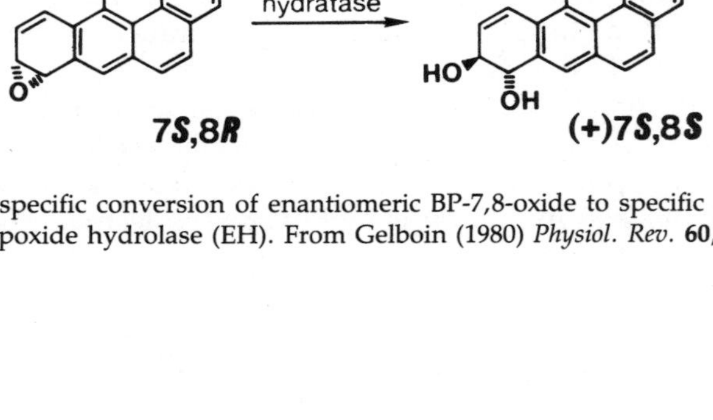

Figure V-8. Stereospecific conversion of enantiomeric BP-7,8-oxide to specific enantiomeric BP-*trans*-7,8-diol by epoxide hydrolase (EH). From Gelboin (1980) *Physiol. Rev.* **60,** 1107.

is either *trans* (BPDE I) or *cis* (BPDE II) to the epoxide oxygen. Since each diastereomer can be resolved into a pair of enantiomers, a total of four isomers may be formed by mammalian enzymes (Figure V-9). The (−)-*trans*-7,8-diol is oxygenated by liver microsomes highly stereoselectively at the 9,10 double bond at *trans* position to the 7-OH, forming mainly (+)-*anti* isomer or 7,8-diol-9,10-epoxide (BPDE I). The (−)-*trans*-7,8-diol is also converted to a smaller extent at the site of the 9,10 double bond, which is *cis* to the 7-OH that forms (−)-*syn* isomer or diol epoxide II (BPDE II). On the other hand, the (+)-*trans*-7,8-diol enantiomer yielded mainly (+)-*syn* diol epoxide and less (−)-*anti* diol epoxide. For example, liver microsomes from 3MC-treated rats converted (−)-*trans* diol into *anti*- and *syn* diol (−) epoxides in a ratio of 6 : 1 while the (+)-*trans* enantiomer yielded these diol epoxides in the ratio 1 : 20.

The unique structural feature of the diol epoxide appears to be that the epoxide is on a saturated angular benzo ring and that it forms part of a "bay region" of the polycyclic hydrocarbons. The simplest example of the "bay

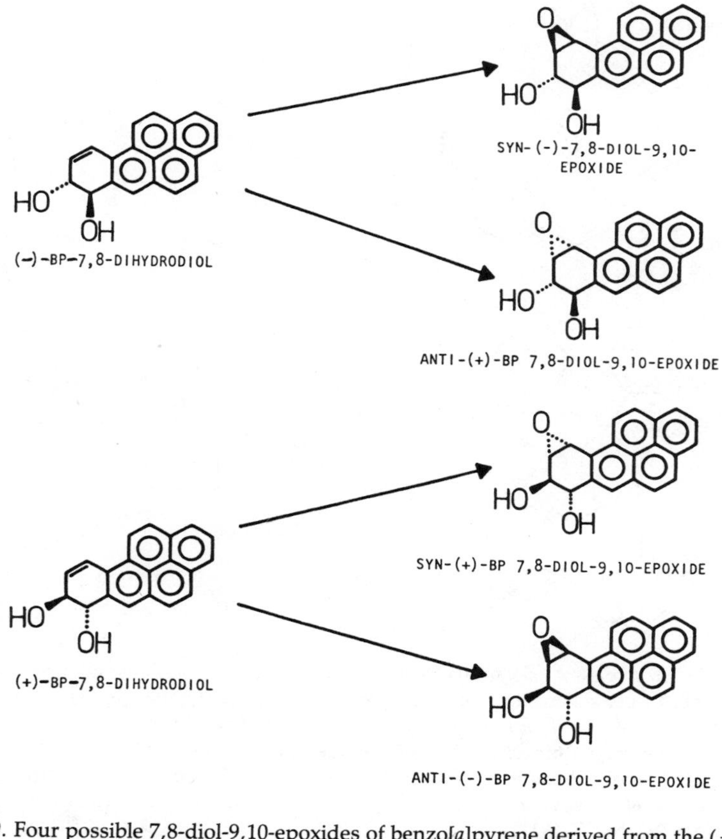

(−)-BP-7,8-DIHYDRODIOL

SYN-(−)-7,8-DIOL-9,10-EPOXIDE

ANTI-(+)-BP 7,8-DIOL-9,10-EPOXIDE

(+)-BP-7,8-DIHYDRODIOL

SYN-(+)-BP 7,8-DIOL-9,10-EPOXIDE

ANTI-(−)-BP 7,8-DIOL-9,10-EPOXIDE

Figure V-9. Four possible 7,8-diol-9,10-epoxides of benzo[*a*]pyrene derived from the (+) and (−) isomers of BP-7,8-dihydrodiol.

region" in a PAH is the sterically hindered region between the 4 and 5 positions in phenanthrene. Examples of "bay region" of various PAHs are shown in Figure V-10A and of dihydrodiol epoxide derivatives in Figure V-10B. The key chemical feature of such epoxide is the ease of carbonium ion formation, which is highly reactive and susceptible to attack by nucleophiles (Figure V-

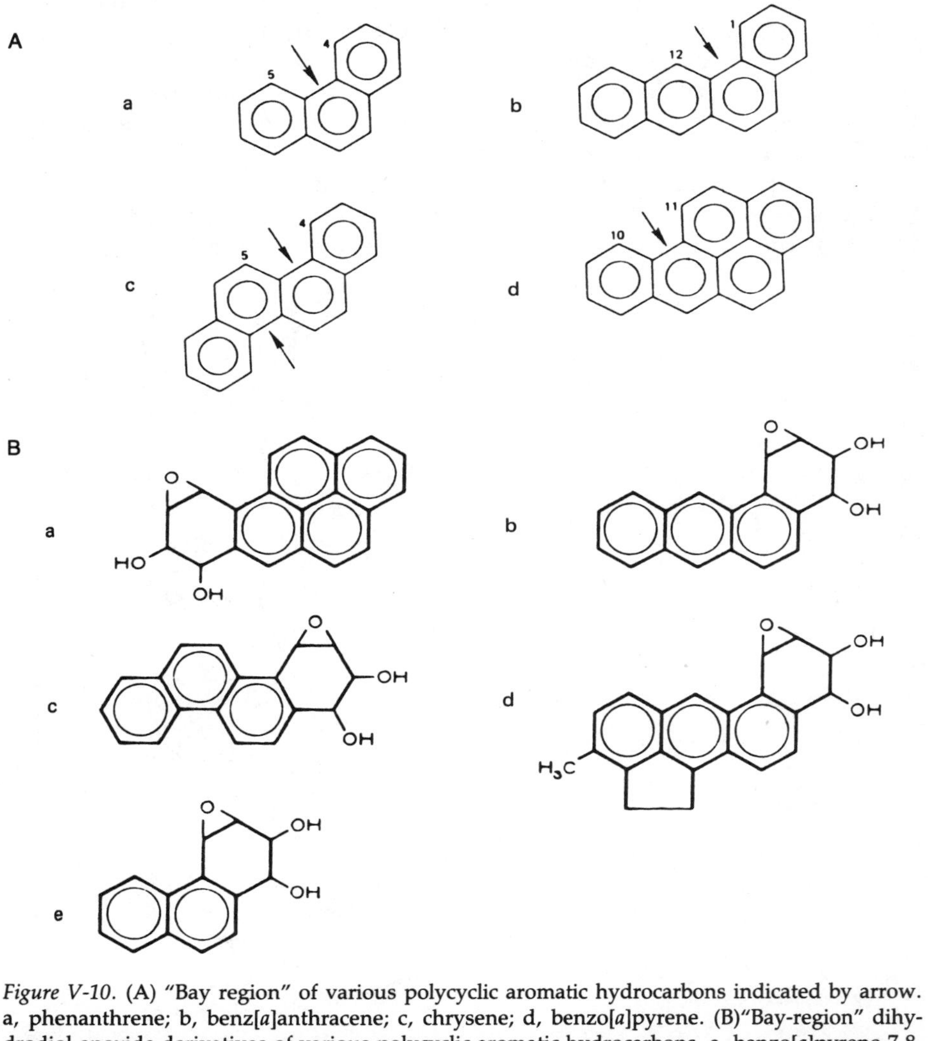

Figure V-10. (A) "Bay region" of various polycyclic aromatic hydrocarbons indicated by arrow. a, phenanthrene; b, benz[*a*]anthracene; c, chrysene; d, benzo[*a*]pyrene. (B)"Bay-region" dihydrodiol epoxide derivatives of various polycyclic aromatic hydrocarbons. a, benzo[*a*]pyrene-7,8-diol-9,10-oxide; b, benz[*a*]anthracene-3,4-diol-1,2-oxide; c, chrysene-1,2-diol-3,4-oxide; d, 3-methylcholanthrene-9,10-diol-7,8-oxide; e, phenanthrene-1,2-diol-3,4-oxide. Adapted from Jerina *et al.* (1977) In *Origins of Human Cancer* (Hiatt, Watson, and Winstein, eds.), Cold Spring Harbor Laboratory, pp. 639–658.

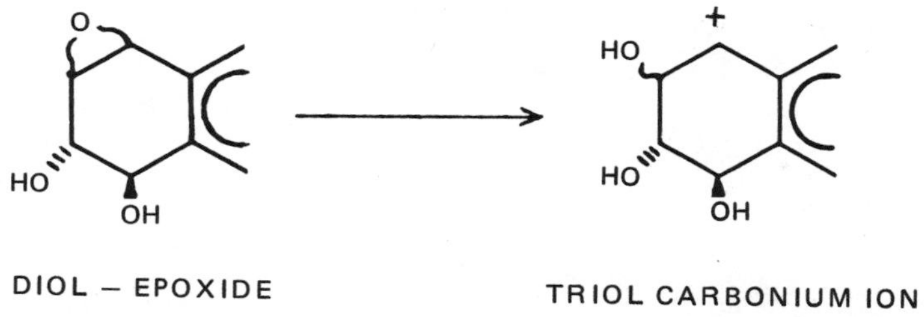

DIOL – EPOXIDE TRIOL CARBONIUM ION

Figure V-11. Formation of carbonium ion from diol epoxide.

11). The high chemical reactivity of "bay-region" diol epoxides can be attributed to their unique electronic properties. Perturbational molecular–orbital calculations performed by Jerina indicate that epoxides on saturated benzo rings that form part of the "bay region" of a hydrocarbon undergo ring opening to a carbonium ion much more easily than the non-bay-region epoxides.

The exceptionally high potency of BP-diol epoxide as a mutagen in bacterial and mammalian systems is in good accord with the postulate that the "bay-region" diol epoxides would be good candidates for ultimate carcinogenic forms of the parent PAHs. The mutagenicity and carcinogenicity of "bay-region" diol epoxides of other polycyclic hydrocarbons such as DMBA, benz[*a*]anthracene, chrysene, and 3MC are also several times higher than their non-bay-region counterparts.

It is also possible that the importance of the "bay region" lies in the fact that it sterically hinders enzymatic detoxification of the epoxide function in this region. Both *anti-* and *syn*-BPDE proved to be resistant to EH-catalyzed hydration (see Section 5), a factor potentially significant with respect to their carcinogenic and mutagenic activity. *Anti*-BPDE generally exhibits greater mutagenic and carcinogenic activity than *syn*-BPDE in most test systems. *Anti*-BPDE has also a greater activity in induction of malignant transformation of mouse fibroblasts and as a carcinogen on mouse skin. Comparison of the tumorigenicities in newborn mouse lung of the BPDE isomers revealed that (+)-*anti*-BPDE is more tumorigenic than (−)-*anti*-BPDE, BP, and (−)- or (+)-*syn*-BPDE. In contrast to 7,8-diol, the electron-rich 4,5-diol, called by Pullman the "K region," undergoes rapid enzymatic hydrolysis and reaction with GSH, thus preventing its binding with nucleic acids.

In summary, Figure V-12 shows the metabolic pathway of formation of the ultimate carcinogen of BP, the 7,8-diol-9,10 epoxide. It proceeds via epoxidation of the aromatic ring of the hydrocarbon by the cytochrome P-450 system to form BP-7,8 oxide, followed by hydration of the arene oxide by EH to form BP-7,8-dihydrodiol and finally by epoxidation of the 9,10 olefinic double bond by the cytochrome P-450 system. All three enzymatic reactions are highly stereoselective.

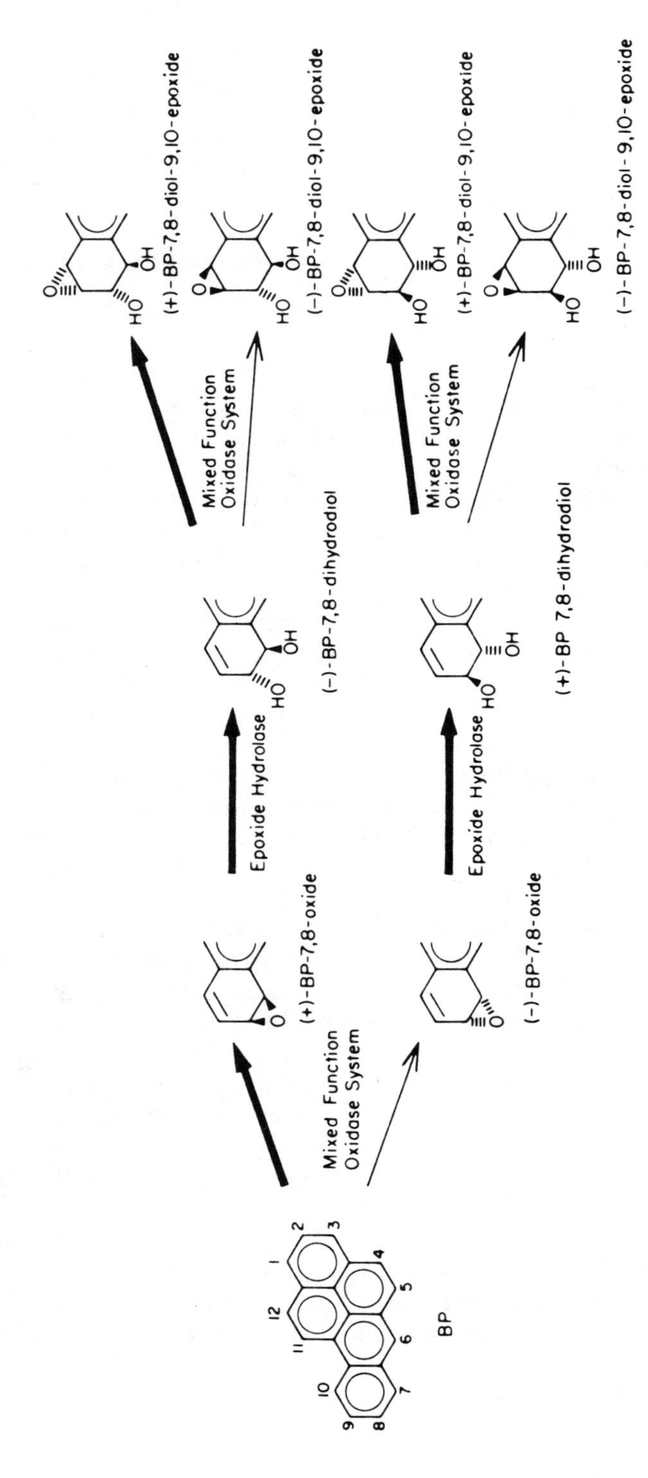

Figure V-12. Formation of benzo[*a*]pyrene metabolites responsible for the carcinogenicity of the parent hydrocarbon. Solid arrows indicate major metabolic pathways. Adapted from Levin *et al.* (1980) *J. Biol. Chem.* **255,** 9067.

2. *Other Polycyclic Aromatic Hydrocarbons*

Studies similar to those performed with BP were carried out with other PAHs such as benz[*a*]anthracene, 7-methylbenz[*a*]anthracene, 7,12-dimethylbenz[*a*]anthracene, dibenz[*a,h*]anthracene, chrysene, and others. The main objectives of these studies were (1) to determine whether the diol epoxide mechanism, described with BP, is a general mechanism of activation and (2) to identify the actual ultimate carcinogenic forms of the parent PAHs. We will discuss here only the metabolic products of the four most studied of these ring compounds.

a. BENZ[A]ANTHRACENE

Benz[*a*]anthracene can form five *trans*-dihydrodiols, the K-region 5,6- and the non-K-region 1,2-, 3,4-, 8,9-, and 10,11-dihydrodiols (Figure V-13). In liver microsomal fractions from normal and 3MC-treated rats and by a purified cytochrome P-448 system, the main metabolites are the 5,6- and 8,9-dihydrodiols, although small amounts of 3,4- and 10,11-dihydrodiols are also formed. Only very small amounts of 1,2-dihydrodiol were detected.

Mouse skin maintained in organ culture with benz[*a*]anthracene formed mainly 3,4-dihydrodiol but no 1,2-dihydrodiol (Figure V-14). By further metabolism of 8,9-dihydrodiol by rat liver microsomes, the *anti* isomer of 8,9-diol-10,11-epoxide was identified. Although no *syn* isomer was found, it is possible that during the metabolism of benz[*a*]anthracene both the *syn* and the *anti* form of the 1,2-diol-3,4-epoxide are formed.

Although benz[*a*]anthracene is generally considered as noncarcinogenic, the synthetically prepared 3,4-dihydrodiol and the "bay-region" 3,4-diol-1,2-epoxide are highly active in inducing mutation in *Salmonella typhimurium* TA100 and in V79 Chinese hamster cell systems. Similarly, the *anti*-3,4-dihydrodiol is 10- to 20-fold more active than the other dihydrodiols in producing tumors on mouse skin. The *anti*-1,2-diol-3,4-epoxide and the *anti*-10,11-diol-8,9-epoxide and the *syn*- and *anti*-8,9-diol-10,11-epoxides were less effective. From these results we can assume that the low carcinogenic activity of the parent compound is probably due to the low level of both metabolism to a "bay-region" diol epoxide and interaction with DNA.

b. 7-METHYLBENZ[A]ANTHRACENE

7-Methylbenz[*a*]anthracene can, in theory, form five *trans* -dihydrodiols, the K-region 5,6-dihydrodiol and the 1,2-, 3,4-, 8,9-, and non-K-10,11- region dihydrodiols (Figure V-15). In addition to hydroxylation on the aromatic ring, hydroxylation of the methyl group can also occur to form 7-hydroxymethylbenz[*a*]anthracene. This derivative can then be further metabolized to ring-hydroxylated products.

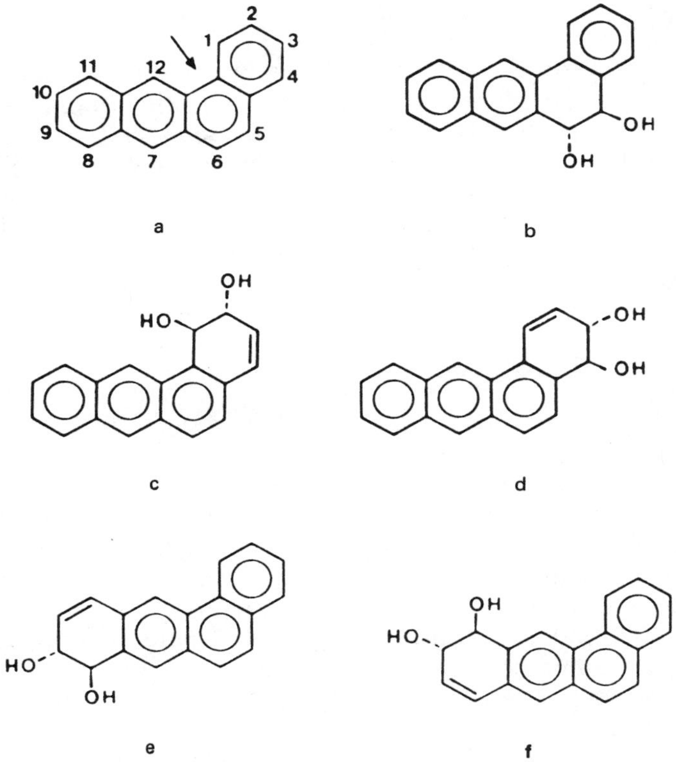

Figure V-13. Structures of *trans*-dihydrodiols of benz[*a*]anthracene. a, benz[*a*]anthracene; b, *trans*-5,6-dihydro-5,6-dihydroxybenz[*a*]anthracene; c, *trans*-1,2-dihydro-1,2-dihydroxybenz-[*a*]anthracene; d, *trans*-3,4-dihydroxybenz[*a*]anthracene; e, *trans*-8,9-dihydro-8,9-dihydroxybenz-[*a*]anthracene; f, *trans*-10,11-dihydro-10,11-dihydroxybenz[*a*]anthracene. Adapted from Sims and Grover (1981) In *Polycyclic Hydrocarbons and Cancer*, Vol. 3 (Gelboin and Ts'o, eds.), Academic Press, pp. 117–181.

All five dihydrodiols were detected in liver homogenates and microsomal fractions prepared from rats pretreated with 3MC. In mouse skin organ culture, mainly the 3,4- and 8,9-dihydrodiols are formed (Figure V-16). The 3,4-dihydrodiol was much more active than the parent hydrocarbon or the four other dihydrodiols when tested for mutagenesis with *S. typhimurium* TA100 as well as in Chinese hamster V79 cells. In induction of malignant transformation in mouse fibroblasts, again the 3,4-dihydrodiol was the most active compound followed by 8,9-dihydrodiol. Both products were more active than 7-methylbenz[*a*]anthracene. Thus, the 3,4-dihydrodiol in these systems is probably converted to vicinal diol epoxides which are one or both isomers of the "bay-region" 3,4-diol-1,2-epoxide.

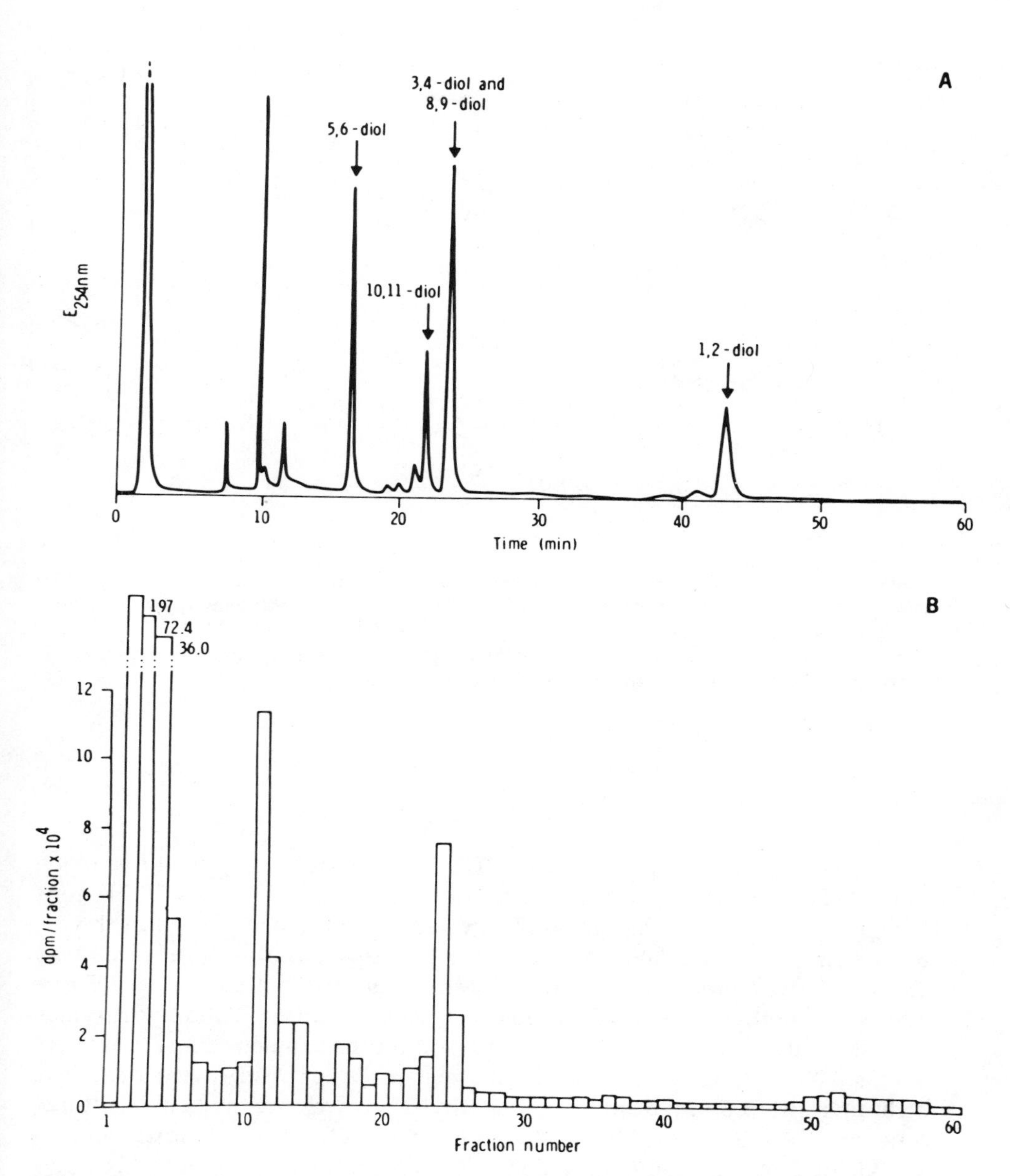

Figure V-14. The formation of dihydrodiols as metabolites of [^{3}H]benz[*a*]anthracene by mouse skin maintained in short-term organ culture. (A) HPLC elution profile obtained for UV absorption of the five reference dihydrodiols of benz[*a*]anthracene; (B) the amounts of radioactive material present in each fraction. From MacNicoll *et al.* (1980) *Chem. Biol. Interact.* **29**, 169.

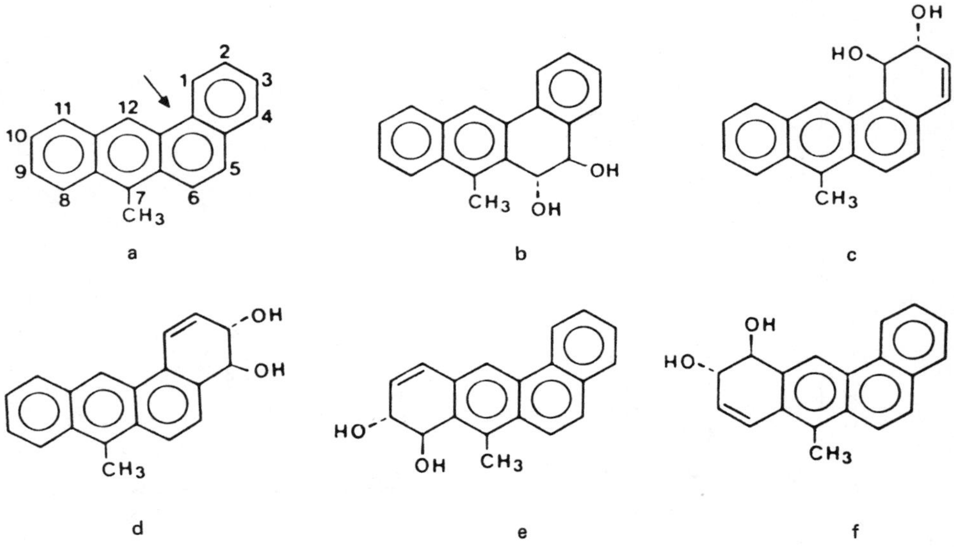

Figure V-15. Structures of *trans*-dihydrodiols of 7-methylbenz[*a*]anthracene. a, 7-methylbenz[*a*]anthracene; b, *trans*-5,6-dihydro-5,6-dihydroxy-7-methylbenz[*a*]anthracene; c, *trans*-1,2-dihydro-1,2-dihydroxy-7-methylbenz[*a*]anthracene; d, *trans*-3,4-dihydro-3,4-dihydroxy-7-methylbenz[*a*]anthracene; e, *trans*-8,9-dihydro-8,9-dihydroxy-7-methylbenz[*a*]anthracene; f, *trans*-10,11-dihydro-10,11-dihydroxy-7-methylbenz[*a*]anthracene.

c. 7,12-DIMETHYLBENZ[A]ANTHRACENE

7,12-Dimethylbenz[*a*]anthracene (DMBA) can, in theory, yield five *trans*-dihydrodiols, the K-region 5,6-dihydrodiol and the 1,2-, 3,4-, 8,9-, 10,11- non-K-region dihydrodiols (Figure V-17). In addition to hydroxylation of the aromatic rings, hydroxylation of the methyl groups can also occur. This was shown in liver preparations where DMBA was metabolized to 7-hydroxymethyl-12-methyl, to 12-hydroxymethyl-7-methyl and to 7,12-dihydroxymethyl derivatives, as well as to hydroxylation on the aromatic rings. In experiments with rat liver microsomes, further metabolites of DMBA were identified, including the 5,6-dihydrodiol of 7-methyl-12-hydroxymethylbenz[*a*]anthracene and the 8,9-dihydrodiol of 7,12-dihydroxymethylbenz[*a*]anthracene.

Using liver microsomal fractions from PB-pretreated rats, four 3,4-dihydrodiols were detected either as metabolites of DMBA itself or of the 7- and 12-hydroxymethyl and of the 7,12-dihydroxymethyl derivatives. In microsomal systems from 3MC-pretreated rats, the 3,4-, 5,6-, 8,9-, and 10,11-dihydrodiols of DMBA were found as metabolites.

The primary DMBA metabolite, 7-hydroxymethyl-12-methylbenz[*a*]anthracene, is further metabolized by rat liver microsomal fractions to the corresponding 3,4-, 5,6-, 8,9-, and 10,11-dihydrodiols. As with the parent

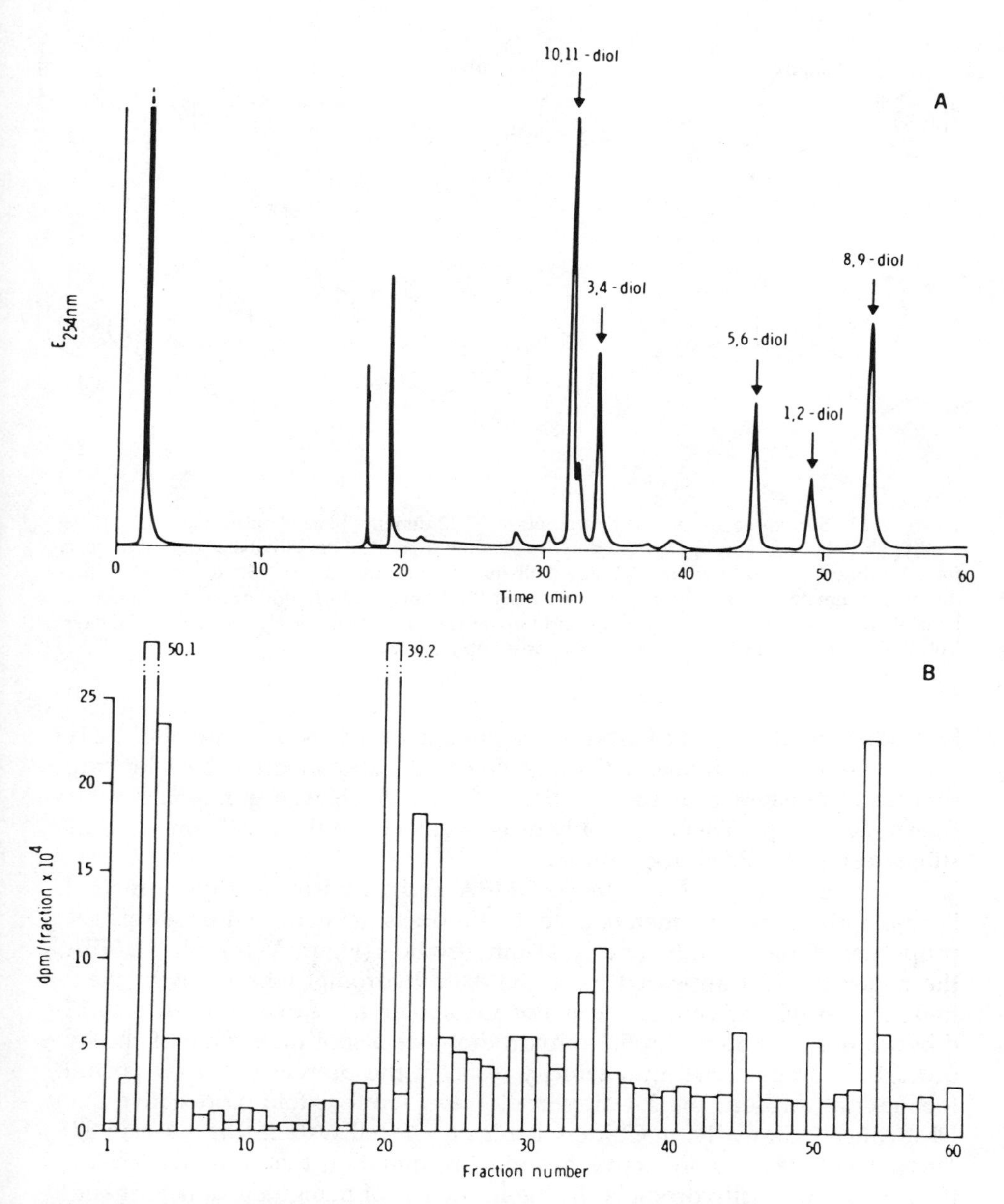

Figure V-16. The formation of dihydrodiols as metabolites of [^{3}H]-7-methylbenz[*a*]anthracene by mouse skin maintained in short-term organ culture. (A) HPLC elution profile obtained for UV absorption of the five reference dihydrodiols of 7-methylbenz[*a*]anthracene; (B) the amounts of radioactive material present in each fraction. From MacNicoll *et al.* (1980) *Chem. Biol. Interact.* **29,** 169.

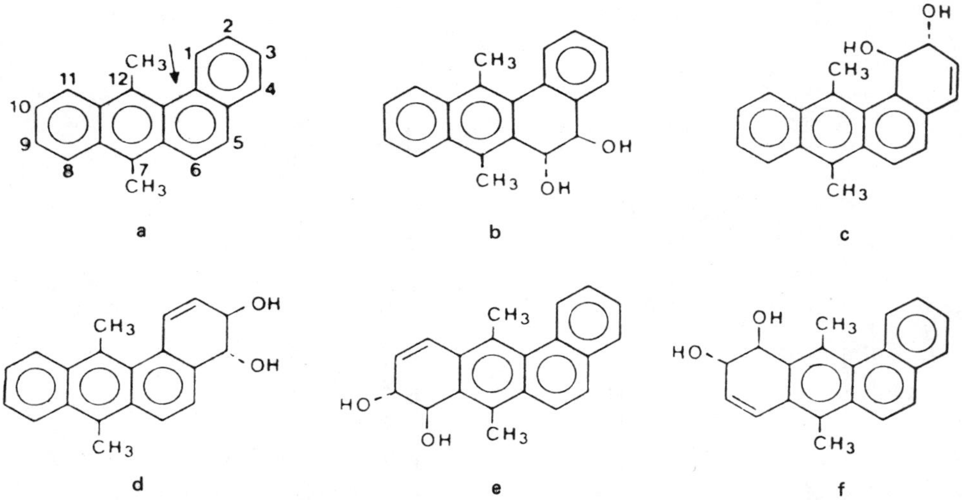

Figure V-17. Structures of *trans*-dihydrodiols of 7,12-dimethylbenz[*a*]anthracene. a, 7,12-dimethylbenz[*a*]anthracene; b, *trans*-5,6-dihydro 5,6-dihydroxy-7,12-dimethylbenz[*a*]anthracene; c, *trans*-1,2-dihydro-1,2-dihydroxy-7,12-dimethylbenz[*a*]anthracene; d, *trans*-3,4-dihydro-3,4-dihydroxy-7,12-dimethylbenz[*a*]anthracene; f, *trans*-10,11-dihydro-10,11-dihydroxy-7,12-dimethylbenz[*a*]anthracene. Adapted from Sims and Grover (1981) In *Polycyclic Hydrocarbons and Cancer*, Vol. 3 (Gelboin and Ts'o, eds.), Academic Press, pp. 117–181.

hydrocarbon, the 1,2-dihydrodiol was not detected as a metabolite. 7-Hydroxymethyl-12-methylbenz[*a*]anthracene was also converted by liver cytosolic sulfotransferase to the reactive sulfate, which was mutagenic in the *Salmonella* system. This is the only direct evidence of the involvement of the sulfate ester of a PAH in cell mutation.

In mouse skin organ culture, DMBA and 7-hydroxymethyl-12-methylbenz[*a*]anthracene are metabolized to products with the chromatographic properties of the 3,4-, 8,9-, and 10,11-dihydrodiols (Figure V-18). With DMBA, the major product appeared to be the 8,9-dihydrodiol whereas with the 7-hydroxymethyl derivative, the major metabolite appeared to be the 10,11-dihydrodiol. The K-region 5,6-dihydrodiol was also a metabolite of the hydrocarbon. In bacterial mutagenicity tests, in the presence of a rat postmitochondrial fraction, the 3,4-dihydrodiol was about sixfold more active than DMBA itself. Similarly, in Chinese hamster V79 cell systems the 3,4- and 8,9-dihydrodiols were more active in inducing mutation than either DMBA or the 5,6- or 10,11-dihydrodiols. In the induction of malignant transformation of mouse fibroblasts, the 3,4- and the 8,9-dihydrodiols proved to be more active than DMBA. In the induction of the mouse skin tumors, the 3,4-dihydrodiol was again more active than the parent hydrocarbon. On the other hand, 8,9- and 5,6-dihydrodiols were essentially inactive. Substitution of methyl groups in the 2 and 3 positions and fluorine in the 1, 2, and 5 positions led to loss of biological activity.

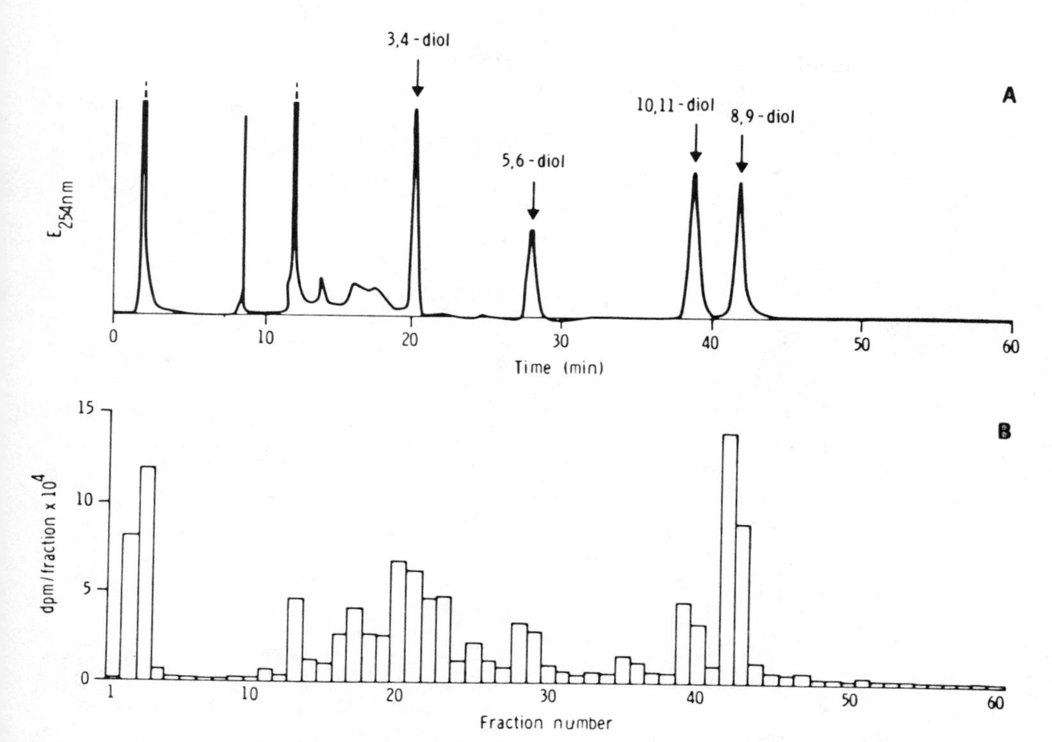

Figure V-18. Formation of dihydrodiols as metabolites of [^{3}H]-7,12 -dimethylbenz[*a*]anthracene by mouse skin maintained in short-term organ culture. (A) HPLC elution profile obtained for UV absorption of the four reference dihydrodiols of 7,12-dimethylbenz[*a*]anthracene; (B) the amounts of radioactive material present in each fraction. From MacNicoll *et al.* (1980) *Chem. Biol. Interact.* **29,** 169.

These results show that in systems where DMBA can be metabolized to diol epoxides, the 3,4-dihydrodiol is the most active diol. This suggests that in the metabolic activation of DMBA, one or possibly both the *anti* and *syn* isomers of the bay-region 3,4-diol-1,2-epoxides are formed and constitute the ultimate carcinogen. The biological role of the active 8,9-dihydrodiol is still uncertain.

d. 3-METHYLCHOLANTHRENE

3MC can yield four *trans*-dihydrodiols: the K-region 11,12-dihydrodiol and the 4,5-, 7,8-, and 8,10- non-K-region dihydrodiols. 3MC also yields dihydroxy derivatives such as the *cis* and *trans* isomers of the 1,2-diol as well as the 1-hydroxy and 2-hydroxy and the 1-keto and 2-keto derivatives (Figure V-19).

The major metabolites of 3MC by liver microsomal preparations from rats pretreated either with 3MC or PB arise through hydroxylation at the 1 and 2

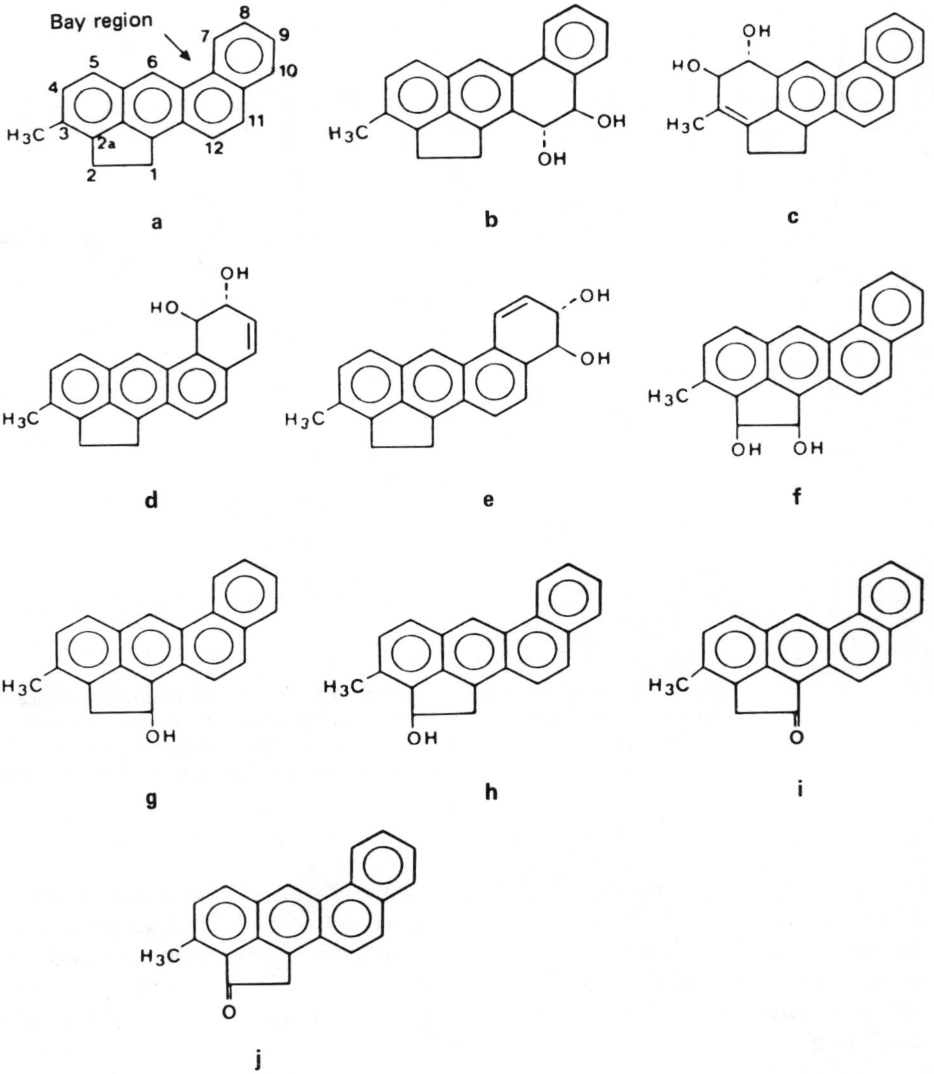

Figure V-19. Structures of 3-methylcholanthrene derivatives. a, 3-methylcholanthrene; b, *trans*-11,12-dihydro-11,12-dihydroxy-3-methylcholanthrene; c, *trans*-4,5-dihydro-4,5-dihydroxy-3-methylcholanthrene; d, *trans*-7,8-dihydro-7,8-dihydroxy-3-methylcholanthrene; e, *trans*-9,10-dihydro-9,10-dihydroxy-3-methylcholanthrene; f, *cis*- and *trans*-1,2-diol-3-methylcholanthrene; g, 1-hydroxy-3-methylcholanthrene; h, 2-hydroxy-3-methylcholanthrene; i, 1-keto-3-methylcholanthrene; j, 2*a*,3-diol-3-methylcholanthrene. Adapted from Sims and Grover (1981) In *Polycyclic Hydrocarbons and Cancer*, Vol. 3 (Gelboin and Ts'o, eds.), Academic Press, pp. 117–181.

positions. The hydroxy metabolites are further metabolized to the 9,10-dihydrodiol of either 1- or 2-hydroxy-3MC. By HPLC it was possible to identify the 4,5-, 7,8-, 9,10-, and 11,12-dihydrodiols. Although the 9,10-dihydrodiol appeared to be the principal dihydrodiol metabolite, its amount was very small compared to the amounts of the 1-hydroxy-3MC. In organ cultures of mouse skin, 3MC was metabolized to all the same diols and dihydrodiols as in liver microsomes (Figure V-20).

When tested for mutagenicity using *S. typhimurium* or Chinese hamster V79 cells, the 9,10-dihydrodiol of 1-hydroxy-3MC was the most active of the compounds tested, although the 1- and 2-hydroxy derivatives also showed some activity. Of the dihydrodiol derivatives the 9,10-dihydrodiol was considerably more active than the 4,5-, 7,8-, and 10,11-dihydrodiols or the hydrocarbon itself.

In the induction of malignant transformation in mouse fibroblasts, 9,10-dihydrodiol was again more active than the hydrocarbon. The 4,5-, 7,8- and 10,11-dihydrodiols were either inactive or weakly active in this system. A comparison of the carcinogenic activities of 3MC and its 1- and 2-hydroxy derivatives in mice indicated that 3MC was the most active. The 9,10-dihydrodiol was also active in inducing tumors on mouse skin, but no more so than 3MC. The evidence from metabolic and biological data suggests that hydroxylation at either the C-1 or the C-2 position, when coupled with the formation of a 9,10-diol-7,8-epoxide, may be essential to the metabolic activation of 3MC.

All the results with PAHs indicate that the general mechanism of the activation into an ultimate carcinogen involves formation of vicinal diol epoxides. The reactive carbonium ions are generated *in vivo* from these epoxides. These are finally involved in interactions with cellular macromolecules such as DNA. As has been predicted by Jerina and co-workers in the "bay region" theory, the diol epoxides of the bay region type should be the most chemically reactive epoxides from which the carbonium ions are formed. For BP, 3MC, DMBA, and 7-methylbenz[*a*]anthracene, much of the available experimental evidence supports this prediction. On the other hand, in addition to "bay-region" diol epoxide reactivity, there are a variety of other, not yet well-understood, factors that will have to be taken into account before a carcinogenic potency of a hydrocarbon can be predicted from the structure. For the time being, however, the "bay-region" hypothesis remains a useful tool to study metabolism and mechanism of carcinogenicity of novel PAHs.

C. ENZYMATIC ACTIVATION OF *N*-SUBSTITUTED AROMATIC COMPOUNDS AND RESULTING METABOLITES

Aromatic compounds that have nitrogen atoms attached to their ring carbons have a potential for eliciting a variety of mutagenic and carcinogenic responses. Although this group of compounds is most often referred to as

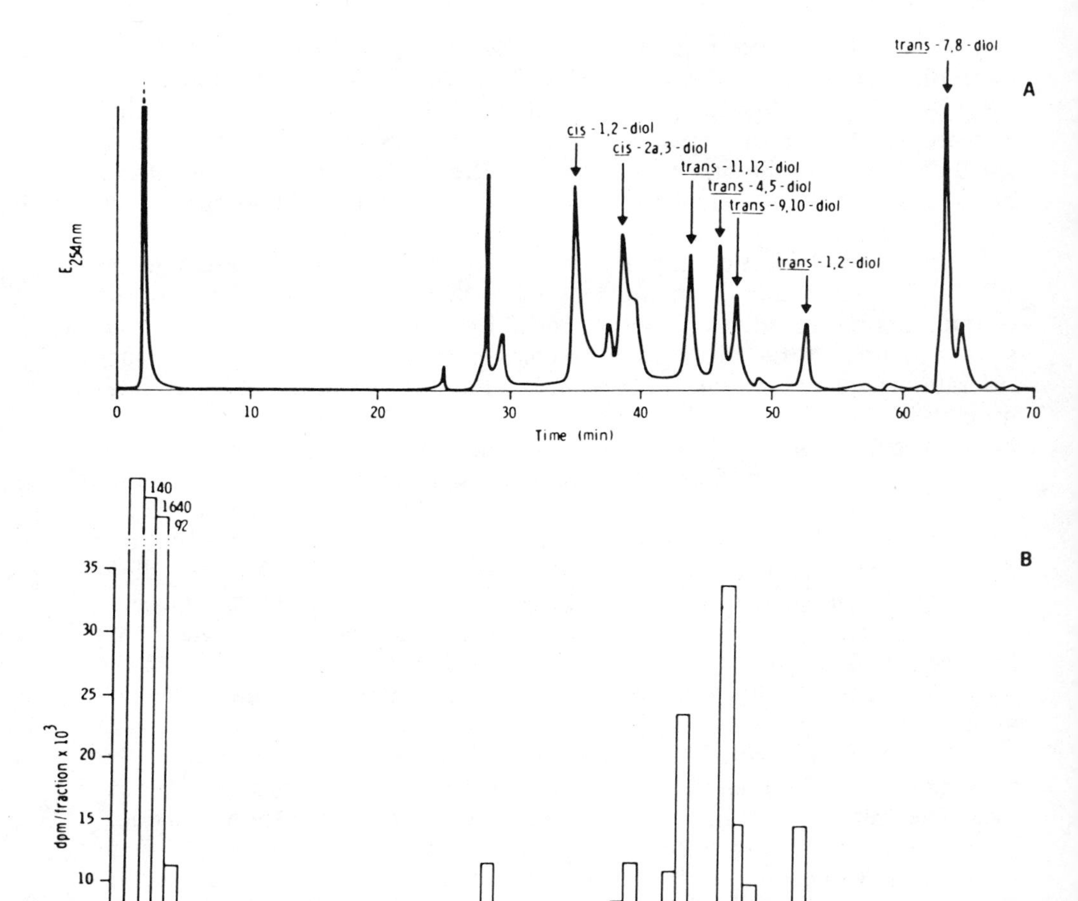

Figure V-20. The formation of diols as metabolites of [^{3}H]-3-methylcholanthrene by mouse skin maintained in short-term organ culture. (A) HPLC elution profile obtained for UV absorption of the seven reference diols of 3-methylcholanthrene; (B) the amounts of radioactive material present in each fraction. From MacNicoll *et al.* (1980) *Chem. Biol. Interact.* **29,** 169.

aromatic amines, the term *N*-substituted aromatic compounds is more appropriate since it is sufficiently broad to include nitrocompounds that may be metabolized to aromatic amines, as well as metabolites of amines, e.g., hydroxamates and amides.

Several compounds of this type are used in many areas of chemical industry especially as intermediates in the production of dyestuffs, plastics, and pharmaceuticals. Some of these compounds, including 2-naphthylamine, 4-aminobiphenyl, and benzidine, proved to be carcinogenic in experimental animals and in humans. Epidemiological data show a high incidence of bladder cancer in people exposed to these agents in the work place. As a result of these findings several of these compounds are on the list of agents termed human carcinogens (see Tables II-1, 2). However, not all *N*-substituted aromatic compounds are potent carcinogens. The carcinogenic potential of these agents depends on their structure, namely on the type of aromatic ring system and on the presence of other substituents in the molecule. Two simple *N*-substituted aromatic compounds, aniline and acetanilide, are considered noncarcinogenic. Methyl substitution on the ring, such as in *o*-toluidine, yields a weak carcinogen. The most active carcinogens of this type contain two or more aromatic rings. These include 2-naphthylamine, 4-aminobiphenyl, 2-aminofluorene, benzidine, and also compounds in which two benzene rings are separated by an azo bond such as 4-aminoazobenzene (Figure V-21). The position of amino groups also affects the carcinogenic activity of the compounds. The highest level of activity is in the *N*-substituted di- and tricyclic

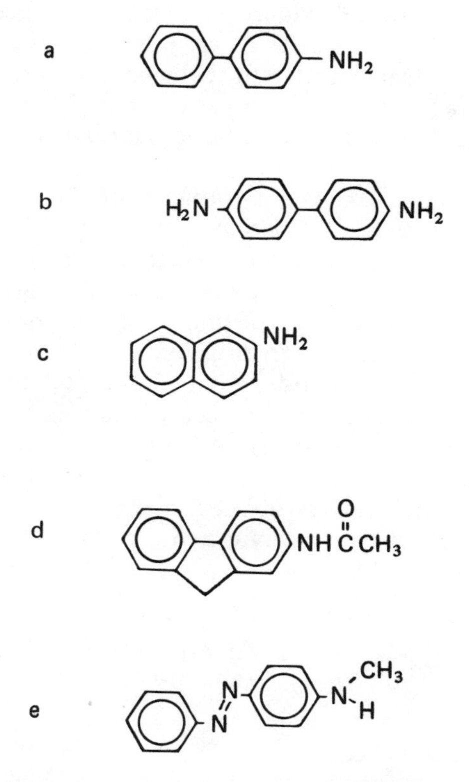

Figure V-21. Structures of some *N*-substituted aromatic compounds. a, 4-aminobiphenyl; b, benzidine; c, 2-naphthylamine; d, 2-acetylaminofluorene; e, *N*-methyl-4-aminoazobenzene.

aromatic compounds when the substitution is at the terminal carbon atom of the longest conjugated chain. In some cases these aromatic compounds are acetylated.

There are great differences among these agents in species and tissue susceptibilities to cancer induction, probably due to the differences in the metabolism of *N*-substituted aromatic compounds. Most of the metabolites are ring-hydroxylated phenolic compounds that are excreted as either sulfate or glucuronide conjugates, all serving as proximate carcinogens for further metabolism to electrophilic ultimate carcinogenic forms.

1. *N-Oxidation to Arylhydroxamic Acids*

All aromatic amines require metabolic conversion into a strong electrophilic reactant to be able to bind chemically to nucleophilic sites in the target cells, particularly to nucleic acids and proteins. Most of the aromatic amines are metabolized primarily in the liver by the microsomal mixed-function oxidase or monooxygenase system, similarly to PAHs. The details of metabolic activation have been worked out primarily with a potent hepatocarcinogen, 2-acetylaminofluorene (AAF) (Figure V-22). Early studies in the Millers' laboratories showed that rats fed AAF converted it into a new metabolite, *N*-hydroxy-AAF, which was found in the urine almost entirely as the glucuronic acid conjugate. Later, *N*-hydroxy derivatives were found in the urine of other species including cat, chicken, rabbit, dog, and hamster in either the free or the conjugated form.

N-Hydroxylation of AAF is considered to be the initial activation step in carcinogenesis, whereas hydroxylation of the aromatic rings appears to provide a detoxification pathway. These reactions are catalyzed by the cytochrome P-450-dependent monooxygenase systems described in Section A of this chapter. Four highly purified forms of rabbit hepatic microsomal cytochrome P-450 catalyze the *N*- and ring hydroxylation of AAF. Form 4, the major form of the cytochrome induced in adult rabbit liver by TCDD or 3MC, is the most active in the *N*-hydroxylation reaction. Form 6, which is induced by TCDD in newborn rabbit liver, and form 3, a constitutive cytochrome species, both catalyze almost exclusively the detoxification pathway, the hydroxylation of AAF in the 7 position. On the other hand, the PB-induced form 2 exhibits practically no catalytic activity with AAF as substrate.

$COCH_3$ N H — rat liver E.R. + NADPH + O_2 → N $COCH_3$ OH

C H_2

2-ACETYLAMINOFLUORENE (AAF) N-HYDROXY-AAF

Figure V-22. Oxidation of *N*-2-acetylaminofluorene by rat liver microsomes.

Although administration of AAF and its *N*-hydroxy derivative yielded nucleic acid- and protein-bound adducts, especially in rat liver, no reactions of these compounds with nucleic acids or proteins occurred *in vitro*. Therefore, further metabolism seemed necessary to convert *N*-hydroxy metabolites as proximate carcinogens to electrophilic reactants that can interact with cellular macromolecules. Similar *N*-oxidation of other aromatic amines and amides was detected mainly in rat liver with 4-acetylaminobiphenyl, 4-acetylaminostilbene, and 2-acetylaminophenanthrene. *N*-Oxidation of all these compounds led to formation of arylhydroxamic acids as proximate carcinogens.

2. *N-Oxidation of Arylamines*

In the case of the above-described aromatic amines such as AAF, the primary activation step involved formation of arylhydroxamic acids through oxidation of the *N*-acetyl derivative. However, there are aromatic amines that are not acetylated and their activation proceeds via *N*-oxidation of the amines. Carcinogenicity of 1- and 2-naphthylamines and 4-aminobiphenyl for the bladder of dogs depends on the *N*-oxidation of the amines. In the urine of treated dogs a conjugate of *N*-hydroxy-4-aminobiphenyl with glucuronic acid was detected. Another aromatic amine, *N*-methyl-4-aminoazobenzene (MAB), is also *N*-oxidized by hepatic microsomes in the presence of NADPH-generating system (Figure V-23). The initial *N*-oxidation product, *N*-OH-MAB, is readily oxidized to a second product that yielded *N*-OH-4-aminoazobenzene (*N*-OH-AB) upon acidic treatment. The secondary *N*-oxidation product may be formed nonenzymatically and is presumed to be N-OH-MAB *N*-oxide or its dehydrated derivative, nitrone. Under the same conditions MAB is also oxidatively *N*-dealkylated to 4-aminoazobenzene (AB), which is further *N*-oxidized to *N*-OH-AB. The microsomal *N*-oxidation of MAB, in contrast to *N*-oxidation of AAF, is cytochrome P-450-independent and therefore it is insensitive to inhibition by 2-[(2,4-dichloro-6-phenyl)phenoxy]ethylamine (DPEA). It is possible that the enzyme system catalyzing the *N*-oxidation of MAB in rat liver microsomes is similar to or identical with the flavoprotein mixed-function amine oxidase purified from pig liver microsomes. The isolation of *N*-OH-MAB as a product of the incubation of MAB with the purified oxidase provided good evidence for the role of a flavoprotein mixed-function amine oxidase in the *N*-oxidation of MAB.

3. *Sulfotransferase*

As was described in the previous section, *N*-hydroxy derivatives of aromatic amines require further metabolism to form the electrophilic reactant that is the ultimate carcinogen. In the case of AAF it is further metabolized after *N*-oxidation by a soluble liver sulfotransferase enzyme that has been prepared in purified form (Figure V-24). In the presence of PAPS and rat liver cytosol, the very reactive AAF-*N*-sulfate ester is formed. There is a good

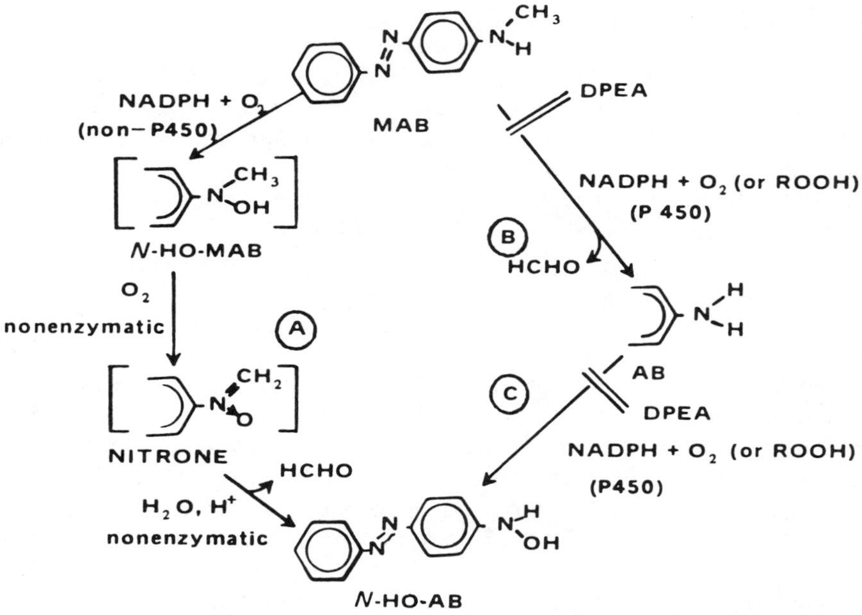

Figure V-23. N-Oxidation of *N*-methyl-4-aminoazobenzene by hepatic microsomes and NADPH. MAB, *N*-methyl-4-aminoazobenzene; *N*-OH-MAB, *N*-hydroxy-MAB; *N*-OH-AB, *N*-hydroxy-4-aminoazobenzene. From Kadlubar *et al.* (1976) *Cancer Res.* **36,** 1196.

correlation between the activity of this metabolic system *in vitro* and the reactivity of the *N*-hydroxy derivative *in vivo,* indicating that this sulfuric acid ester is a major reactive ultimate carcinogen in rat liver. The high carcinogenicity of AAF and *N*-hydroxy-AAF in male and female Fisher rats is in good correlation with their ability to sulfonate *N*-hydroxy-AAF. The reactivity and carcinogenicity of *N*-hydroxy-AAF can be lowered by limitation of the sulfate ions *in vivo* through the administration of *p*-hydroxyacetanilide, which is excreted in the urine as a stable sulfuric acid ester. Similarly, by administration of acetanilide, which is oxidized to *p*-hydroxyacetanilide *in vivo,* the carcinogenicity of *N*-hydroxy-AAF could be inhibited. This inhibition was partially removed by the simultaneous administration of sulfate ions. The sulfuric acid

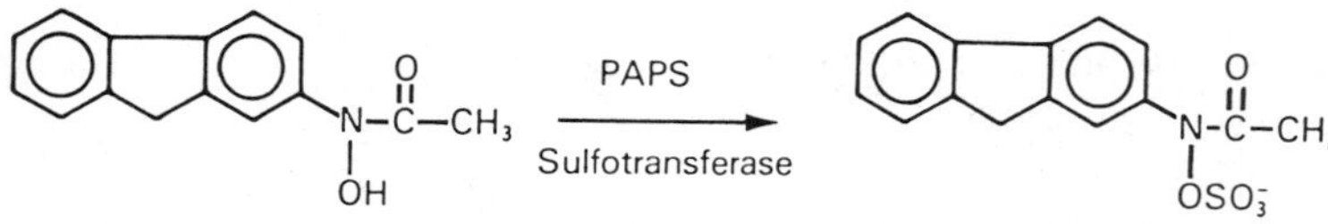

Figure V-24. Esterification of *N*-hydroxy-AAF by sulfotransferase in the presence of 3′-phosphoadenosine-5′-phosphosulfate (PAPS).

ester is very unstable, highly reactive, and has a half-life in water of less than 1 min. Because of the short half-life and high reactivity, isolation of sulfate esters from tissues or from urine of animals receiving aromatic amines has not been possible. The activation pathway through *O*-esterification by sulfuric acids seems to be restricted exclusively to liver. No sulfotransferase activity has been detected in Zymbal's gland or in mammary gland. However, in each of these tissues *N*-hydroxy-AAF is strongly carcinogenic and therefore other activation reactions must exist in these tissues. This aspect will be discussed later.

In contrast to AAF-*N*-sulfate, which is very unstable and difficult to obtain in pure form, the sulfuric acid esters of other aromatic amines such as *N*-hydroxy-4-acetylaminobiphenyl, *N*-hydroxy-4-acetylamino-4′-fluorophenyl, and *N*-hydroxy-2-acetylaminophenanthrene have been chemically synthesized and characterized.

N-Hydroxy-MAB could also be converted into a reactive sulfoester in the presence of microsomes, NADPH, cytosol, and PAPS. The suggestion that MAB-*N*-sulfate is the sole ultimate carcinogenic species of MAB is supported by the observation that addition of sodium sulfate to the diet increased the carcinogenicity of 3′-methyl-4-dimethylaminoazobenzene.

4. *N,O*-Acyltransferase

Although the enzymatic conjugation of *N*-oxidized derivatives of aromatic amines with sulfate may be an important ultimate carcinogenic form, there is increasing evidence suggesting that enzymatic *N,O*-acyl transfer from arylhydroxamic acids may be another important pathway leading to carcinogenic compounds. Activation of *N*-OH-AAF by *N,O*-acyltransferase has been demonstrated in the cytosol fraction from the liver of a number of species and in several extrahepatic tissues in the rat.

Arylhydroxamic acid *N,O*-acyltransferase has been purified from rat and rabbit liver cytosols. It is a sulfhydryl-dependent enzyme and therefore inhibited by additional *p*-chloromercuribenzoic acid or iodoacetamide but it is protected by dithiothreitol. The enzyme from rat liver has a molecular weight of 28,000 and that from rabbit, 33,000.

In the *N,O*-acyltransferase-catalyzed reaction two successive steps are involved, i.e., *N*-deacylation and *O*-acylation. The mechanism of *N,O*-acyltransferase-catalyzed reaction of *N*-OH-AAF with guanosine is depicted in Figure V-25. In the first step the *N*-acyl moiety of the hydroxamine acid is removed, and in the second step the acyl group is transferred to an amine to form a stable amide or to the oxygen of the hydroxylamine to yield a reactive *N*-acyloxyarylamine. Due to extreme reactivity of *N*-acyloxyarylamines the synthesis of *N*-acetoxy-*N*-2-aminofluorene has not been achieved. Its existence has been deduced from its reactivity with guanosine residues in nucleic acids as seen in Figure V-25.

The role of *N,O*-acyltransferase in mutagenesis has been shown in the

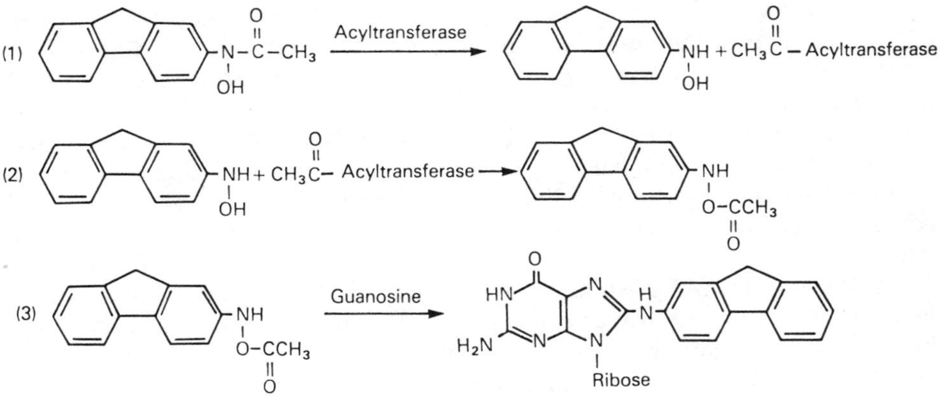

Figure V-25. Mechanism of *N,O*-acetyltransferase-catalyzed reaction of *N*-OH-AAF with guanosine.

Salmonella mutagenesis system by the activation of *N*-OH-AAF, which is a weak mutagen. The transferase-catalyzed metabolic activation of *N*-OH-AAF can be inhibited by 2-aminofluorene, which competes with the initial product of deacetylation (*N*-OH-AF) for the acetyl group, thereby inhibiting the formation of *N*-acetoxy-AF. On the other hand, no increase of revertants was observed by AAF, which is not a substrate for the enzyme.

The *N,O*-acyltransferase-catalyzed activation of arylhydroxamic acids may be responsible for the induction of some aromatic amine-incuded tumors. The strongest evidence for the role of this enzyme in the carcinogenic process comes from mammary gland tumor induction experiments in the rat. A single administration of *N*-OH-AAF to the mammary gland of the female rat resulted in the formation of tumors at the site of application. Tumors were not induced by AAF or *N*-2-OH-AF. The enzyme is widely distributed in several tissues of the rat, hamster, mouse, guinea pig, and rabbit that are targets for the induction of tumors by AAF derivatives. The enzyme was also detected in human liver, small intestine, colon, and lung. In contrast, no detectable acyltransferase was found in dogs, although bladder and liver tumors develop on treatment with AAF. Therefore, it seems that while acyltransferase may be involved in the production of some tumors by aromatic amines, it cannot be responsible for the development of all lesions that are ascribed to aromatic amines.

5. *Deacetylases*

Arylhydroxylamines are formed from arylhydroxamic acids by enzymatic removal of the *N*-acetyl group from the *N*-hydroxy derivatives (Figure V-26). Two types of deacetylases were detected. One enzyme is located in the en-

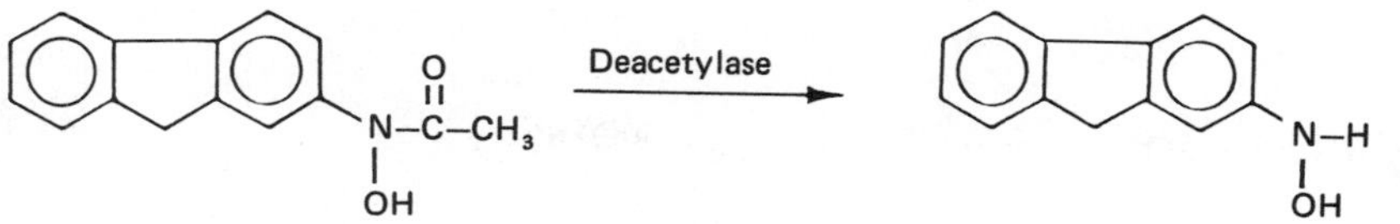

Figure V-26. Formation of *N*-hydroxy-2-aminofluorene from *N*-hydroxy-2-acetylaminofluorene by microsomal deacetylase.

doplasmic reticulum and the other is a soluble enzyme. The activity of deacetylases varies with the species. In the liver of guinea pig and hamster the activity of the microsomal enzyme to deacetylate *N*-hydroxy-AAF is very high in comparison with that of rat liver. The deacetylated product, *N*-hydroxy-AF, is able to react with guanosine residues in nucleic acids *in vitro* at acidic pH. Similarly, *N*-hydroxy-AF generated by incubation of *N*-hydroxy-AAF with soluble rat liver deacetylase forms reaction products with guanosine and with tRNA. In this reaction, however, there is a possibility that the binding occurred through transfer of the *N*-acetyl group of *N*-hydroxy-AAF followed by electrophilic attack of the *O*-acetyl derivative as was described in the previous section.

The importance of microsomal deacetylation in the mutagenic activation of *N*-OH-AAF was clearly demonstrated using the deacetylase inhibitor diethyl-*p*-nitrophenylphosphate (paraoxon). In the presence of this inhibitor, complete inhibition of both mutagenesis and deacetylation of *N*-OH-AAF was observed in *Salmonella* test systems containing liver and/or kidney microsomes. These data indicate that the endoplasmic reticulum may be an important site for the metabolic activation, not only of AAF but also of *N*-OH-AAF. Liver cell nuclei from guinea pig, rat, and mouse have also been shown to deacetylate *N*-OH-AAF. The same nuclear preparations are capable of mutagenic activation of *N*-OH-AAF, which could be completely blocked by paraoxon.

It is reasonable to suggest that *N*-OH-AAF deacetylase plays a role not only in the *in vitro* mutagenic activation of *N*-OH-AAF and other arylhydroxamic acids but also in the *in vivo* carcinogenic process of those compounds.

6. *Free Radical Activation*

One route of activation for arylhydroxamic acids is by peroxidases and H_2O_2, which form a free radical from the *N*-hydroxy-AAF (Figure V-27). It was demonstrated that the chemical or enzymatic one-electron oxidation of *N*-hydroxy-AAF gives a nitroxide free radical that easily dismutates to 2-nitrosofluorene and N-acetoxy-AAF. Both of the products are highly carcinogenic and mutagenic. Other *N*-hydroxy arylamines undergo a similar reac-

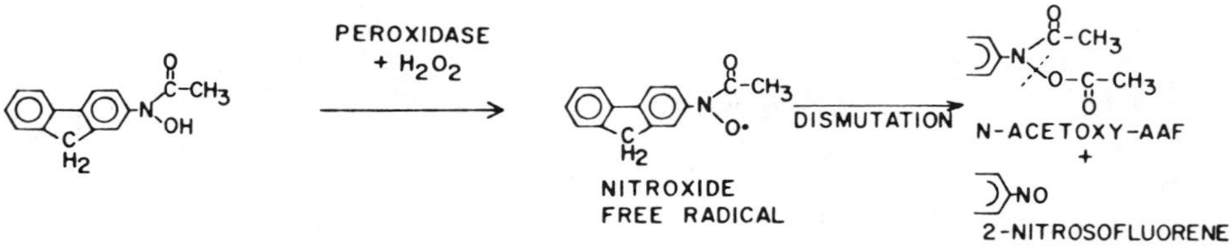

Figure V-27. Peroxidase-catalyzed oxidation of N-hydroxy-AAF to the nitroxide free radical after dismutation results in 2-nitrosofluorene and N-2-acetoxy-AAF. Based on Bartsch and Hecker (1971) *Biochim. Biophys. Acta* **237,** 567.

tion. The activation of *N*-hydroxy-AAF in microsomal systems can be inhibited by ascorbate. It is interesting that linoleic acid hydroperoxide, which is a component of the microsomal membranes, catalyzes the formation of 2-nitrosofluorene and *N*-acetoxy-AAF from *N*-hydroxy-AAF in the presence of hematin or methemoglobin. It is possible that the effect of dietary fat upon the carcinogenic potency of AAF in rats is related to radical activation of *N*-hydroxy-AAF. Generally, rats fed a diet containing high levels of unsaturated fats are more susceptible to AAF than rats on a diet with more highly saturated components.

7. *O*-Glucuronides of Arylhydroxamic Acids

The formation of *O*-glucuronides is quantitatively the most important type of reaction involved in excretion of *N*-hydroxy-AAF. The conjugation reaction of *N*-hydroxy derivatives with glucuronic acid is catalyzed by glucuronyl transferase contained in the liver of all species investigated, with the exception of cats (Figure V-28). The *O*-glucuronide is also electrophilic but much weaker than the other esters previously described. This is reflected in a considerably lower rate of interaction with nucleophilic sites in macromolecules, *in vitro*, than with *N*-acetoxy-AAF. Therefore, it is assumed that the

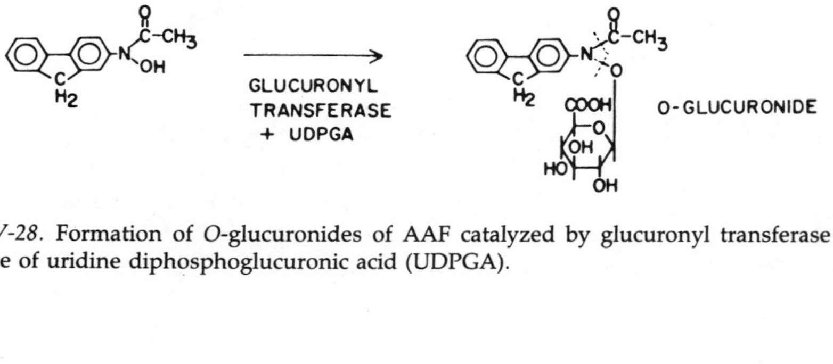

Figure V-28. Formation of *O*-glucuronides of AAF catalyzed by glucuronyl transferase in the presence of uridine diphosphoglucuronic acid (UDPGA).

importance of glucuronides is that they represent the transport form of these carcinogens. After their formation in liver, the glucuronides are transported to other organs where they may be subjected to various enzymatic reactions.

8. *N-Glucuronides of Arylhydroxylamines*

Microsomes of rat, dog, and human supplemented with UDPGA metabolize *N*-hydroxy-2-naphthylamine to the corresponding water-soluble *N*-glucuronide (Figure V-29). Presumably the same reaction takes place with other aromatic amines or amides, such as 4-aminobiphenyl, 2-aminofluorene, and 4-aminoazobenzene. The *N*-glucuronides appear to be relatively stable and nonreactive at neutral pH. However, at pH 5 they are rapidly hydrolyzed back to the corresponding hydroxylamines, which are then converted to highly reactive intermediates, probably arylnitrenium ions.

This supports the concept that arylamine bladder carcinogens first are *N*-oxidized and *N*-glucuronidated in the liver. Then the *N*-glucuronides are transported to the bladder. In the acidic urine of dogs and human the *N*-glucuronides are hydrolyzed to *N*-hydroxyarylamines, which are converted to electrophilic arylnitrenium ions capable of reaction with nucleic acids and which play a critical role in tumor induction in the bladder.

D. SUMMARY

Some of the possible metabolic pathways involved in the metabolism of *N*-acetylarylamines are illustrated in Figure V-30. The first step in the metabolic activation of these compounds is *N*-hydroxylation catalyzed by some specific forms of the cytochrome P-450 system. In contrast to the *N*-hydroxylation, oxidation of various carbon atoms in the ring system are considered as detoxification pathways.

The *N*-hydroxy derivatives serve as substrate for several metabolic pathways that can convert this hydroxamic acid into reactive electrophiles capable of covalent interaction with cellular macromolecules. Sulfotransferase, *N,O*-acetyltransferase and deacetylase enzymes are involved mainly in activation.

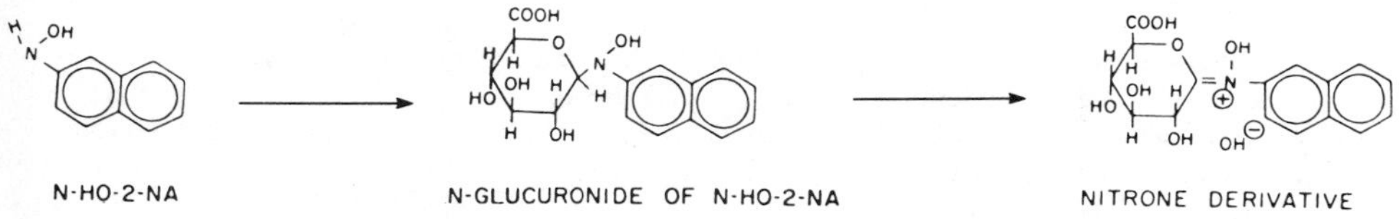

N-HO-2-NA → N-GLUCURONIDE OF N-HO-2-NA → NITRONE DERIVATIVE

Figure V-29. Formation of *N*-glucuronides and arylnitrenium ion from 2-naphthylamine.

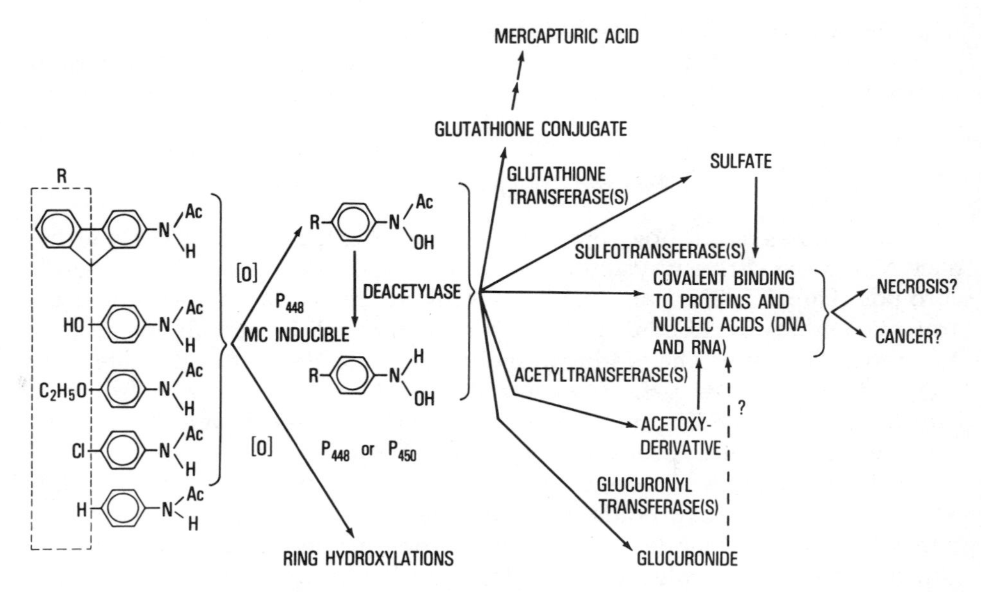

Figure V-30. Scheme of possible metabolic pathways of *N*-acetylarylamines. From Thorgeirsson *et al.* (1983) In *Reviews in Biochemical Toxicology*, Vol. 5 (Hodgson, Bent, and Philpot, eds.), Elsevier Publ. Co., pp. 349–389.

On the other hand, GSH-S-transferase and glucuronyl transferases are involved mainly in detoxification mechanisms.

Selected References

DiGiovanni, J., Sina, J. F., Ashurst, S. W., Singer, J. M., and Diamond, L. (1983) Benzo(*a*)pyrene and 7,12-dimethylbenz(*a*)anthracene metabolism and DNA adduct formation in primary cultures of hamster epidermal cells. *Cancer Res.* **43,** 163–170.

Gelboin, H. V. (1980) Benzo(*a*)pyrene metabolism, activation, and carcinogenesis: Role and regulation of mixed-function oxidases and related enzymes. *Physiol. Rev.* **60,** 1107–1166.

Gozukara, E. M., Guengerich, F. P., Miller, H., and Gelboin, H. V. (1982) Different patterns of benzo(*a*)pyrene metabolism of purified cytochrome P-450 from methylcholanthrene, β-naftoflavone and phenobarbital treated rats. *Carcinogenesis* **3,** 129–133.

Fagan, J. B., Pastewka, J. V., Park, S. S., Guengerich, F. P., and Gelboin, H. V. (1982) Identification and quantitation of a 2.0-kilobase messenger ribonucleic acid coding for 3-methylcholanthrene-induced cytochrome P-450 using cloned cytochrome P-450 complementary deoxyribonucleic acid. *Biochemistry* **21,** 6574–6580.

Jakoby, W. B., and Habig, W. H. (1980) Glutathione transferases. In *Enzymatic Basis of Detoxification* (W. B. Jakoby, ed.), Academic Press, New York, pp. 63–94.

Jakoby, W. B., Sekura, R. D., Lyon, E. S., Marcus, C. J., and Wang, J.-L. (1980) Sulfotransferases. In *Enzymatic Basis of Detoxification* (W. B. Jakoby, ed.), Academic Press, New York, pp. 199–223.

Jernström, S. H., Moldens, M. M. P., Christodoulides, L., and Ketterer, B. (1982) Inactivation of DNA-binding metabolites of benzo(*a*)pyrene and benzo(*a*)pyrene-7,8-dihydrodiol by glutathione and glutathione S-transferases. *Carcinogenesis* **3,** 757–760.

Kalinyak, J. E., and Taylor, J. M. (1982) Rat glutathione S-transferase cloning of double stranded cDNA and induction of its mRNA. *J. Biol. Chem.* **257,** 523–530.

Keller, G. M., Turner, C. R., and Jefcoate, C. R. (1982) Kinetic determination of benzo(*a*)pyrene metabolism to dihydrodiol epoxides by 3-methylcholanthrene-induced rat liver microsomes. *Mol. Pharmacol.* **22,** 451–458.

King, C. M., and Allaben, W. T. (1980) Arylhydroxamic acid acyltransferase. In *Enzymatic Basis of Detoxification,* Vol. II (W. B. Jakoby, ed.), Academic Press, New York, pp. 187–197.

Kriek, E., and Westra, J. G. (1979) Metabolic activation of aromatic amines and amides and interactions with nucleic acids. In *Chemical Carcinogens and DNA,* Vol. II (P. L. Grover, ed.), CRC Press, Boca Raton, Fla., pp. 1–28.

Lu, A. Y. H., and West. S. B. (1980) Multiplicity of mammalian microsomal cytochromes P-450. *Pharmacol. Rev.* **31,** 277–295.

Miller, E. C. (1978) Some current perspectives on chemical carcinogenesis in humans and experimental animals: Presidential address. *Cancer Res.* **38,** 1479–1496.

Miller, E. C., and Miller, J. A. (1981) Mechanisms of chemical carcinogenesis. *Cancer* **47,** 1055–1064.

Miller, J. A., and Miller, E. C. (1979) Perspectives on the metabolism of chemical carcinogens. In *Environmental Carcinogenesis* (P. Emmelot and E. Kriek, eds.), Elsevier/North-Holland, Amsterdam, pp. 25–50.

Morgenstern, R., Guthenberg, C., and Depierre, J. W. (1982) Microsomal glutathione S-transferase. Purification, initial characterization and demonstration that it is not identical to the cytosolic glutathione S-transferases A, B and C. *Eur. J. Biochem.* **128,** 243–248.

Nemoto, N. (1981) Glutathione, glucuronide and sulfate transferase in polycyclic aromatic hydrocarbon metabolism. In *Polycyclic Hydrocarbons and Cancer,* Vol. 3 (H. V. Gelboin and P. O. P. Ts'o, eds.), Academic Press, New York, pp. 213–258.

Oesch, F. (1979) Epoxide hydratase. In *Progress in Drug Metabolism,* Vol. 3 (J. W. Bridges and L. F. Chasseaud, eds.), Wiley, New York, pp. 253–300.

Pelkonen, O., and Nebert, D. W. (1982) Metabolism of polycyclic aromatic hydrocarbons: Etiologic role in carcinogenesis. *Pharmacol. Rev.* **34,** 190–222.

Sims, P., and Grover, P. L. (1981) Involvement of dihydrodiols and diol epoxides in the metabolic activation of polycyclic hydrocarbons other than benzo(*a*)pyrene. In *Polycyclic Hydrocarbons and Cancer,* Vol. 3 (H. V. Gelboin and P. O. P. Ts'o, eds.), Academic Press, New York, pp. 117–181.

Thakker, D. R., Yagi, H., Akagi, H., Koreeda, M., Lu, A. Y. H., Levin, W., Wood, A. W., Conney, A. H., and Jerina, D. M. (1977) Metabolism of benzo(*a*)pyrene. VI. Stereoselective metabolism of benzo(*a*)pyrene and benzo(*a*)pyrene 7,8-dihydrodiol to diol epoxides. *Chem. Biol. Interact.* **16,** 281–300.

Thakker, D. R., Levin, W., Yagi, H., Tada, M., Conney, A. H., and Jerina, D. M. (1980) Comparative metabolism of dihydrodiols of polycyclic aromatic hydrocarbons to bay-region diol epoxides. In *Polynuclear Aromatic Hydrocarbons: Fourth International Symposium on Analysis, Chemistry and Biology,* Battelle Press, Columbus, Ohio, pp. 267–286.

Thorgeirsson, S. S., Wirth, P. J., Staiano, N., and Smith, C. L. (1982) Metabolic activation of 2-acetylaminofluorene. In *Chemical Mechanisms and Biological Effects, Adv. Exp. Med. Biol.* Vol. 136B, pp. 897–919.

Wattenberg, L. W. (1979) Inhibitors of chemical carcinogens. In *Environmental Carcinogenesis* (P. Emmelot and E. Kriek, eds.), Elsevier/North-Holland, Amsterdam, pp. 214–263.

VI

Reactions of Metabolically Activated Carcinogens with Nucleic Acids

A. Detection of DNA–Carcinogen Adducts

Interactions of chemical carcinogens with macromolecules were first noted over 30 years ago with findings that aminoazodyes and benzo[*a*]pyrene are bound to proteins in tissues in which they produce tumors. Later, covalent binding of residues of many chemical carcinogens with both nucleic acids and proteins in tumor-susceptible tissues was found. Although the extent of binding to DNA and RNA, but not to proteins, generally showed reasonable correlation with the carcinogenic potency, this was not always true. Although all adequately studied carcinogens bind covalently to cellular macromolecules *in vivo*, some exceptions have been noted thus far: one is the carcinogenic and mutagenic antitumor agent, adriamycin, which binds *in vivo* to DNA in a noncovalent fashion.

After examination of the binding of a large number of chemical carcinogens to nucleic acids, E. C. and J. A. Miller proposed the theory that the active carcinogenic agents are highly reactive electrophiles that form covalent linkages with nucleophilic residues in macromolecules. Compounds that are not electrophilic per se undergo metabolism by mixed-function oxidase system to ultimate carcinogenic forms. The metabolism of this type of carcinogens was described in Chapter V. Since the formation of nucleic acid–carcinogen adducts is likely to be the key factor in the initiation stage of carcinogenesis, the elucidation of the chemical structures of these adducts is of extreme importance. Therefore, it is desirable to describe a general experimental approach used to obtain information on the types of adducts formed in cells exposed to metabolically activated carcinogens.

1. *Characterization of DNA–Carcinogen Adducts*

To obtain preliminary evidence for the formation of the DNA adducts, the cells are exposed to the agent for about 24 hr and the DNA is then extracted and extensively purified to remove all noncovalently bound materials. If the carcinogen is fluorescent, then the analysis of the fluorescence spectra of this DNA may indicate the types of chromophores but not the site of the attachment of the carcinogen to DNA. This was the case with benzo[*a*]pyrene (BP) where the level of detection of modified DNA is on the order of one adduct per 10^5–10;6 bases. The sensitivity of the fluorescence technique could be increased by measuring the spectra at $-70°C$. Figure VI-1 shows the fluorescence spectra of BP-treated DNA.

When the carcinogen is not fluorescent, e.g., *N*-2-acetylaminofluorene (AAF), UV spectra are helpful in the detection of the types of the DNA adducts formed.

As a next step of DNA–carcinogen adduct characterization, it is useful to react deoxyhomopolymers *in vitro* with the synthetically prepared ultimate

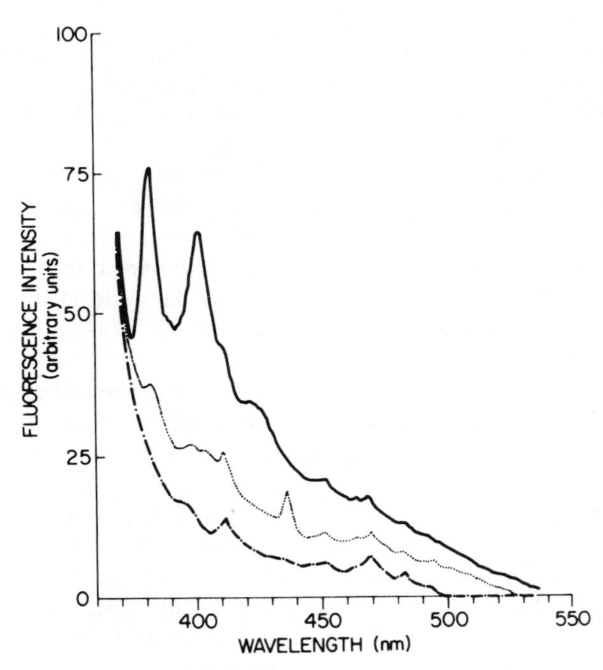

Figure VI-1. Effect of temperature on fluorescence emission spectra. ——,16 A_{260} units of DNA modified with 1 residue of BP/45,000 nucleotide residue measured at 77°K. · · · · ·, The same sample measured at 300°K. – · – · – · –, Control DNA measured at 77°K. From Ivanovic *et al.* (1976) *Biochem. Biophys. Res. Commun.* **70,** 1172.

carcinogenic form of the parent agent. Reaction conditions for some specific carcinogens will be described later in this chapter.

After thorough extraction of the noncovalently bound derivatives with organic solvents, the modified homopolymers are digested to deoxyribonucleosides using a combination of both endonucleases and exonucleases. Good results in the treatment of the polymers were first achieved with DNase I, followed by snake venom and/or spleen phosphodiesterase and alkaline phosphatase. Since the modified deoxyribonucleosides are hydrophobic, they can be separated from unmodified nucleosides by Sephadex LH-20 column chromatography (Figure VI-2). The nucleoside–carcinogen adducts elute later from the column. Further resolution of these products can be achieved by reverse-phase HPLC. For the final chemical identification of the purified nucleoside adduct, a variety of spectral techniques are used, including NMR spectra, field desorption, and electron impact mass spectrometers, as well as circular dichroism and Fourier-transformed infrared spectra. Once the structure of such an adduct is determined it can be used as a marker for comparisons with the DNA products formed *in vivo* upon reaction with radioactively labeled parent carcinogens. The radioactive DNA isolated from these cells is enzymatically hydrolyzed as described previously. Finally the similarity between the marker and the labeled modified nucleosides isolated from the *in vivo* experiment is compared by HPLC (Figure VI-3). If the radioactivity coelutes with the marker which is detected by UV or fluorescence spectra, the identity of the nucleoside adduct formed *in vivo* can be regarded as established. Some specific cases will be presented in this chapter. Sometimes further derivatization of the marker and labeled adducts is necessary to obtain definite identification of the product.

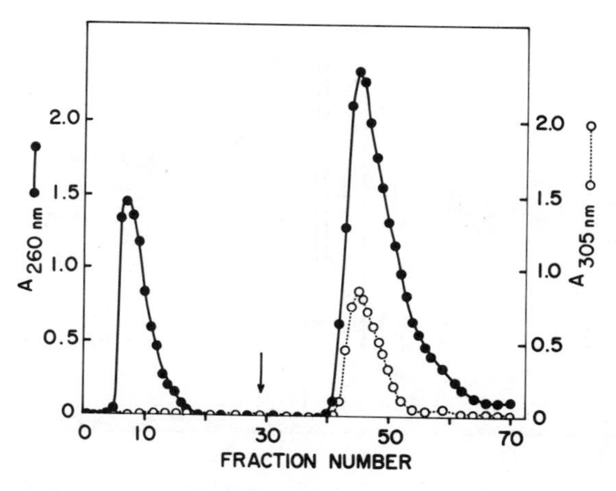

Figure VI-2. Separation of AAF-modified deoxyguanosine by Sephadex LH-20 chromatography. Elution with 0.01 M ammonium bicarbonate and (at arrow) elution with ethanol–0.02 M ammonium bicarbonate. ●–●, absorbance at 260 nm; ○- -○, at 305 nm (AAF residues absorb).

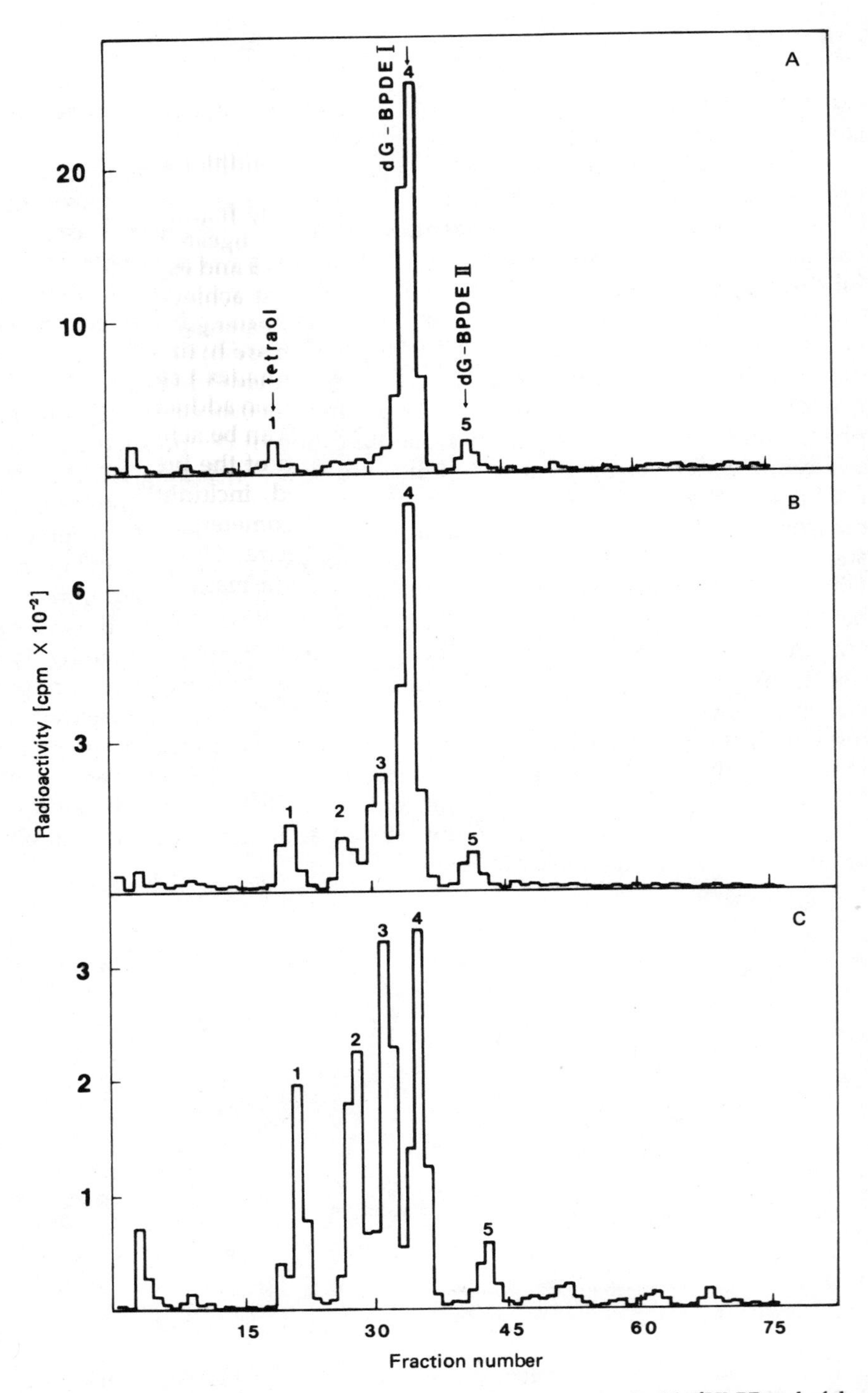

Figure VI-3. HPLC profiles of DNA from human keratinocytes treated with [^{3}H]-BP and of deoxyguanosine adducts synthesized *in vitro* with BPDE I and II. (A) Synthetic markers; (B, C) two keratinocyte cultures. From Theall *et al.* (1981) *Carcinogenesis* **2,** 581.

2. *Postlabeling Methods to Detect DNA–Carcinogen Adducts*

The methods for detection of DNA–carcinogen adducts thus far described require compounds in radioactive form or possessing some specific fluorescence or UV spectral characteristics. It is also possible to use a chemical test for detecting the binding of agents to DNA by a combination of ^{32}P-postlabeling and chromatographic techniques.

The scheme of ^{32}P-postlabeling of DNA adducts according to Randerath is presented in Figure VI-4. The ^{32}P label can be introduced into DNA constituents after exposure of the DNA to a nonradioactive mutagen or carcinogen. Following treatment of DNA with the agent, the unlabeled DNA is isolated and digested to nucleotides by the mixture of DNase I or micrococcal endonuclease and spleen phosphodiesterase. The mixture of digested DNA containing deoxynucleotide 3′-monophosphates is reacted with [γ-^{32}P]-ATP and T4 polynucleotide kinase. After removal of unreacted [^{32}P]-ATP, mixture of authentic dpGp, dpAp, dpCp, and dpTp and the modified unknown ^{32}P-labeled deoxynucleoside 3′,5′-diphosphates are chromatographed on PEIcellulose thin layers. Another technique is to treat the modified diphosphate with T4 polynucleotide kinase containing 3′-phosphatase activity. In this case the modified deoxynucleoside 5′-monophosphates can be separated by HPLC using a reverse-phase column. In this separation all four 5′-nucleotides are resolved in addition to new peaks that result from modification of nucleotides. Given the high specific activity of [γ^{32}P]-ATP available and the high resolution of HPLC, infrequent base modification at a rate of one in 10^4–10^5 should be detectable.

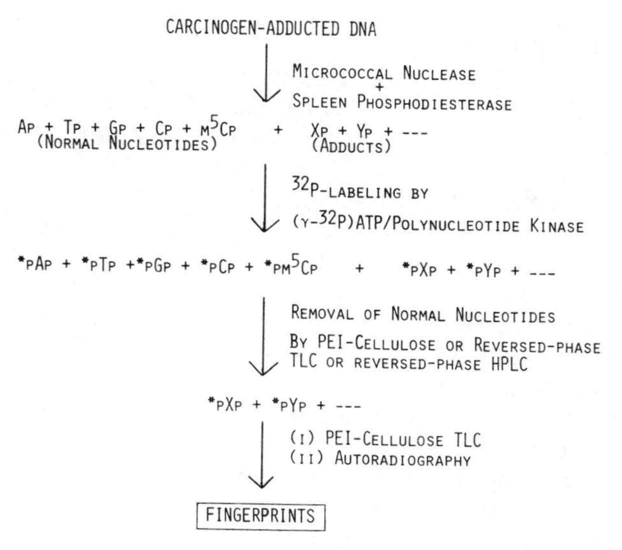

Figure VI-4. Scheme of ^{32}P-postlabeling of DNA adducts. Based on Randerath *et al.* (1981) *Proc. Natl. Acad. Sci. USA* **78,** 6126 and *Carcinogenesis* **3,** 1081 (1982).

These rapid and sensitive assays should facilitate direct detection of the covalent binding of chemicals to DNA in order to assess possible correlations between DNA damage *in vivo* and carcinogenesis. Such approach may also be suitable to detect DNA modifications in human tissues exposed to environmental agents.

3. *Radioimmunoassay of DNA–Carcinogen Adducts*

Recently antisera have been raised in rabbits against the DNA adducts of aromatic amines, polycyclic aromatic hydrocarbons, and alkylating carcinogens. It was then possible to use these specific antibodies to assay for chemical carcinogens bound to DNA. Examples of adducts quantitated by this technique include the major deoxyguanosine adduct formed between BP and DNA, between AAF and DNA, and between aflatoxin B_1 and DNA.

The determination of carcinogen–DNA adducts by such immunological procedures has certain advantages over other techniques. The sensitivity is usually higher than that obtained with radioactive carcinogens. The specificity of the antibodies can be very high since carcinogen–DNA antibodies do not cross-react appreciably with structurally dissimilar adducts of the same carcinogen, the carcinogen alone, unmodified nucleosides, or unmodified DNA. The assays are rapid, highly reproducible, and can be used in cases when the radiolabeled compound is not available. The high sensitivity of detecting nonradioactive carcinogen–DNA adducts may be of great value in monitoring human tissues. The immunological techniques can also be applied in combination with morphologic procedures (immunofluorescence and electron microscopy) to localize adducts in particular cells, subcellular fractions, and DNA molecules *in vitro* and *in vivo*.

The conventional RIA can be used to detect the presence of a carcinogen–DNA adduct. In this assay two chemically identical haptens compete for the same antibody binding site. When the concentration of one hapten is constant, the concentration of the other is reflected in the degree to which it inhibits binding of the constant hapten to the antibody. An example of an RIA standard curve by Poirier and co-workers in which the binding of [^{3}H]dG-BPDE adduct to specific antiserum was measured in the presence of the immunogen BPDE-DNA or unmodified DNA is presented in Figure VI-5.

The limitation of RIA in using radioactive isotopes was eliminated by solid-phase immunoassay (also known as enzyme-linked immunosorbent assay or ELISA). In this system an enzyme such as alkaline phosphatase is used as the immunoglobulin marker, instead of a radioactive isotope. This enzyme–antibody conjugate is bound to the solid phase by a series of antibody–antigen reactions and essentially converts a substrate, e.g., *p*-nitrophenylphosphate, to products with a visible yellow color that can be measured spectrophotometrically. Since the sensitivity of the ELISA system has not significantly exceeded that of RIA, the combination of the advantages of both

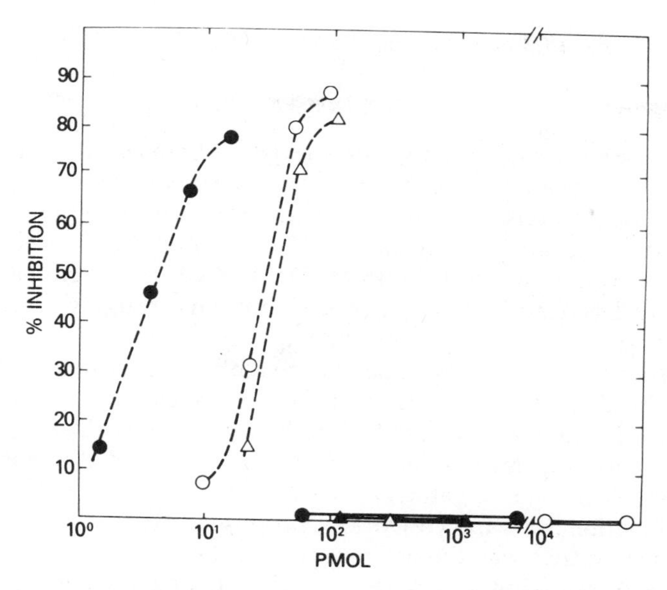

Figure VI-5. Radioimmunoassay standard curves in which the binding of BPDE I-[^{3}H]dG to specific antiserum was measured in the presence of increasing amounts of the immunogen BPDE I-DNA (●- -●), BPDE I-dG (○- -○), BPDE II-dG (△- -△), unmodified DNA (○–○), BPDE I-tetrol (●–●). From Poirier *et al.* (1980) *Cancer Res.* **40,** 412.

systems led Harris and co-workers to the development of an ultrasensitive enzymatic radioimmunoassay (USERIA). This system uses [^{3}H]-AMP as substrate, which is converted by the antibody-bound alkaline phosphatase into [^{3}H]adenosine. USERIA is approximately 500-fold more sensitive than RIA and 5-fold more sensitive than ELISA for detection of DP-DNA adducts. As little as 1 BP-DNA adduct per 7×10^6 nucleotides can be detected.

Even more sensitive techniques were achieved by using the selected hybridomas that produce monoclonal antibodies specific for aflatoxin B_1-DNA adducts. Using this antibody it is possible by USERIA to measure 15 fmoles of the adducts in 10 mg of DNA. This method is at least 100-fold more sensitive than the standard ELISA procedure. With these monoclonal antibodies the ELISA assay reliably quantitated 1 aflatoxin B_1 residue per 250,000 nucleotides in DNA. It should be noted that simple alkyl modified nucleosides can also be detected with great sensitivity (Table IV-13).

Generally, the immunoassays using either antibodies raised in rabbits or monoclonal antibodies against carcinogen–DNA adducts serve as useful analytical tools for studying not only the carcinogen–DNA adduct interactions but also the occurrence of these adducts in biological specimens from people exposed to different environmental carcinogens. However, this method can be used only with chemically well-characterized adducts and does not serve for detection of unknown DNA products.

B. Polycyclic Aromatic Hydrocarbons

1. *Benzo[a]pyrene (BP)*

The first strong evidence for the correlation between the binding of different polycyclic aromatic hydrocarbons to cellular macromolecules and carcinogenicity was an observation of Brookes and Lawley. When they applied ^3H-labeled hydrocarbons to the skin of mice, a small fraction of the administered dose became covalently bound to DNA, RNA, and proteins. The extent of binding to DNA and RNA, but not to proteins, roughly correlated with carcinogenic potency of the compounds applied.

In subsequent studies it became clear that BP requires metabolic activation in order to bind to macromolecules. This was achieved, *in vitro,* in a microsomal system in the presence of NADPH and O_2, as described in Chapter V. After it was found in microsomal systems that the 7,8-diol metabolite of BP bound to DNA to a far greater extent that BP or any other known diol or phenol metabolite, Sims and co-workers proposed that the diol was converted to a diol epoxide that was the active form that binds to DNA.

Chapter V contains a detailed description of enzymatic formation of (−)-*trans*-7,8-diol of BP and its further metabolism to a pair of diastereomeric 7,8-diol-9,10-epoxides in which the 7-OH is either *anti* (diol epoxide I, BPDE I) or *syn* (diol epoxide II, BPDE II) to the oxirane ring. Both diastereomeric diol epoxides of BP are among the most mutagenic compounds tested. The diol epoxide I is also highly tumorigenic on mouse skin. On the other hand, chemical synthesis of diol epoxides yields a racemic mixture of the 7*R* or (+) and 7*S* or (−) enantiomers of BPDE I.

The strongest evidence of diol-epoxides as ultimate carcinogenic metabolites comes from nucleic acid binding studies. The major product of nonenzymatic reaction (±) BPDE I with DNA, following degradation to the nucleoside level, was compared chromatographically with the DNA adducts formed in rodent, bovine, and human cells exposed to BP (Figure VI-3). In this way it was possible to confirm that the major nucleoside product formed from reactions of BPDE with polydeoxyguanosine *in vitro* were identical to those formed in DNA as a result of metabolic activation of BP by several cell systems as human bronchial segments, human epidermal keratinocytes, 10 T½ cells, etc.

One of the most important steps in identifying *in vivo* products has been to unambiguously determine the structures and stereochemistry of these modified nucleosides. Using a variety of techniques, including NMR spectra, high-resolution mass spectra, and circular dichroism spectra, Weinstein and co-workers were able to conclude that when the chemically synthesized BPDE I reacted with polydeoxyguanosine, the carbon of the 10 position of BPDE I is linked to the 2-amino group of the guanine residue (Figure VI-6). In addition, when BPDE I reacts with native DNA, the ratio of reactivity of the 7*R* and 7*S* enantiomers with G residues is about 19 : 1. Interestingly, when BP

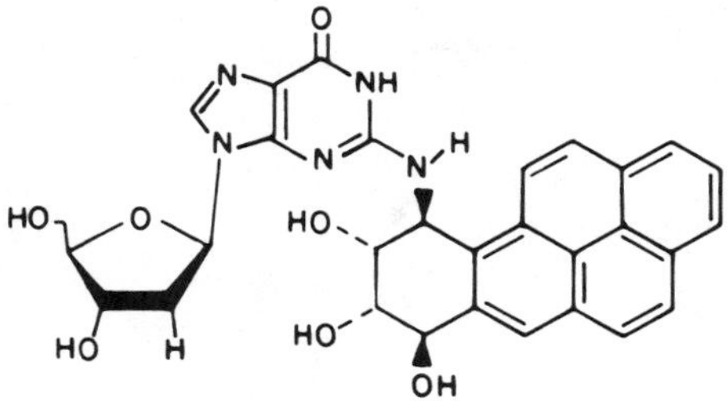

Figure VI-6. Structure of the major dG-BPDE I adduct. The N^2 of G is bound to the C-10 position of the BP moiety. The G residue is *trans* to the 9-OH, and the 8- and 7-OH are *trans*. Based on Weinstein *et al.* (1976) *Science* **193,** 592.

is used, in cellular systems, the major adduct is also derived from the 7*R* enantiomer of BPDE I. The product is formed by *trans* addition of the amino group of guanine to the 10 position of the diol epoxide. Similar adducts are also formed between BPDE II and guanine residues. Although the ratio of BPDE I to BPDE II adducts may vary in particular tissues and species, the major product in all cases studied is that formed with BPDE I. This is in good agreement with the "bay region" theory described in Chapter V.

Besides this predominant adduct, it was also indirectly observed that BPDE reacts with the N-7 position of guanine. This adduct has been difficult to characterize since the glycosyl bond of the adduct in DNA is extraordinarily labile. The proposed structure of this adduct is shown in Figure VI-7.

In DNA reacted with BPDE, guanine is not the only base involved in interaction. When DNA was reacted with BPDE *in vitro*, adenine as well as cytosine products were detected. Structure determination by Jeffrey and coworkers showed that the adenine adduct is derived by the *trans* addition of the 6-amino group of adenine to the 10 position of the diol epoxide (Figure VI-8). It is interesting that investigation of the dA-BPDE adduct formed in plasmid DNA showed that the 7*S* adduct predominated, which is in contrast with results found with dG. Although the precise structure of BPDE-C adducts has not yet been determined, it was proposed from indirect measurements that the exocyclic amino group of cytosine is involved in binding to the 10 position of the diol epoxide. Therefore, we can conclude that when BPDE

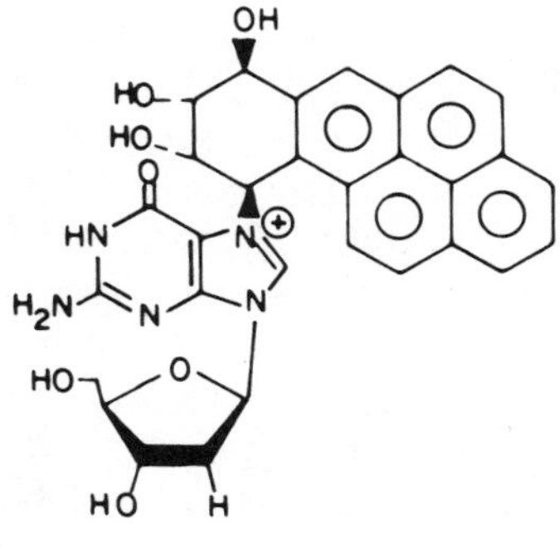

Figure VI-7. Proposed structure of guanine-N^7-BPDE I adduct. Based on Osborne *et al.* (1978) *Chem. Biol. Interact.* **20,** 123.

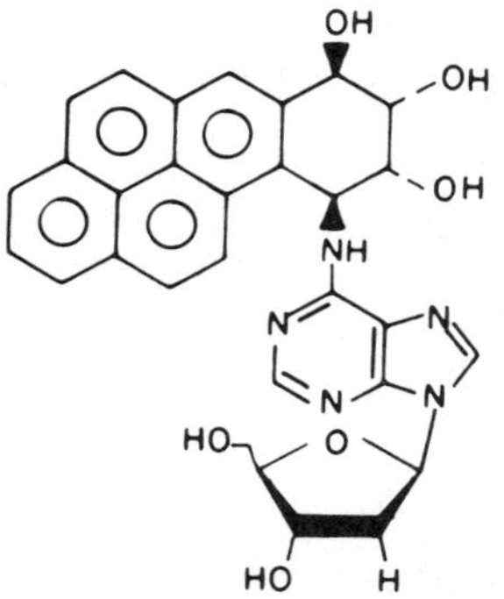

Figure VI-8. Structure of adenosine adduct with BPDE I. The N^6 of adenine is bound to the C-10 position of the BPDE residue. The A residue is *trans* to the 9-OH, and the 8- and 7-OH are *trans*. Based on Jeffrey *et al.* (1979) *Science* **206,** 1309.

reacts with DNA the exocyclic amino groups of all the bases bind to the 10 position of the diol epoxide.

Although the predominant DNA-binding metabolite of BP is the diol epoxide, several other metabolities also bind to DNA. In the early studies the electron-rich "K-region" BP-4,5-oxide was to be the most likely reactive intermediate capable of alkylating cellular macromolecules. Chemical synthesis of BP-4,5-oxide made it possible to react it with DNA in aqueous solution. The products of the reaction were degraded enzymatically to the nucleoside level and separated by a Sephadex LH-20 column chromatography. In the elution volume, the position of resulting BP–nucleoside products failed to correspond to those of BP–DNA adducts isolated from mouse embryo cells treated with [^{3}H]-BP. These experiments represent the strongest evidence against BP-4,5-oxide as ultimate carcinogenic metabolite of BP.

In the microsome-catalyzed binding of BP to DNA, an additional adduct was detected. It was suggested that by further metabolism of 9-OH-BP metabolite, 4,5-epoxy-9-OH-BP was formed that in turn bound to DNA. This pathway may be of some importance since 9-OH-BP is a major metabolic intermediate bound to DNA in hepatocytes grown in culture. Studies with isolated rat liver nuclei have also shown that in addition to diol epoxides, both 9-OH-BP and 3-OH-BP when further metabolized were covalently bound to DNA. The efficiency of this binding, however, was much lower than that with diol epoxides. The structures of the suggested products are not yet known.

The participation of the 6-oxy-BP radical in DNA binding has also been indicated. The possible role of free radicals in DNA binding and carcinogenicity is not well understood.

Taken together, results with binding BP metabolites to DNA strongly suggest that the BP-diol epoxides are the most important metabolites involved in binding to DNA and to result in mutagenesis as well as in carcinogenesis. These results, however, do not exclude the possibility that other BP metabolites that can bind to DNA may have some importance in carcinogenesis, especially in specific species or tissues.

2. *Benz[a]anthracene (BA)*

The "bay-region" 3,4-diol-1,2-epoxide of BA is highly active both as mutagen and as tumor initiator (Chapter V). A non-bay-region 8,9-diol-10,11-epoxide is also involved in metabolic activation of BA in hamster embryo cells. When DNA was isolated by Sims and co-workers from either hamster embryo cells or mouse skin treated with BA, hydrolyzed to nucleosides, and chromatographed on Sephadex LH-20 and HPLC, two adducts were obtained. Figure VI-9 shows that a nonlabeled deoxyribonucleoside–hydrocarbon adduct arising from *anti*-BA-8,9-diol-10,11-oxide is indistinguishable from a major ^{3}H-labeled product prepared from DNA of the skin of mice or hamster embryo cells treated with ^{3}H-labeled BA. Thus, BA-8,9-diol is also a metabolite of BA and can be converted into the *anti* isomer of BA-8,9-diol-10,11-oxide, which in turn reacts with DNA. The second adduct probably arises from the reaction of the "bay-region" *anti*-3,4-diol-1,2-epoxide with DNA. Hence, both a bay-region and a non bay-region diol epoxide appear to be involved in the metabolic activation of BA and in binding to nucleic acids.

Concerning the structures of DNA-BA adducts there is some evidence that when *anti*-BA-8,9-diol-10,11-oxide and *anti*-BA-3,4-diol-1,2-oxide are incubated with DNA, the hydrocarbon becomes attached to the exocyclic amino group of guanine in a similar way as described for BP-7,8-diol-9,10-epoxide. Figure VI-10 shows the proposed structure of guanosine-*anti*-BA-8,9-diol-10,11-oxide adduct. It is evident that the 2-amino group of guanosine becomes attached to the benzylic 11 position rather than to the non benzylic 10 position.

When *syn*- and *anti*-BP-7,8-diol-9,10-oxide were incubated with poly(G), the orientation of the amino group of the guanine to the 10 position of the hydrocarbon was mainly *trans,* relative to the epoxide oxygen. By analogy, the linkage of the amino group of G to the 11 position of *anti*-BA-8,9-diol-10,11-oxide was tentatively assigned as resulting also from *trans* addition.

Both the *anti*-3,4-diol-1,2-oxide and the anti-8,9-diol-10,11-oxide reacted equally well with double- and with single-stranded DNA. It appeared that the 3,4-diol-1,2-oxide also reacted with the 2-amino group of guanine.

Unlike BP adducts, other BA–DNA adducts were not well characterized. Although the *anti*-BA-8,9-diol-10,11-oxide reacted to a high extent with dAdo, only traces of an adenosine adduct were detected in DNA hydrolysates that had been reacted with this diol epoxide.

In conclusion, the "bay-region" 3,4-diol-1,2-epoxide binds covalently to nucleic acids and is highly active as a mutagen and carcinogen. On the other hand, the significance of the formation of DNA–BA adducts derived from the non-bay-region 8,9-diol-10,11-oxide in BA-treated cells is not yet clear. The biological activity of this diol oxide is lower than that of BA and much lower than that of "bay-region" oxide. This is a case in which the "bay region" theory failed to predict correctly all the metabolites involved in the binding of polycyclic aromatic hydrocarbons to DNA. Such a failure in prediction is expected since the "bay region" theory, as proposed, does not allow for differences in the formation and further metabolism of different diol epoxides.

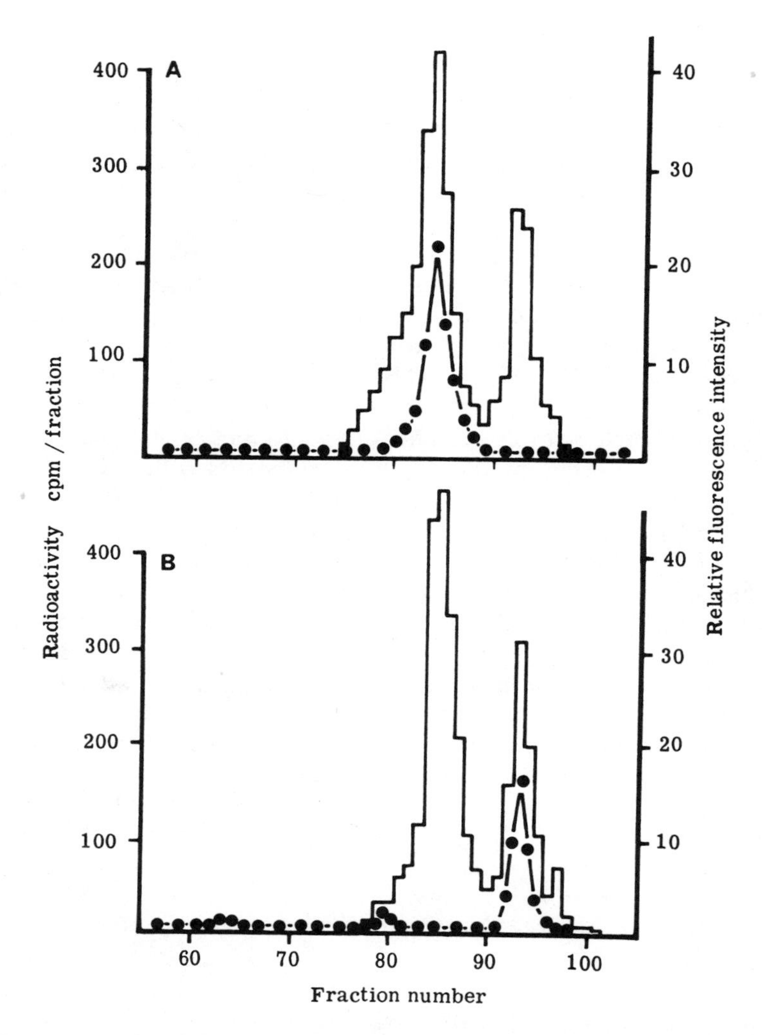

Figure VI-9. HPLC of ^{3}H-labeled acetylated nucleoside–BA adducts present in hydrolysates of DNA from hamster embryo cells treated with [^{3}H]-BA and nonlabeled adducts isolated from DNA that had been incubated with BA-diol epoxides *in vitro.* (A) *Anti*-BA-8,9-diol-10,11-oxide; (B) *anti*-BA-3,4-diol-1,2-oxide. From Cooper *et al.* (1980) *Carcinogenesis* **1,** 233.

Therefore, other factors will have to be taken into consideration if the ultimate carcinogenic forms of polycyclic aromatic hydrocarbons are to be predicted reliably.

3. *7-Methylbenz[a]anthracene (MBA)*

The metabolism of MBA in rat liver homogenate or skin organ culture yields five dihydrodiols. This is described in Chapter V. Fluorescence spectra

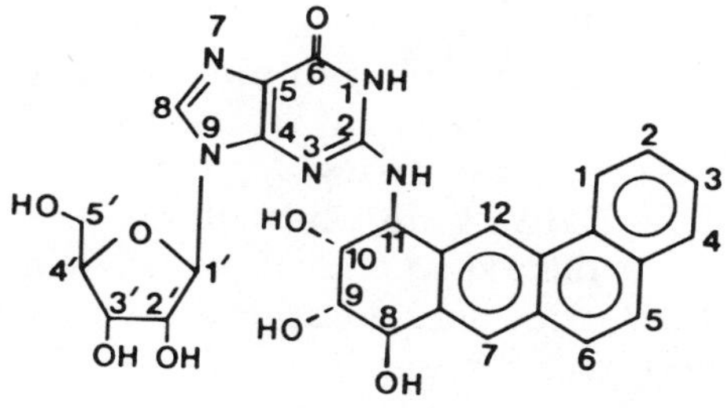

Figure VI-10. Structure of guanosine-*anti*-BA-8,9-diol 10,11-oxide adduct. From Carey *et al.* (1980) *Carcinogenesis* **1,** 505.

of DNA isolated from either mouse skin or hamster embryo cells treated with MBA showed anthracene-type of spectra, suggesting that reaction with DNA involved the 1,2,3,4 ring of the hydrocarbon. The 8,9,10,11 ring involvement in DNA binding would have resulted in fluorescence spectra characteristic of phenanthrene. The spectrum of the nucleoside adduct was similar to that of the adduct obtained in the reaction of DNA with the "bay-region" 3,4-diol-1,2-epoxide and not with 8,9-diol-10,11-oxide (Figure VI-11).

The chromatographic properties of the nucleoside–MBA adducts obtained from DNA isolated from hamster embryo cells or mouse skin treated

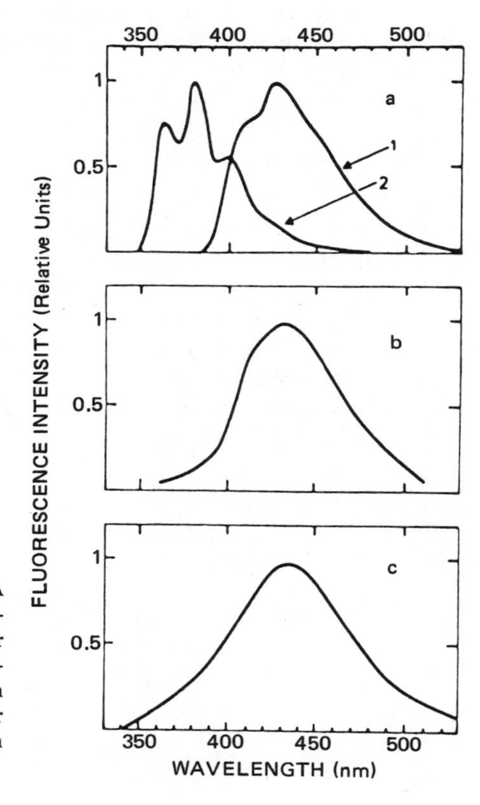

Figure VI-11. Fluorescence spectra of DNA (a) treated with 3,4-dihydro-3,4-dihydroxy-7-methyl-benzanthracene-1,2-oxide (curve 1) or with 8,9-dihydro-8,9-dihydroxy-7-methyl-benzanthracene-10,11-oxide (curve 2) or isolated from (b) hamster embryo cells or (c) mouse skin after treatment with 7-methylbenzanthracene. From Vigny *et al.* (1977) *FEBS Lett.* **75,** 9.

with [^{3}H]-MBA were similar only to the adducts formed from mouse skin that had been treated with the 3,4-dihydrodiol. Moreover, the major nucleoside adduct detected in hydrolysates of DNA reacted with the 3,4-diol-1,2-epoxide possessed chromatographic properties similar to those of the adduct isolated from DNA of MBA-treated mouse skin.

Unlike DNA adducts with BA, in the binding of MBA the 8,9-dihydrodiol-10,11-oxide does not appear to be involved. Based on the fluorescence spectral and chromatographic evidence, the ultimate carcinogenic form of MBA is likely to be the "bay-region" 3,4-diol-1,2-epoxide, which does form adducts with DNA. The site of binding of the activated MBA in DNA remains to be determined.

4. *7,12-Dimethylbenz[a]anthracene (DMBA)*

The metabolism of DMBA (described in Chapter V, Section B) yields several dihydrodiols and also metabolites with hydroxylated methyl group(s). Therefore, determination of the ultimate carcinogenic form that binds to DNA and the structure assignment of the nucleoside–DMBA adducts are more difficult than with BP. Evidence of the involvement of all four rings of DMBA in the reaction with DNA came from fluorescence spectra. DNA isolated from hamster embryo cells incubated with DMBA exhibits fluorescence excitation and emission spectra similar to those of 9,10-dimethylanthracene (Figure VI-12). The major hydrocarbon–deoxyribonucleoside adducts present in the hydrolysate of DNA isolated from mouse skin treated with DMBA also possessed anthracene-like and not phenanthrene-like spectra. Thus, activation on the 1, 2, 3, 4 positions of DMBA appear to be implicated in the reaction of DMBA with DNA. Direct evidence that DMBA is metabolically activated in the "bay region" to 3, 4-diol1,2-oxide that then binds to DNA is not yet available, since this compound has not been prepared. Jeffrey and co-workers have compared DNA adducts prepared by microsomal or *m*-chloroperbenzoic acid oxidation of DMBA-3,4-dihydrodiol. The products from both reactions were similar. Enzyme digestion of DNA isolated from hamster embryo cells or human bronchial or esophageal samples exposed to [^{3}H]-DMBA yielded on HPLC a major component identical with the adduct found in the microsomal or chemically oxidized 3,4-diol system (Figure VI-13).

Structural determination of this adduct has come from its field desorption mass spectrum and fluorescence, UV, and CD spectra. From these results it was suggested that epoxidation was in the 1, 3 positions and binding occurred to the N^2-amino group of guanine as described for BPDE (Figure VI-14). This major component represented about 50% of the cell adducts detected by HPLC. The characterization of the minor components, of which there were at least five, has not yet been accomplished.

Although 8,9-dihydrodiol is highly mutagenic, no adducts corresponding to products from 10,11-epoxidation of this dihydrodiol were detected. Similarly no DNA adducts coming from 7- or 12-hydroxymethyl derivative have thus far been characterized.

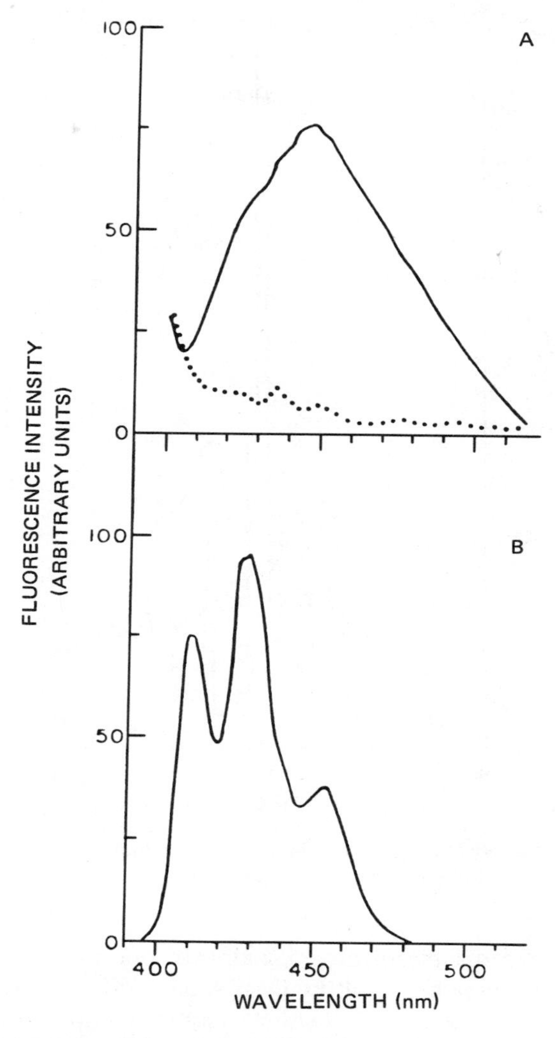

Figure VI-12. Fluorescence emission spectra of DNA modified by *in vivo* exposure of hamster embryo cells to DMBA compared to 9,10-dimethylanthracene: (A)——, DNA modified by 1 residue of DMBA per 10^4 nucleotides; · · · · ·, control DNA. (B) Spectrum of 9,10-dimethylanthracene. From Ivanovic *et al.* (1978) *Cancer Lett.* **4**, 131.

When Dipple and co-workers treated DNA with [^{3}H]-DMBA in rat hepatic microsomal systems, they found that the hydrocarbon–nucleoside adducts differed from those arising from DMBA treatment of mouse skin or mouse embryo cells in culture. In the microsomal system, especially at higher DMBA doses, the products arise mainly through the reaction of the K-region 5,6-epoxide with DNA. Structures of four guanosine adducts formed by the reaction of DMBA-5,6-oxide with poly(G) have been determined. In all four compounds the hydrocarbon residue is attached to the N^2 group of guanine.

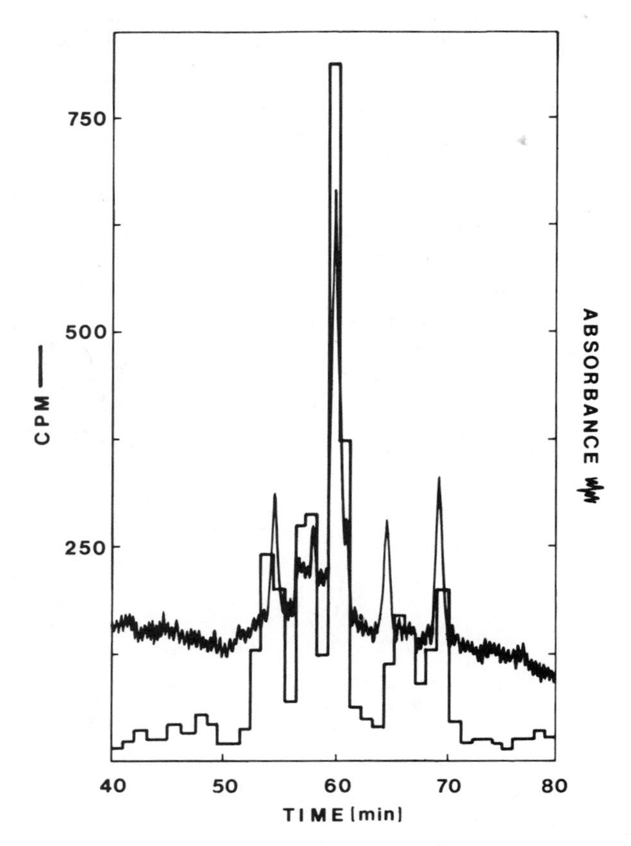

Figure VI-13. HPLC of enzymatic digest of DNA isolated from human bronchial explants exposed to [^{3}H]-DMBA. Courtesy of A. M. Jeffrey.

They constitute two diastereoisomeric pairs. In one pair the adduct results from addition at the 6 position of the DMBA-5,6-oxide, whereas in the other pair the attachment is at the 5 position (Figure VI-15). Three minor RNA adducts representing about 5% of total adducts were also identified in rat liver cells treated with [^{3}H]-DMBA (Figure VI-16). In one product the 2′-hydroxy group of the ribose moiety of guanosine was linked to the C-5 (1a,

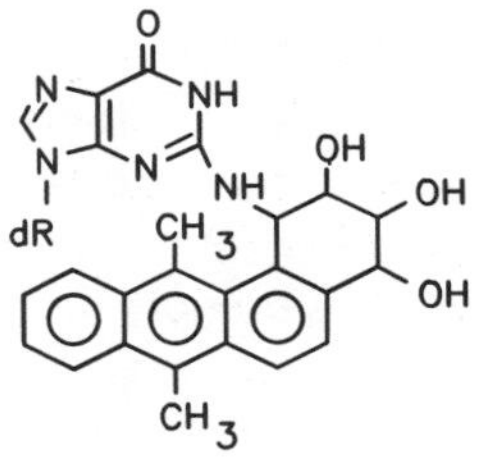

Figure VI-14. Proposed structure of DMBA-G adduct. Courtesy of A. M. Jeffrey.

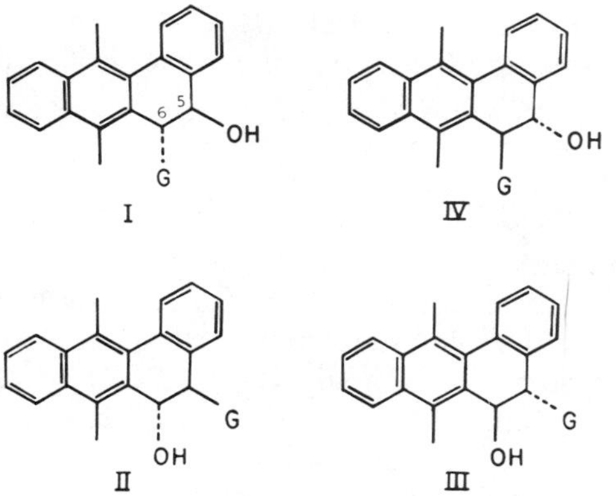

Figure VI-15. Structures of DMBA-5,6-oxide-guanosine adduct. In all four adducts DMBA is linked to the N^2 of G. In diastereoisomeric pair I and IV the N^2 is linked to the 6 position, and in II and III to the 5 position of the DMBA residue. From Jeffrey *et al.* (1976) *Proc. Natl. Acad. Sci. USA* **73,** 2311.

Figure VI-15) and in the second to the C-6 position of DMBA-5,6-oxide (1b). In the third product the C-8 position of guanine was linked to the C-5 of the 5,6-oxide (2).

In summary, it seems that the types of adducts formed between DMBA metabolites and nucleic acids are different in microsomal systems and in whole cell systems. In microsomal systems, especially at higher doses, the K-region epoxide is mainly involved in the binding to nucleic acids. At lower doses of DMBA, the diol epoxide adducts are also formed. On the other hand, in cellular systems, such as mouse skin, mouse embryo cells, hamster embryo cells, and human skin, bronchial, and esophageal segments, the "bay-region" diol epoxide type adducts are mainly formed, independent of the doses used.

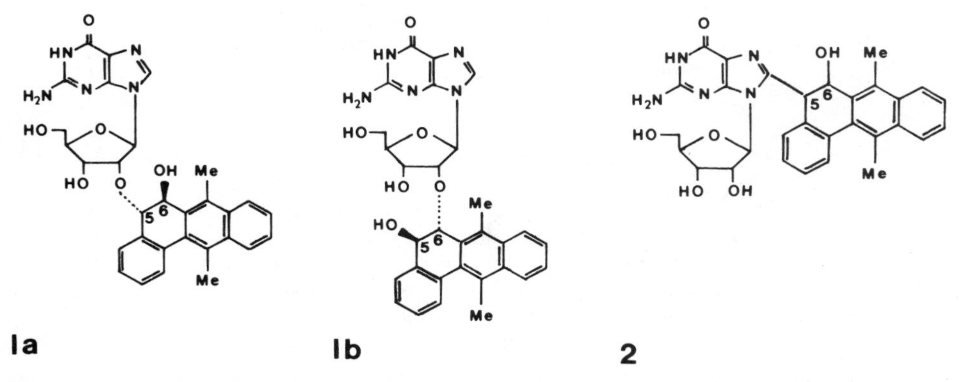

Figure VI-16. Structures of DMBA-5,6 oxide-rG adducts detected in RNA of rat liver cells treated with DMBA. From Frenkel *et al.* (1981) *Biochemistry* **20,** 4377.

At present, all the adducts characterized are reaction products with the guanine residues. However there is a possibility that binding of DMBA derivatives to other nucleosides will also be detected.

5. 3-Methylcholanthrene (3MC)

Since metabolism of 3MC yields not only diol epoxides but also several hydroxylated derivatives, formation of multiple DNA adducts could be expected. At least eight major deoxyribonucleoside adducts and one minor adduct were resolved by HPLC analysis of a DNA hydrolysate obtained after treatment of mouse skin with ^{3}H-labeled 3MC. The DNA hydrolysate subjected to chromatography on Sephadex LH-20 resolved into three peaks in the region expected for nucleosides–hydrocarbon adducts (Figure VI-17). Product I was further resolved by HPLC into three major components (Ia, Ib, Ic, Figure VI-18). Product II from Sephadex LH-20 was similarly resolved into three components and product III was resolved into two major components.

The fluorescence spectra of most of the resolved products are anthracene-like (Figure VI-19). These observations indicate that the metabolites of 3MC involved in the formation of DNA adducts in mouse skin are most probably derivatives containing saturated 7,8,9,10 ring. This is consistent with the hypothesis that the metabolic activation occurs through the formation of the "bay-region" diol epoxide, 3MC-9,10-diol-7,8-oxide, and other derivatives of diol epoxide, that retain the intact anthracene ring system. Similarly, a DNA hydrolysate from hamster embryo cells treated with 3MC was resolved into five nucleoside–hydrocarbon adducts. The fluorescence spectra of these adducts were similar to that of 7,8,9,10-tetrahydro-3-methylcholanthrene. Calf thymus DNA incubated with [^{3}H]-3MC and liver microsomes from Aroclor-pretreated rats also had anthracene-like fluorescence spectra.

From all these data it seems likely that some of the adducts are formed by the reaction of 3MC-9,10-diol-7,8-oxide with DNA, and some are formed

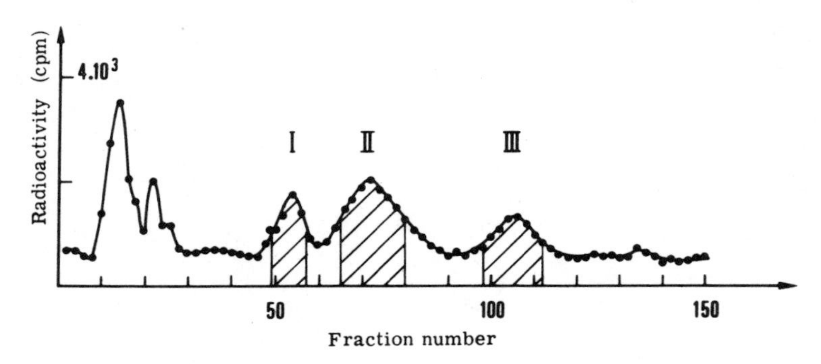

Figure VI-17. Sephadex LH-20 chromatography of DNA hydrolysate of [^{3}H]-3MC treated mouse skin. The three ^{3}H-labeled deoxyribonucleoside–hydrocarbon adducts are denoted by the hatched areas. From Cooper *et al.* (1980) *Carcinogenesis* **1,** 855.

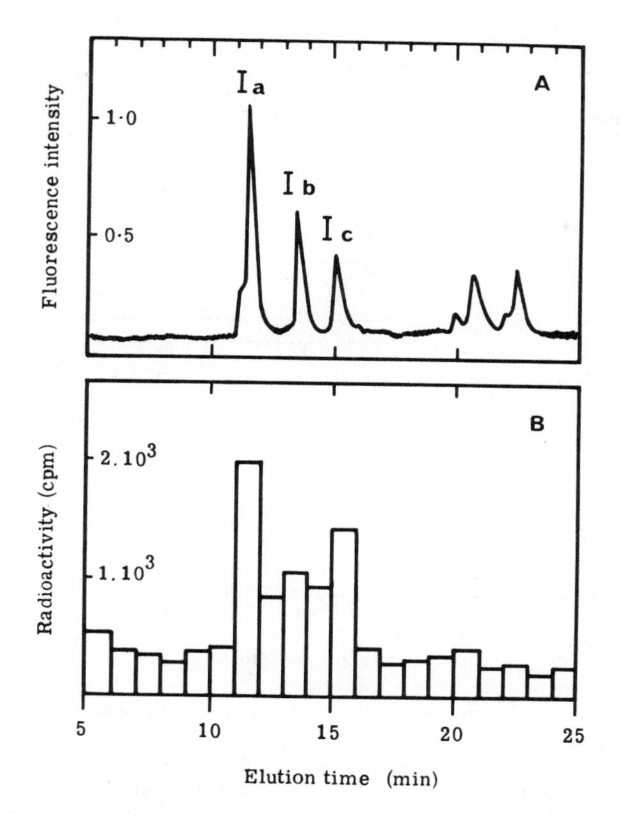

Figure VI-18. HPLC of deoxyribonucleoside–hydrocarbon product I from Sephadex LH-20 column of Figure VI-17. (A) Fluorescence and (B) radioactivity. From Cooper *et al.* (1980) *Carcinogenesis* **1,** 855.

by the reaction of the hydroxylated derivatives of this diol epoxide with DNA. Since 3MC-9,10-diol and 1-hydroxy- and 2-hydroxy-3MC, which can be metabolized into "bay-region" epoxide, are active in mutagenicity and carcinogenicity tests, it is possible to conclude that the "bay-region" diol epoxide, 3MC-9,10-diol-7,8-oxide, and one or more of its hydroxylated derivatives contribute to the biological activity of this carcinogen. Structural determination of 3MC–deoxynucleoside adduct has not yet been accomplished.

C. *N*-SUBSTITUTED AROMATIC COMPOUNDS

1. *N-2-Acetylaminofluorene (AAF)*

As described in the previous section on polycyclic aromatic hydrocarbons, binding of the ultimate carcinogen to DNA is now well established. One of the best studied interactions of carcinogens with macromolecules is that of

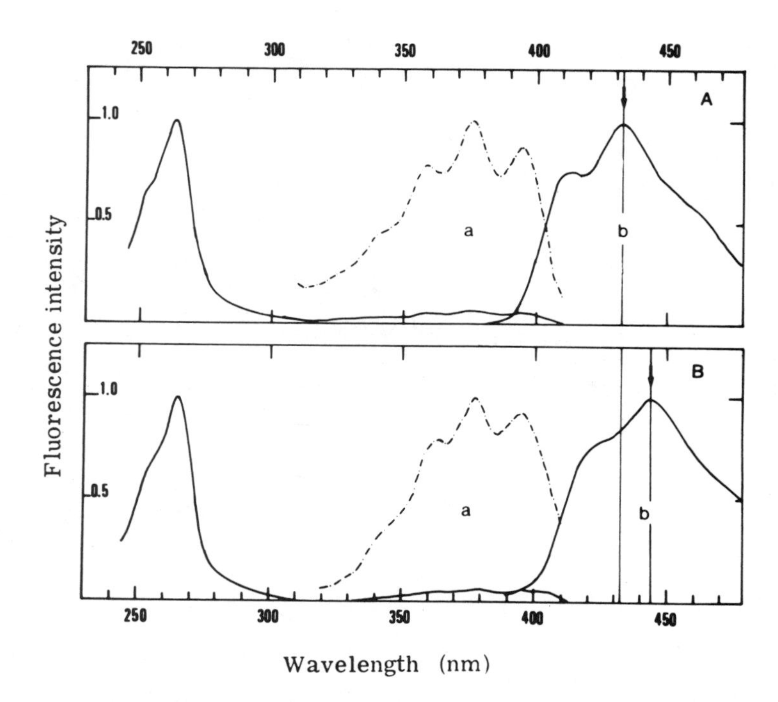

Figure VI-19. Fluorescence excitation (a) and emission (b) spectra of (A) a hydrocarbon–deoxyribonucleoside adduct (product 1a, Figure VI-18) prepared from mouse skin treated with MC and (B) hydrocarbon–deoxyribonucleoside adducts prepared from DNA treated *in vitro* with 7-methylbenz[*a*]anthracene-3,4-diol-1,2-oxide. From Cooper *et al.* (1980) *Carcinogenesis* **1,** 855.

covalent attachment of the reactive metabolites of *N*-substituted aromatic compounds to nucleic acids. Early in 1960, with the recognition of proximate and ultimate carcinogenic forms of carcinogens by the Millers, the ultimate carcinogen AAF was used to study its nonenzymatic reaction with nucleic acids. The major products formed in these reactions were characterized and shown to be identical with covalent adducts formed *in vivo* when the parent carcinogen or the proximate carcinogen *N*-OH-AAF was administered to rats. At least three types of DNA adducts were determined (Figure VI-20) by using Sephadex LH-20 column chromatography in combination with HPLC. In order to prepare these nucleoside derivatives as marker compounds, *N*-acetoxy-AAF was reacted with nucleosides or homopolymers. It was then shown that the major product (80%) obtained from hydrolysates of the modified DNA is *N*-(deoxyguanosin-8-yl)-AAF (Figure VI-21b). The remainng fraction of the AAF residues (20%) was identified by Kriek and co-workers as 3-(deoxyguanosin-N^2-yl)-AAF (Figure VI-21a). In contrast to the C-8 adduct of G, the N^2-G adduct was detected only in native DNA. When RNA or denatured DNA was reacted with *N*-acetoxy-AAF *in vitro,* only one product was formed that

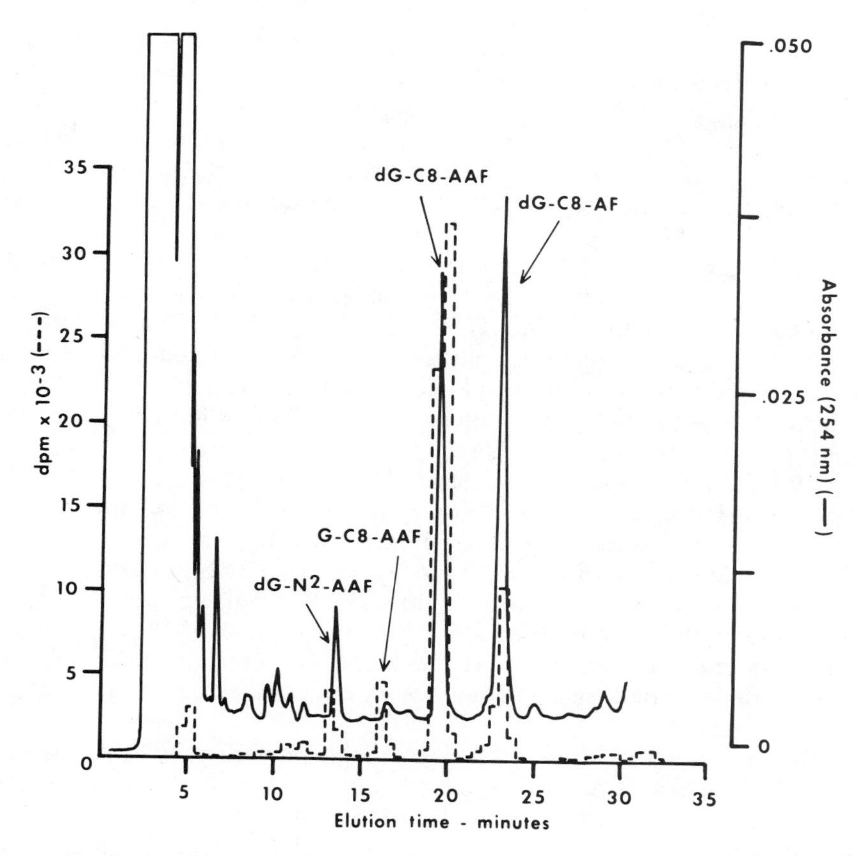

Figure VI-20. HPLC of DNA hydrolysate of rat liver cells treated with [^{3}H]-*N*-hydroxy-2-acetyl-aminofluorene. Adapted from Howard *et al.* (1981) *Carcinogenesis* **2**, 97.

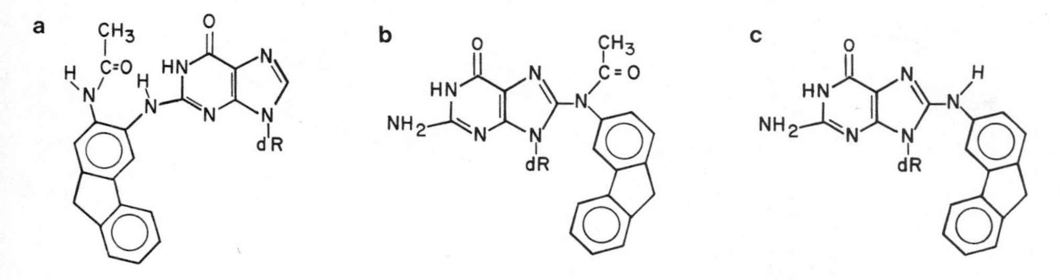

Figure VI-21. Structures of deoxyguanosine-*N*-2-acetylaminofluorene and -*N*-2-aminofluorene adducts. (a) 3-(Deoxyguanosin-N^2-yl)-AAF; (b) *N*-(deoxyguanosin-8-yl)-AAF; (c)*N*-(deoxyguanosin-8-yl)-AF.

bound to the C-8 of G. After administration of AAF to rats *in vivo,* a third DNA adduct was also detected. It represented a deacetylated form, *N*-(deoxyguanosin-8-yl)-AF (Figure VI-21c). A similar DNA adduct is also formed after application of *N*-OH-AAF in primary rat hepatocyte culture. The formation of the deacetylated arylamine substituted product can be the result of a deacetylation reaction or alternatively, of *N,O*-acyltransferase-catalyzed formation of *N*-acetoxy-AF.

The deacetylated product *N*-(deoxyguanosin-8-yl)-AF can be formed *in vitro* upon alkaline deacetylation of *N*-(deoxyguanosin-8-yl)-AAF or by treatment of DNA with *N*-OH-AF at pH 5. The AF-modified DNA is unstable and decomposes rapidly in aqueous solution. When AF-modified DNA is hydrolyzed with phosphodiesterases and alkaline phosphatase at pH 9.0, the dG-AF derivative decomposes, but hydrolysis of AF modified DNA in trifluoroacetic acid avoids decomposition.

Structural determination by Kriek and Westra showed that alkaline treatment of dG-AF residues results in two pyrimidine derivatives. From chemical and spectral analyses these pyrimidine derivatives have been identified as 1-[6-(2,5-diamino-4-oxopyrimidinyl-N^6-deoxyriboside)]-3-(2-fluorenyl) ureas, which probably are stereoisomers (Figure VI-22). Similar products were isolated from enzymatic hydrolysates of DNA reacted with *N*-OH-AF at pH 5.0 and subsequent treatment with 0.1 N NaOH. The presence or possible formation of ring-opened forms of dG-AF products in cellular DNA *in vivo* is not yet established.

Although reaction of *N*-acetoxy-AAF with polyadenylic acid was demonstrated, all attempts to show any reaction with adenine residues in DNA after treatment with *N*-acetoxy-AAF failed. Similarly, no reaction products of AAF with adenine have been detected in studies *in vivo.* Therefore, it is possible to conclude that administration of AAF *in vivo,* or interaction of active esters of AAF with nucleic acids *in vitro,* probably yields no products other than with guanosine residues.

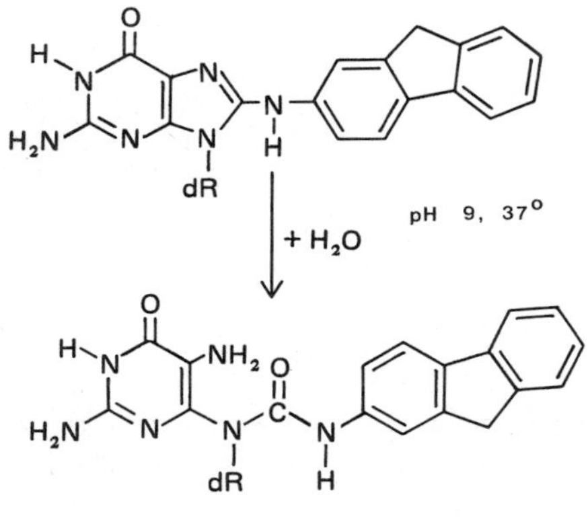

$+ H_2O$ pH 9, 37°

Figure VI-22. Structure of the pyrimidine derivative formed from *N*-(deoxyguanosin-8-yl)-AF at alkaline pH. Adapted from Kriek and Westra (1980) *Carcinogenesis* **1,** 459.

2. *N-Methyl-4-Aminoazobenzene (MAB)*

This compound represents one of the best-studied examples of the group of carcinogenic azo dyes. Determination of the products formed with nucleic acids became possible when the synthetic ester *N*-benzoyloxy-MAB became available. Interaction of this synthetic ultimate carcinogen with DNA *in vitro* yielded approximately 1 bound carcinogen residue per 10^3 nucleotides. After enzymatic hydrolysis of the DNA and HPLC analysis, at least six MAB adducts were detected (Figure VI-23). Only two products cochromatographed with MAB-DNA adducts formed in rat liver *in vivo* following oral administration of the radioactive precarcinogen MAB. Identification of these two products by mass, UV, and NMR spectroscopy presented evidence that the major product ("4," Figure VI-23) is formed by binding of the MAB residue to C-8 of dGuo (Figure VI-24a). This adduct was predominant after 8 hr treatment, but after 24 hr it represented only 50% of the original radioactivity. The adduct continued to decrease and after 7 days, it could not be detected. This is very similar to the formation and removal of C-8-dG-AAF adduct in DNA *in vivo*. The second minor adduct ("2," Figure VI-23) formed in rat liver *in vivo* and cochromatographed with one of the markers prepared *in vitro* was charac-

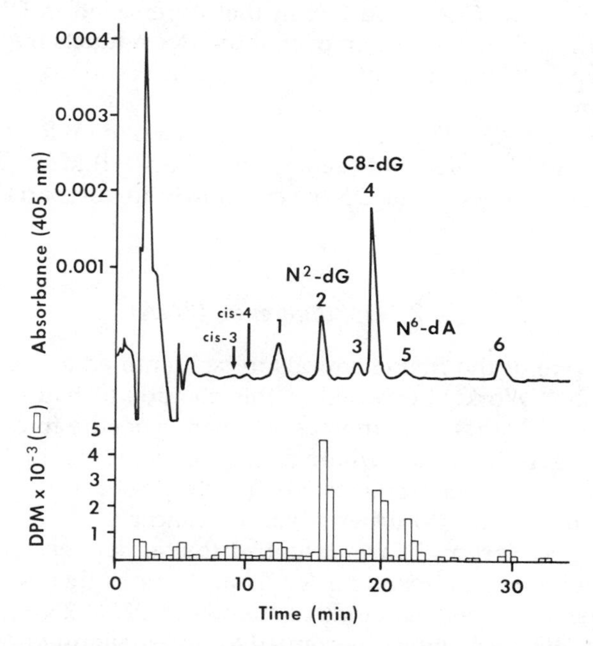

Figure VI-23. HPLC of hydrolysate of DNA of rat liver treated with [^{3}H]-MAB. The solid line is the absorbance at 405 nm of adducts from calf thymus DNA added as marker; the histogram represents radioactivity from [^{3}H]-MAB that is associated with rat liver DNA. From Tullis *et al.* (1981) *Chem. Biol. Interact.* **38,** 15.

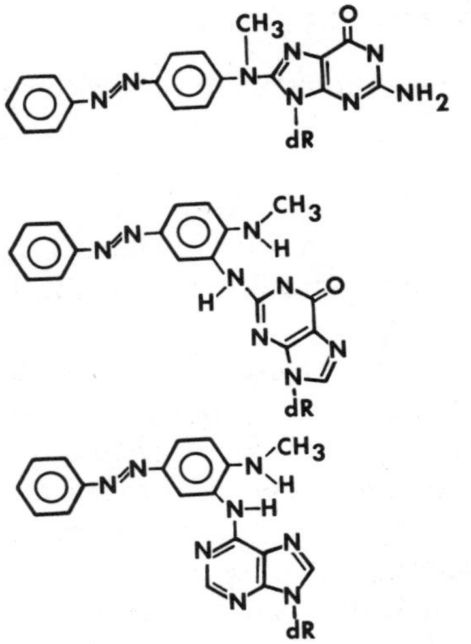

Figure VI-24. Structures of deoxyribonucleoside–MAB adducts. From Beland *et al.* (1980) *Chem. Biol. Interact.* **31,** 1.

terized by a number of spectroscopic techniques and established to be 3-(deoxyguanosin-N^2-yl)-MAB (Figure VI-24b). This product also behaved similarly to the minor N^2-G-AAF adduct in that it persisted in DNA. As in the case of AAF, the differences in removal of the two adducts may be explained by different types of distortion of DNA double helix caused by the two types of modification.

A third adduct ("5," Figure VI-23) was detected in rat liver after multiple doses of MAB and identified as 3-(deoxyadenosin-N^6-yl)-MAB (Figure VI-24c). This product is removed from DNA very slowly in comparison to the C-8-dG adduct.

3. 2-Naphthylamine (2NA)

2NA was one of the first chemicals to be identified as a causative agent in human cancer. Workers exposed to this compound had an increased incidence of urinary bladder carcinomas. Human exposure to 2NA also occurs in the form of cigarette smoke, where nanogram quantities of 2NA have been detected, and it has beeen suggested that its presence may contribute to increased risk of smokers to urinary bladder cancer.

In Chapter V, Section B, we described that arylamines are metabolized through *N*-oxidation. The resulting *N*-OH-arylamines then serve as the ultimate carcinogenic species that binds covalently to DNA. 2NA is metabolically activated to *N*-OH-2NA, which is converted under slightly acidic conditions in urine to an arylnitrenium ion–carbocation electrophile that is capable of binding covalently to cellular macromolecules. When Kadlubar and coworkers reacted *N* -OH-2NA with DNA *in vitro* at pH 5.0 and hydrolyzed the DNA, they obtained three nucleoside–arylamine adducts. The adducts were iden-

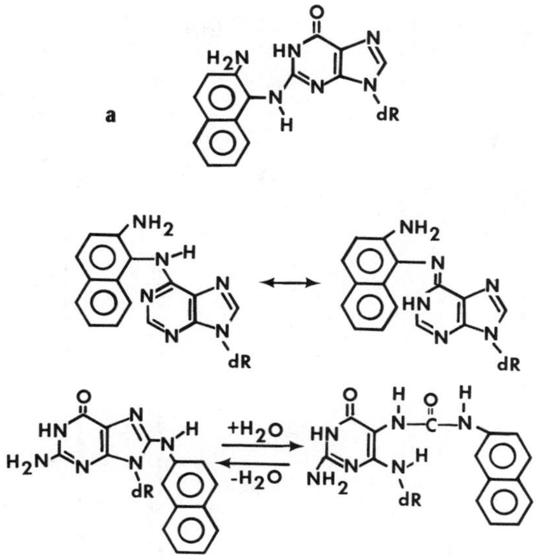

Figure VI-25. Structures of 2-naphthylamine-DNA adducts. (a) 1-(Deoxyguanosin-N^2-yl)-2-naphthylamine; (b) 1-(deoxyadenosin-N^6-yl)-2-naphthylamine; (c) purine ring-opened derivative of *N*-(deoxyguanosin-8-yl)-2-naphthylamine, tentatively identified as 1-(5-(2,6-diamino-4-oxopyrimidinyl-N^6-deoxyriboside))-3-(2-naphthyl) urea. From Kadlubar *et al.* (1980) *Carcinogenesis* **1,** 139.

tified by chemical and spectrometric analyses as 1-(deoxyguanosin-N^2-yl)-2NA (Figure VI-25a), 1-(deoxyadenosin-N^6-yl)-2NA (Figure VI-25b), and a purine ring-opened derivative of *N*-(deoxyguanosin-8-yl)-2NA (Figure VI-25c). The same three adducts were detected in the DNA of liver and urothelium of male beagle dogs treated with [^{3}H]-2NA and sacrificed after 2 or 7 days. The predominant adduct *in vivo* was then one modified on the N^6 adenine, although there was a substantial amount of the C-8-guanine derivative and lesser amount of the N^2-guanine adduct (Figure VI-26). The major difference detected between target (urothelium) and nontarget (liver) tissue was in the total level of binding to DNA, which was fourfold higher in the urothelium at 2 days and eight fold higher at 7 days after 2NA application. This appeared to be a result of the relatively greater persistence of the C-8-G adduct in the urothelium as compared to the liver. Thus, the susceptibility of the bladder to 2NA induced tumors may be a direct consequence of the higher binding levels and greater persistence of 2NA-DNA adducts in this tissue.

4. *1-Naphthylamine (1NA)*

Although 1NA is suspected as a human carcinogen, it has not been proven to be carcinogenic in dog or rodent bioassays. The marked contrast in carcinogenicity between 1NA and 2NA is probably the lack of metabolic

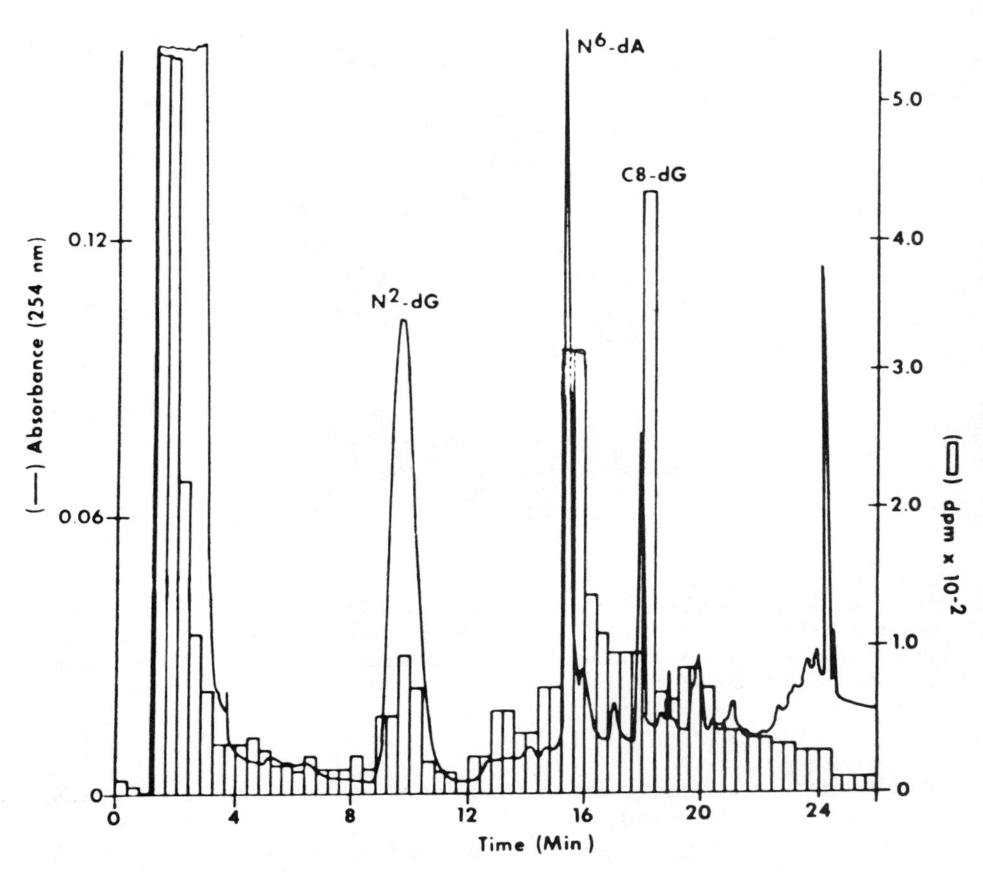

Figure VI-26. HPLC profiles of DNA hydrolysates of liver from 2NA-treated dog. The UV-absorbing components are shown by solid lines and the radioactivity associated with each is plotted as histograms. N^2-dG, 1-(deoxyguanosin-N^2-yl)-2-NA; N^6-dA, 1-(deoxyadenosin-N^6-yl)-2NA; C8-dG, ring-opened derivative of *N*-(deoxyguanosin-8-yl)-2NA. Adapted from Kadlubar *et al.* (1981) *Carcinogenesis* **2**, 467.

activation of 1NA. Only very small amounts of *N*-oxidation products could be detected in the urine of dogs treated for many years with high doses of 1NA. Similarly, no urothelial DNA adducts could be detected in these animals.

On the other hand, *N*-OH-1NA is a direct-acting, highly carcinogenic compound, and it reacts with DNA *in vitro* under mild acidic conditions to yield two DNA–carcinogen adducts. The major adduct, which accounted for up to 60% of the 1NA-bound DNA, was identified by chemical and spectrometric analyses as *N*-(deoxyguanosin-O^6-yl)-1NA and the minor adduct as 2-(deoxyguanosin-O^6-yl)-1NA (Figure VI-27).

The mechanism by which the activated carcinogen is inserted into the DNA at the O^6 position of the guanine base is similar to that of several other

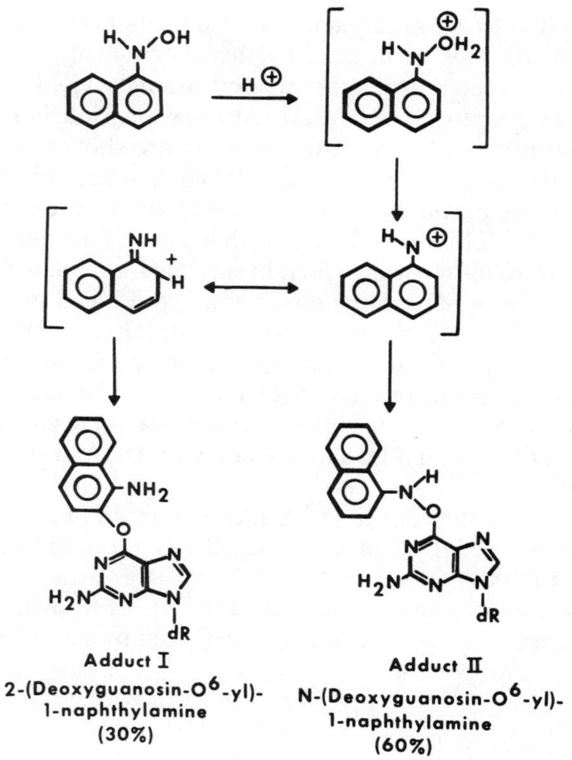

Figure VI-27. Structures of the DNA adducts formed by reaction with *N*-OH-1NA. From Kadlubar *et al.* (1981) *Cancer Res.* **41,** 2168.

N-OH-arylamines. An acid-dependent formation of an electrophilic arylnitrenium ion may be involved in the interaction with guanine residues in DNA as shown in Figure VI-27. The binding of 1-OH-1NA to the O^6 of guanine represents an unusual modification of DNA by *N*-substituted aromatic compounds.

5. *Benzidine (BZ)*

On the basis of epidemiological evidence, BZ has been identified as a human urinary bladder carcinogen and has been shown to induce tumors in a variety of experimental animals.

Metabolic activation of BZ is less well understood than with other aromatic amines. There is indication that acetylation of BZ may be a prerequisite for its activation. When [ring-^{14}C]*N*-acetylbenzidine (ABZ) was injected into rats, the major DNA adduct isolated was shown to coelute on HPLC with that obtained from BZ-treated animals. This suggested that either ABZ was deacetylated or that BZ was acetylated prior to further activation. Involvement of acetylation of BZ was confirmed by the injection of [*acetyl*-^{3}H]-ABZ into rats. Liver DNA isolated from the treated animals yielded a radiolabeled product chromatographically identical with that previously obtained.

Possible pathways for this activation of BZ are shown in Figure VI-28. In the first stage BZ may be *N*-acetylated to yield ABZ, which in turn is *N*-hydroxylated on the primary amine function to yield a proximate carcinogen, *N*-OH-*N*′-ABZ. The other possibility is that after *N*-acetylation, ABZ is *N*-hydroxylated to *N*-OH-*N*′-ABZ, which in turn is *N*′-acetylated to yield *N*-OH-DABZ. Either of these *N*-hydroxy metabolites could lead to DNA-bound adducts. The *N*-OH-*N*′-ABZ can react directly with DNA or might become further activated by hepatic sulfotransferase-catalyzed *O*-esterification, by seryltranserase, or via intermolecular *N,O*-acyltransfer to yield a DNA adduct that retains one acetyl group. However, sulfotransferase-catalyzed esterification of *N*-OH-DABZ would lead to reaction with DNA to yield a deacetylated derivative.

The chemical identity of the DNA adduct formed *in vivo* was established by the comparison on HPLC of the synthetically prepared markers with the materials found *in vivo*.

Liver DNA isolated from mice fed [2,2′-^{3}H]-BZ in drinking water for a week was enzymatically hydrolyzed and analyzed by HPLC. Only one major

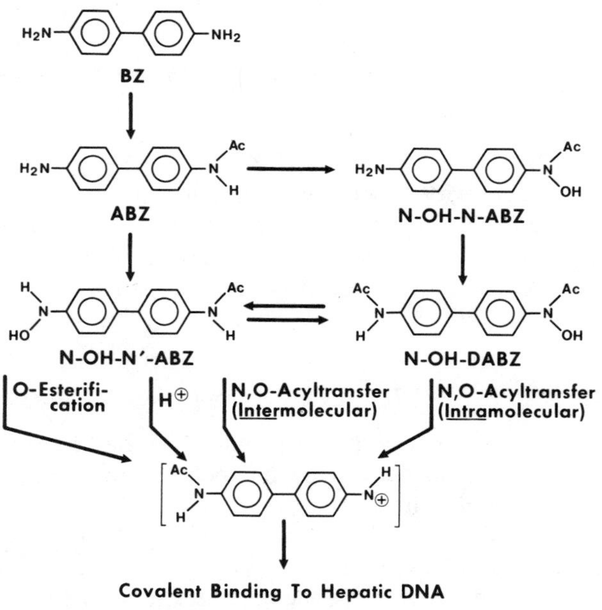

Figure VI-28. Possible pathways for the metabolic activation of benzidine to an ultimate hepatocarcinogen. From Martin *et al.* (1982) *Cancer Res.* **42,** 2678.

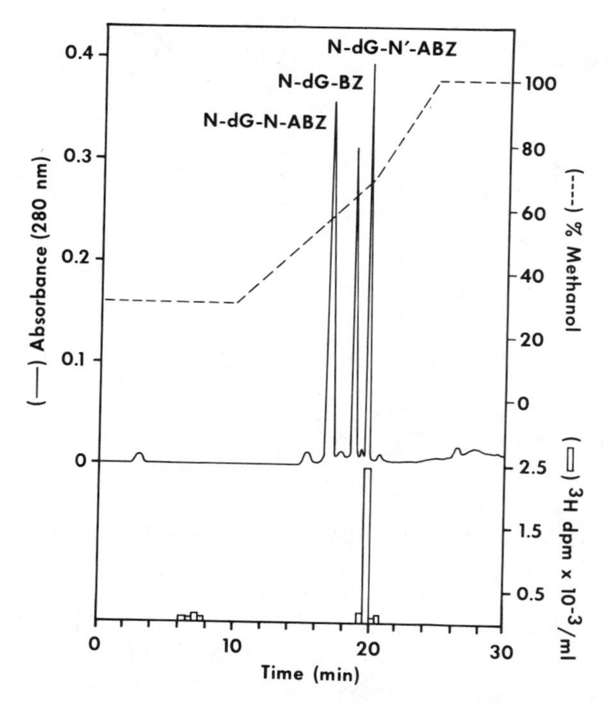

Figure VI-29. HPLC profile of radioactivity associated with liver DNA hydrolyzated from mice administered [2,2'-^{3}H]-BZ in drinking water. The histogram represents the radioactivity associated with the liver DNA, the solid line is the UV absorbance of the synthetic markers, and the dashed line is the methanol–water gradient. *N*-dG-BZ, *N*-(deoxyguanosin-8-yl)-BZ; *N*-dG-*N*-ABZ, *N*-(deoxyguanosin-8-yl)-*N*-ABZ; *N*-dG*N'*-ABZ, *N*-(deoxyguanosin-8-yl)-*N'*-ABZ. From Martin *et al.* (1982) *Cancer Res.* **42,** 2678.

radioactive peak was detected (Figure VI-29). It cochromatographed with the synthetic marker, which was identified by NMR and mass spectra as *N*-(deoxyguanosin-8-yl)-*N*-ABZ (Figure VI-30). No *N* -(deoxyguanosin-8-yl) derivatives of DABZ or BZ were detected. This confirmed that the most probable proximate carcinogen form of BZ is *N*-OH-*N'*-ABZ.

It is interesting that the ABZ residue bound to C-8 of G, in contrast to AAF bound to the same site of G, is not readily repaired in mice or rats after injection of BZ. The persistence of ABZ lesion in DNA may indicate that this modification does not distort the conformation of DNA as it does in the case of AAF (see Chapter VIII).

6. *4-Nitroquinoline-1-Oxide (4NQO)*

4NQO is a potent carcinogen in mammalian cells and a mutagen in microorganisms. It requires metabolic activation for binding to DNA. Homogenates of liver, kidney, spleen, and lung of rats are capable of converting

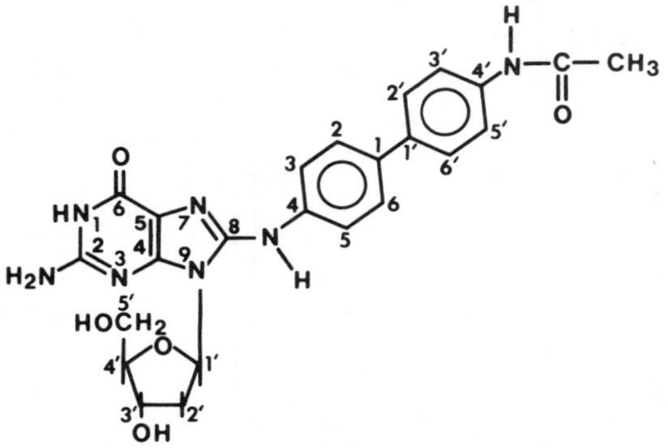

Figure VI-30. Structure of the major DNA adduct with *N*-acetylbenzidine, *N*-(deoxyguanosin-8-yl)-*N*′-ABZ. From Martin *et al.* (1982) *Cancer Res.* **42,** 2678.

4NQO to 4-hydroxyaminoquinoline-1-oxide (HAQO) and 4-aminoquinoline-1-oxide (Figure VI-31). The enzyme was partially purified from a rat liver supernatant. $NADH_2$ and $NADPH_2$ serve as hydrogen donors and 4NQO could act as hydrogen acceptor in this system. Since HAQO is a more potent carcinogen than the nitro derivative in rats and mice, it is considered to be the proximate carcinogen. It serves as an intermediate in the formation of DNA-bound derivative *in vivo* as well as *in vitro*. In order to interact with DNA *in vitro,* conversion of HAQO to an ultimate carcinogenic form is required. One enzymatic system responsible for the activation of HAQO *in vivo* could be seryl-tRNA synthetase, as shown by Tada and Tada. This enzyme requires ATP, serine, and Mg^{2+} for its activity. According to the proposed mechanism, for the first step of the reaction seryl-AMP is formed from serine, ATP, and the enzyme. If tRNA is present, it accepts the serine residue on the enzyme to form seryl-tRNA. This is the physiological function of this enzyme. On the other hand, if HAQO is present in this system, it may be acylated by the seryl-AMP enzyme complex. Seryl-HAQO may then introduce quinoline groups into nucleic acids. With rat hepatoma cell preparations,

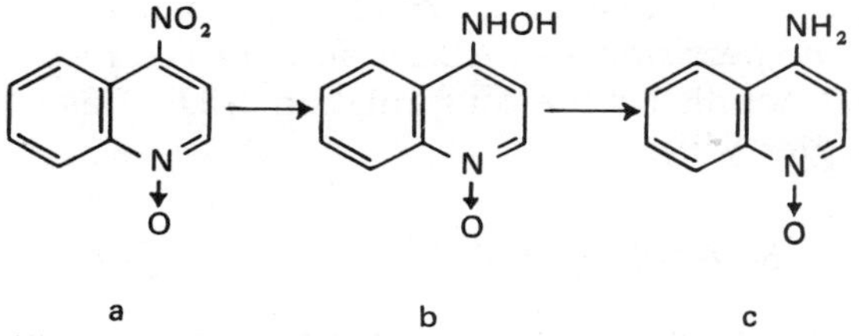

Figure VI-31. Metabolic conversion of 4-nitroquinoline-oxide. (a) 4-Nitroquinoline-1-oxide; (b) 4-hydroxyaminoquinoline-1-oxide; (c) 4-aminoquinoline-1-oxide.

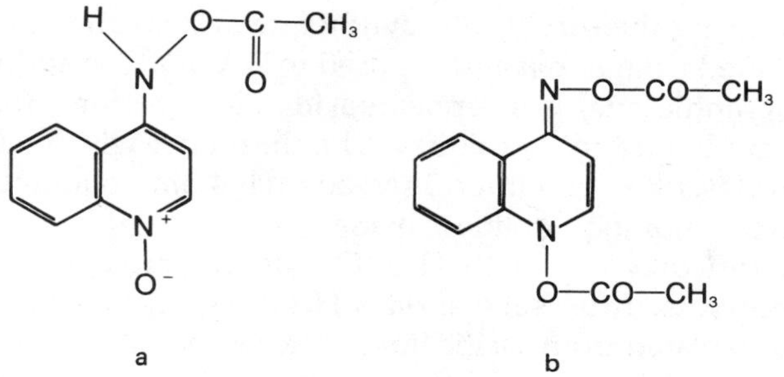

Figure VI-32. Structures of acetyl esters of 4-hydroxyaminoquinoline-1-oxide. (a) 4-Acetoxyaminoquinoline-1-oxide; (b) O,O′-diacetyl derivative.

proline is also active in this reaction. This suggests that HAQO may be activated by both seryl-and prolyl-tRNA synthetases.

Acetyl esters of HAQO also react with cellular macromolecules *in vitro*. When the two esters of HAQO, the 4-acetoxyaminoquinoline-1-oxide (AcHAQO) and the diacetyl derivative of HAQO (diAcHAQO) (Figure VI-32), reacted with purine nucleosides, five adducts were obtained. In both cases, two products were formed with adenine and three with guanine residues. The structure of the major adduct was identified as *N*-(deoxyguanosin-C^8-yl)-4-aminoquinoline-1-oxide (Figure VI-33). The structure of the other G adducts has not been determined.

Comparisons of the fluorescence spectrum of DNA modified *in vitro* by diAcHAQO with that of DNA modified *in vivo* by HAQO showed that both spectra exhibited the principal maximum at the same wavelength. Thus, similar fluorescence spectra support the proposal that diAcHAQO constitutes a valid model for the ultimate carcinogen of 4NQO.

A comparison was made between the adducts obtained after hydrolysis of DNAs modified in two different ways to exposure to HAQO. One DNA

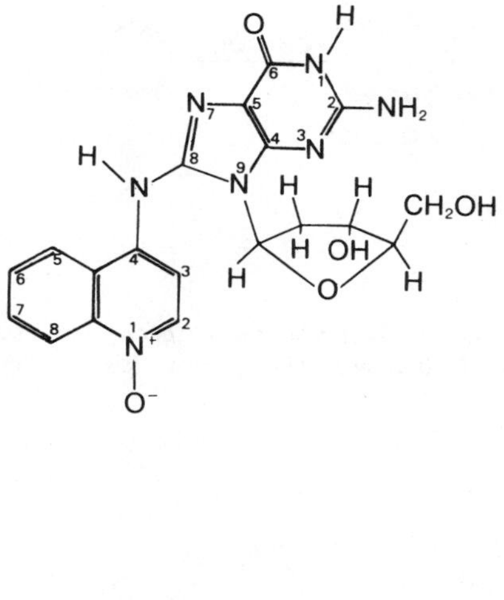

Figure VI-33. Structure of the major adduct, *N*-(deoxyguanosin-8-yl)-4-aminoquinoline-1-oxide. From Bailleul *et al.* (1981) *Cancer Res.* **41,** 4559.

was modified using the seryl-tRNA synthetase system and another modified DNA was isolated from *E. coli* and exposed to HAQO. The adducts, identified by chromatographic analyses, appeared identical. Among the isolated adducts, two were bound to G residues. The third one, which was bound to A residues was described as either 3-(N^6-adenyl)-4-aminoquinoline-1-oxide or 3-(*N* -1-adenyl)-4-aminoquinoline-1-oxide.

DNA modified *in vivo* with HAQO was prepared by injecting ascites tumor cells into rats. After several days HA[2-^{3}H]QO was injected and after 2 hr DNA was isolated from ascitic fluid. After enzymatic hydrolysis of DNA, the chromatographic analysis of the digest revealed a series of tritiated derivatives. Among the different compounds, one was identified as the dG adduct that was prepared *in vitro.* Since only about 20% of the total radioactivity was recovered in this product, this seems to indicate that the C-8 guanine adduct does not constitute a major product in the total DNA modification induced by 4NQO.

There is some analogy between the function of diAcHAQO as the ultimate carcinogen and AAF. In a manner similar to *N*-OH-AAF, HAQO would, by loss of its acetyl group, give an arylnitrenium ion that would react with DNA. Hypothetical mechanisms of these reactions are shown in Figure VI-34. The existence of a 4-quinolinium ion is in agreement with the dG adduct formation. On the other hand, the binding of the carcinogen with adenine through the C-3 of the quinoline ring supports the existence of the 3-quinoline carbocation form.

Since both compounds, seryl-HAQO and AcHAQO, represent monofunctional forms of HAQO and produce similar adducts upon interaction with DNA, it is reasonable to suggest that both are ultimate carcinogenic forms of 4NQO.

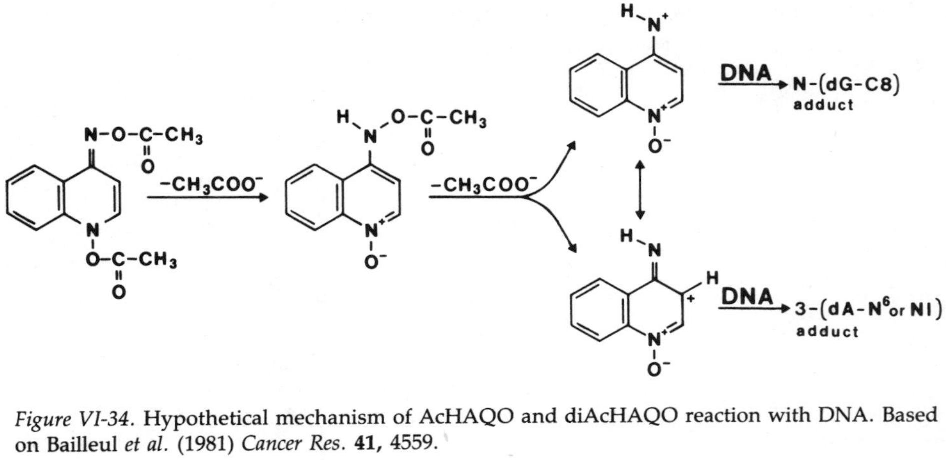

Figure VI-34. Hypothetical mechanism of AcHAQO and diAcHAQO reaction with DNA. Based on Bailleul *et al.* (1981) *Cancer Res.* **41,** 4559.

7. *Heterocyclic Amines from Amino Acid and Protein Pyrolysates*

a. CHARACTERIZATION

In the past few years a number of new mutagens have been identified in cooked foods in Japan. Extracts of the charred surfaces of broiled fish and meat showed very high mutagenic activities in histidine-requiring strains of *Salmonella typhimurium*. Mutagenic activity in this system was dependent on the presence of the postmitochondrial supernatant from homogenized livers of rats pretreated with Aroclor. Thus, substances in the extract required metabolic activation to exhibit mutagenic activity. To determine what constituents in the food were contributing to the mutagenic activity produced by cooking, Sugimura and co-workers isolated several previously unknown compounds from the pyrolysates of amino acids, broiled fish, and beef.

The mutagenic products resulting from pyrolysis of tryptophan were identified as two previously unknown amino-q-carbolines: 3-amino-1,4-dimethyl-5*H*-pyrido[4,3-*b*]indole (named Trp-P-1, for "trytophan pyrolysate 1") and 3-amino-1-methyl-5*H*-pyrido[4,3-*b*]indole (Trp-P-2) (Figure VI-35).

The mutagenic activity resulting from pyrolysis of glutamic acid was shown to be due to the formation of 2-amino-6-methyldipyrido[1,2-*a*:3′,2′-*d*]imidazole (Glu-P-1) and 2-amino-dipryrido[1,2-*a*:3′,2-′*d*]imidazole (Glu-P-2) (Figure VI-35).

Another heterocyclic mutagen has been isolated from the pyrolysis of phenylalanine and identified as 2-amino-5-phenylpyridine (Phe-P-1) (Figure VI-36). The mutagenic substance isolated from lysine pyrolysis was identified as 3,4-cyclopentenopyrido[3,2-*a*] carbazole (Lys-P-1) (Figure VI-36.)

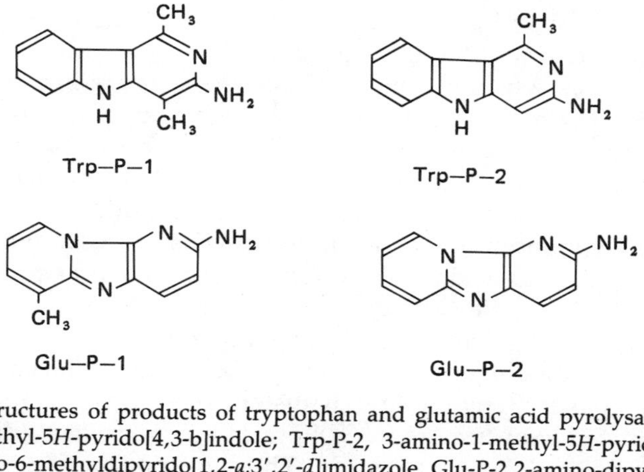

Trp—P—1 Trp—P—2

Glu—P—1 Glu—P—2

Figure VI-35. Structures of products of tryptophan and glutamic acid pyrolysates. Trp-P-1, 3-amino-1,4-dimethyl-5*H*-pyrido[4,3-b]indole; Trp-P-2, 3-amino-1-methyl-5*H*-pyrido[4,3-b]indole; Glu-P-1, 2-amino-6-methyldipyrido[1,2-*a*:3′,2′-*d*]imidazole. Glu-P-2,2-amino-dipyrido[1,2-*a*:3′-2′-*d*]imidazole. Based on Yamamoto *et al.* (1978) *Proc. Jpn. Acad.* **54,** 248, and Sugimura *et al.* (1977) *Proc. Jpn. Acad.* **53,** 58.

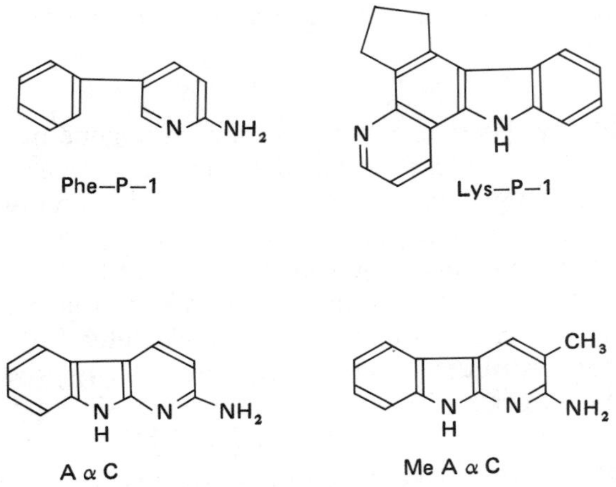

Figure VI-36. Structures of products of phenylalanine, lysine, and soybean globulin pyrolysates. Phe-P-1,2-amino-5-phenylpyridine; Lys-P-1, 3,4-cyclopentenopyrido[3,2-*a*]carbazole; AαC, 2-amino-9*H*-pyrido[2,3-*b*]indole; MeAαC, 2-amino-3-methyl-*q*,*H*-pyrido[2,3-*b*]indole.

Two additional mutagenic compounds not previously identified as pyrolysis products of any amino acid have been isolated from soybean globulin pyrolysates and identified as 2-amino-9*H*-pyrido[2,3-*b*]indole and 2-amino-3methyl-α,*H*-pyrido-[2,3-*b*]indole (MeAαC) (Figure VI-36).

From broiled sardines, two previously unknown mutagens were further isolated by Kasai and co-workers. These are 2-amino-3-methylimidazo[4,5-f]quinoline (IQ) and 2-amino-3,4-dimethylimidazo[4,5-*f*]quinoline (MeIQ) (Figure VI-37). From fried beef, a third new potent mutagen was isolated. The chemical structure of this mutagen was determined as 2-amino-3,8-dimethylimidazo[4,5-*f*] quinoxaline (MeIQx) (Figure VI-37).

It was also shown that some of these products are formed in normally cooked foods. For example, Trp-P-1 has been found in "very well done" broiled beef and Glu-P-2 in broiled fish. Similarly, sardines broiled to a dark brown color contain Trp-P-1, Trp-P-2, and Phe-P-1. In pyrolysates of casein and gluten, Trp-P-1, Trp-P-2, and Glu-P-2 were found.

Table VI-1 summarizes the new mutagenic heterocyclic amines produced by pyrolysis of amino acids and proteins, their source of preparation, and presence in food according to Sugimura and co-workers.

Several of the mutagenic pyrolysates of amino acids and proteins have been tested for carcinogenicity *in vivo*. Subcutaneous injection of Trp-P-1 induced sarcomas in Syrian golden hamsters and in Fisher rats. Trp-P-1 and Trp-P-2 produced liver tumors in CDF_1 mice that were fed a diet containing 0.02% of either of these compounds. Thus, it appears that the mutagenic substances identified in pyrolysates of proteins and amino acids may also have carcinogenic activity. However, the presence of a carcinogenic chemical

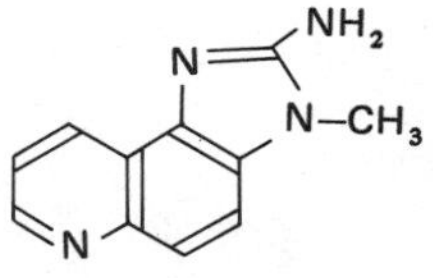

Figure VI-37. Structures of mutagens isolated from fried beef. IQ, 2-amino-3-methylimidazo[4,5-*f*]quinoline; MeIQ, 2-amino-3,4-dimethylimidazo[4,5-*f*]quinoline; MeIQx, 2-amino-3,8-dimethylimidazo[4,5-*f*]quinoxaline. Based on Kasai *et al.* (1981) *Chems. Lett.* **1981.** 485.

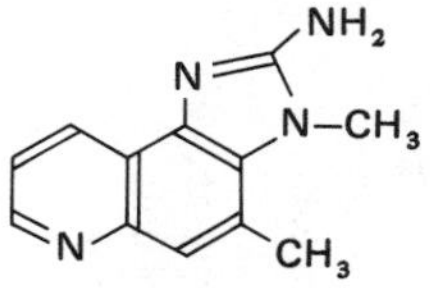

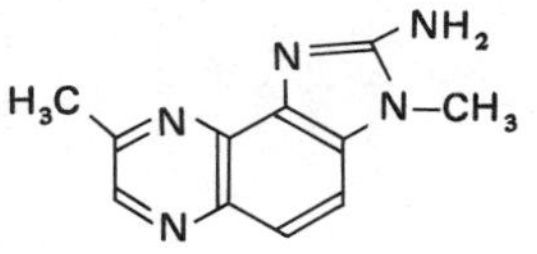

in pyrolyzed proteins does not necessarily imply that more than traces of this compound are formed in normally cooked food. More work has to be done before the effect of cooking procedures on the formation of such chemicals in food can be established.

b. METABOLIC ACTIVATION AND BINDING TO DNA

All the pyrolysis products of amino acids and proteins require metabolic activation to be transformed into highly mutagenic intermediates. This was achieved by liver subcellular fractions containing the cytochrome P-450-mediated monooxygenase system.

In the presence of liver fractions from genetically "responsive" C57BL/6N mice, the mutagenicity of all pyrolytic products was 1000-fold greater than with fractions from "nonresponsive" DBA/2N mice. This is very similar to the situation with BP and AAF, where the mutagenicity in the presence of the P-450 system from "responsive" mice was also much higher. The cytochrome P-450 system in these mice appears to be controlled by the regulatory gene of the [*AH*] locus described in Chapter V. From all polycyclic aromatic-inducible forms of P-450, the cytochrome P_1-450 subclass is most closely associated with polycyclic aromatic-induced aryl hydrocarbon hydroxylase activity. Therefore, it is possible to conclude that P_1-450 induced by BP and controlled by the regulatory gene of the [*AH*] locus is far more efficient than other forms of cytochrome P-450 in metabolizing all of these products of pyrolysis to their active metabolic forms.

In contrast to the polycyclic aromatic hydrocarbons, the pyrolytic products are not effective inducers of P_1-450 and aryl hydrocarbon hydroxylase

Table VI-1. New Heterocyclic Amines Produced by Pyrolysis of Amino Acids, Proteins, and Proteinous Food

Abbreviation	Full name	Source	Existing in
Trp-P-1	3-Amino-1,4-dimethyl-5*H*-pyrido[4,3-*b*]indole	Tryptophan pyrolysate	Broiled sardine, pyrolysates of casein, gluten, globulin, albumin, chicken meat, horse mackerel
Trp-P-2	3-Amino-1-methyl-5*H*-pyrido[4,3-*b*]indole	Tryptophan pyrolysate	Broiled sardine, pyrolysates of casein, gluten, globulin, albumin, chicken meat, horse mackerel
Glu-P-1	2-Amino-6-methyldipyrido[1,2-*a* : 3′,2′-*d*]imidazole	Glutamic acid pyrolysate	
Glu-P-2	2-Amino-6-dipyrido[1,2-*a* : 3′,2′-*d*]imidazole	Glutamic acid pyrolysate	Broiled dried-squid, pyrolysate of casein
Phe-P-1	2-Amino-5-phenylpyridine	Phenylalanine pyrolysate	
Lys-P-1	3,4-Cyclopentenopyrido[3,2-*a*]carbazole	Lysine pyrolysate	
AαC	2-Amino-α-carboline	Soybean globulin pyrolysate	Pyrolysates of tryptophan, casein, gluten, globulin, albumin, chicken meat, horse mackerel
MeAαC	2-Amino-3-methyl-α-carboline	Soybean globulin pyrolysate	Pyrolysates of tryptophan, casein, gluten, globulin, albumin, chicken meat, horse mackerel
IQ	2-Amino-3-methylimidazo[4,5-*f*]quinoline	Broiled sardine	Heated beef extract
MeIQ	2-Amino-3,4-dimethylimidazo[4,5-*f*]quinoline	Broiled sardine	
MeIQx	2-Amino-3,8-dimethylimidazo[4,5-*f*]quinoxaline	Broiled beef	

activity. Therefore, BP and other polycyclic aromatic hydrocarbons might have a synergistic action with the *N*-containing heterocyclic products in pyrolysates from cigarettes or certain foods in such a way that the polycylic aromatic hydrocarbon acts as a potent inducer of cytochrome P_1-450, which then converts the inert pyrolytic products to highly reactive mutagenic and perhaps carcinogenic intermediates. In order to characterize metabolites of Glu-P-1 and Glu-P-2, the compounds were activated in a reconstituted cytochrome P-450 system. The system contained cytochrome P-448 isolated from liver microsomes of rats treated with a polychlorinated biphenyl mixture and 3MC.

HPLC analysis revealed the presence of three metabolites from Glu-P-1 or Glu-P-2 (Figure VI-38). Among the metabolities of Glu-P-1, two were mutegenic without further activation. The major active metabolite of Glu-P-1 was characterized as *N*-hydroxy-Glu-P-1 (Figure VI-39).

From all the pyrolytic products thus far identified, only the binding of

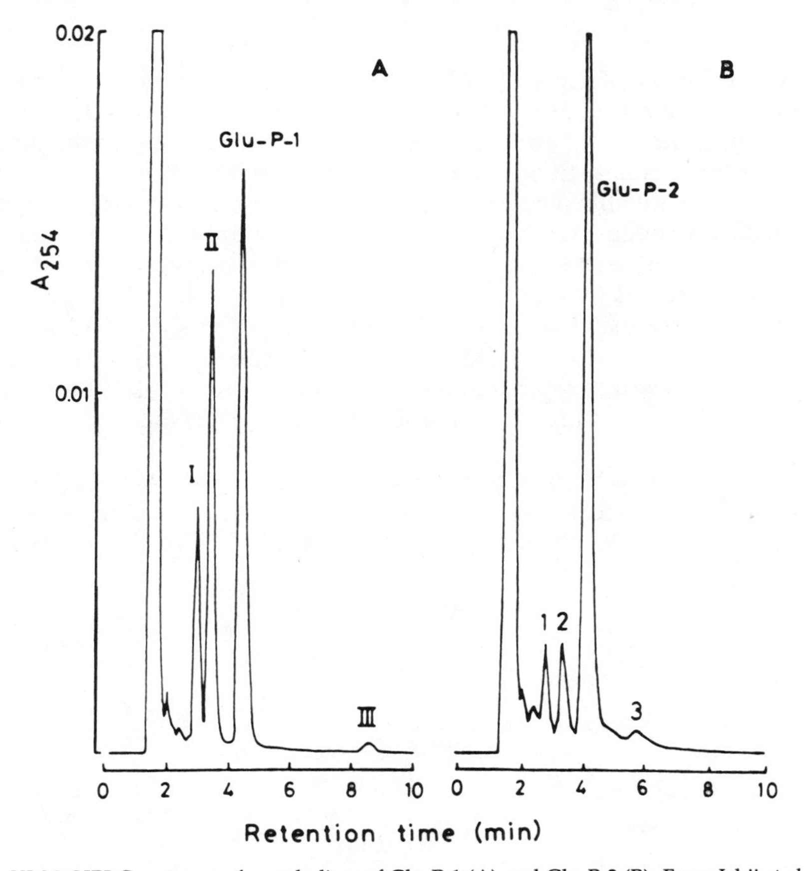

Figure VI-38. HPLC patterns of metabolites of Glu-P-1 (A) and Glu-P-2 (B). From Ishii *et al.* (1981) *Chem. Biol. Interact.* **38,** 1.

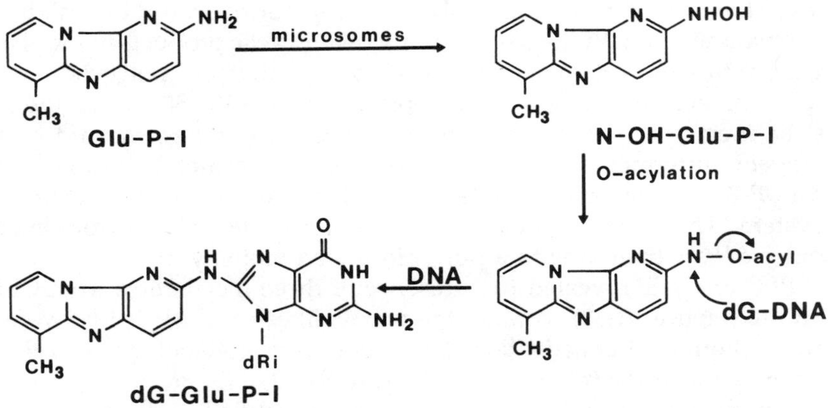

Figure VI-39. Scheme of Glu-P-1 metabolic activation and binding to dG in DNA. Based on Hashimoto *et al.* (1980) *Biochem. Biophys. Res. Commun.* **92,** 971.

the activated form of Glu-P-1 and Trp-P-2 to DNA has been described. In both cases, the binding of Glu-P-1 and Trp-P-2 occurred only in the presence of microsomal fractions. When DNA was reacted with Glu-P-1 in the presence of rat liver microsomes, the major modified adduct isolated from the DNA hydrolysate was identified as 2-(C-8-guanyl)-amino-6-methyldipyrido[1,2-*a*:3′,2′-*d*]imidazole (Gua-Glu-P-1) (Figure VI-40). This type of binding suggested that the activated form leading to the modified base is probably the *N*-hydroxy derivative of Glu-P-1 (*N*-OH-Glu-P-1), because *N*-OH-Glu-P-1 can be an electrophile and can attack the nucleophilic 8 position of guanine. Indeed, when Glu-P-1 was incubated in the microsomal system, one of the metabolites in the ethylacetate extract separated by HPLC was identified as *N*-OH-Glu-P-1. The scheme of Glu-P-1 activation and binding to DNA is illustrated in Figure VI-39.

On incubation of *N*-OH-Glu-P-1 with DNA under neutral or slightly acidic conditions, no significant binding was observed. After O-acetylation of *N*-OH-Glu-P-1, about 40 times higher binding of *N*-OH-Glu-P-1 to DNA was

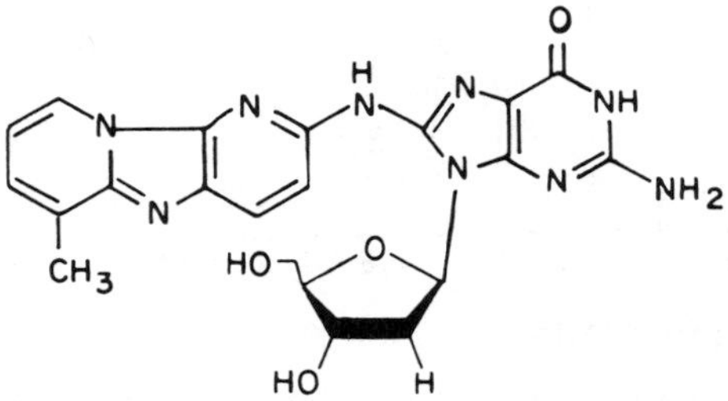

Figure VI-40. Proposed structure of Glu-P-1 and deoxyguanosine adduct. Based on Hashimoto *et al.* (1980) *Biochem. Biophys. Res. Commun.* **92,** 971.

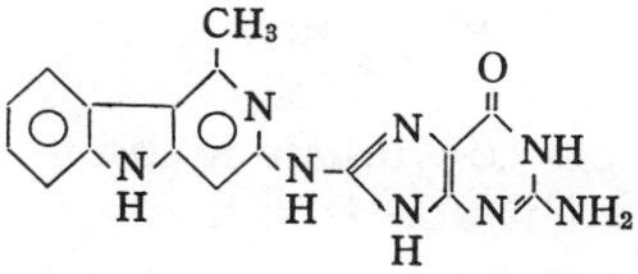

Figure VI-41. Structure of Trp-P-2 and guanine adduct. Based on Hashimoto *et al.* (1979) *Chem. Pharm. Bull.* **27**, 1058.

obtained than in the microsome-mediated binding of Glu-P-1 to DNA. However, in both systems the same C-8 guanine adduct was formed.

The binding of Trp-P-2 to DNA also occurred in rat liver microsomal system. In the hydrolysate of the modified DNA, one of the products was identified as 3-(*C*-8-guanyl)-amino-1-methyl-5*H*-pyrido[4,3-*b*]indole (Gua-Trp-P-2) (Figure VI-41). Since the binding site of activated Trp-P-2 to DNA is the 8 position of guanine, similar to that of AAF and Glu-P-1, it is quite possible that the activated form of Trp-P-2 is also the corresponding hydroxylamine and its ester. Hence, the proximate form of both pyrolytic products, Glu-P-1 and Trp-P-2, is possibly their *N*-OH derivative. Consequently, the ultimate electrophilic form could be either the sulfo ester, as with other arylhydroxylamines, or, as in the case of HAQO, the activation could involve a characteristic O-acylation.

Metabolic activation and binding to DNA of the other products from amino acid and protein pyrolysates have not been described yet. It is possible that the formation of the electrophilic forms and their interaction with DNA will follow a similar pattern to that described with Glu-P-1 and Trp-P-2, respectively.

D. NATURALLY OCCURRING CARCINOGENS

In this part of the chapter we will describe interactions of DNA with naturally occurring carcinogenic compounds. Although some of these substances occur in certain common human foods, they have been found primarily in foods contaminated by certain fungi.

1. *Aflatoxins*

a. BIOLOGICAL EFFECTS

Concern about the biological hazard of fungal metabolites largely stems from the finding in the early 1960s of an epidemic in England killing thousands of chicks, ducklings, and turkeys with acute hepatic necrosis. At about the same time, many of the trout in fish hatcheries developed liver cancers. The epidemic was quickly traced to food sources, especially to peanuts from Brazil and Africa. Soon it was shown that peanuts were harmless unless the nut became the medium for growth of the mold *Aspergillus flavus*. This common mold elaborated compounds that demonstrated blue and green fluorescence under UV. Following extraction of the peanut meals with solvents, Wogan and co-workers found four compounds in the contaminated food, each with a characteristic fluorescence under far-UV light and each having different chromatographic properties. They were named collectively "aflatoxins"and given a letter (B or G) to indicate their blue or green fluorescence (Figure VI-42). There was a perfect correlation between content of aflatoxin in the fish or poultry diets and carcinogenicity in these animals. Aflatoxin promptly induced hepatomas in trout and, in fact, the trout hepatomas were the first tumors produced by the defined chemical. Moreover, in Africa and in Southeast Asia, where human hepatomas are frequent, there is an excellent correlation between their frequency and the aflatoxin content in the food.

Species-, sex-, and strain-related differences to aflatoxin B_1 (AFB_1) carcinogenicity have been detected. Female rats are less susceptible than male rats. At a dose of 0.2 μg/day and a total dose of approximately 100 μg, all of the rats in the experiments develop liver tumors in an average of 476 days. On the other hand, both mouse and hamster livers are resistant to the carcinogenic effects of AFB_1. Orally administered AFG_1 appears to be less carcinogenic in rat liver than AFB_1, but it appears to be more active than AFB_1 in inducing kidney tumors in the rat.

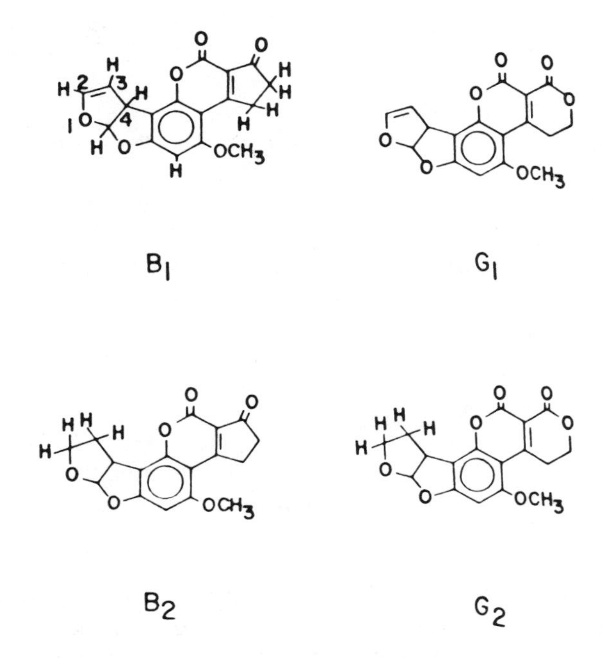

Figure VI-42. Structures of aflatoxins.

b. METABOLISM AND ACTIVATION OF AFB_1

Several studies led to the conclusion that AFB_1 requires metabolic activation in order to be carcinogenic and mutagenic. In *Salmonella* assay system it is mutagenic only in the presence of the mixed-function oxidase system. AFB_1 binds covalently to cellular macromolecules both *in vivo* and only after activation *in vitro* by liver chromosomes. Early work done mainly in the E. C. and J. A. Miller laboratories suggested that oxidative metabolism at the 2,3- double bond at the terminal furan ring was likely to be the critical reaction in the metabolic activation of AFB_1. It was later shown that the predominant metabolite of AFB_1 that binds to DNA in animal and human tissues is the AFB_1-2,3-oxide (Figure VI-43). This metabolite possesses such a high degree of chemical reactivity that it has never been isolated. Its structure was established mainly on the basis of indirect evidence, including identification of the AFB_1-2,3-dihydrodiol as an acid degradation product of AFB_1-modified DNA and RNA (Figure VI-43). In more recent studies, it was shown that AFB_1 can also be metabolically activated by at least two other metabolic routes, both of which probably also involve epoxidation. Known metabolic activations of AFB_1 in liver are summarized in Figure VI-44. It shows that part of total DNA binding apparently is due to the activation of its hydroxylated metabolites, AFM_1 (4-hydroxy-AFB_1) and AFP_1 (the O-demethylated derivative of AFB_1).

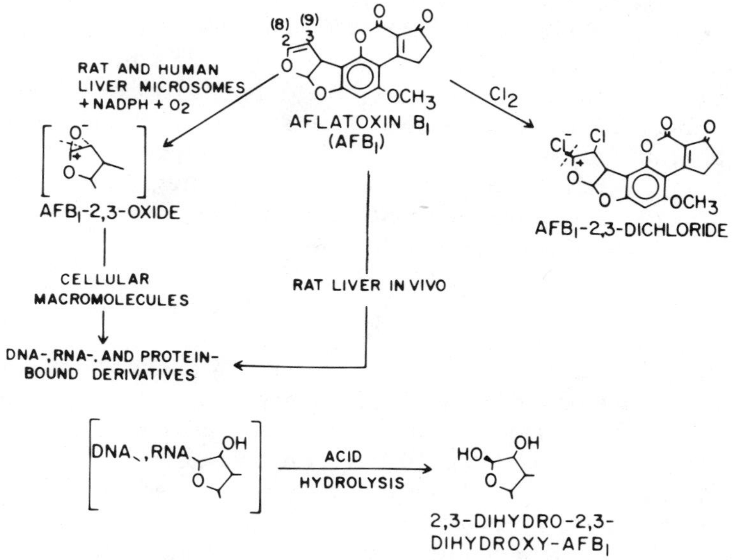

Figure VI-43. Metabolic activation of aflatoxin B_1 by liver microsomes to form nucleic acid-bound derivatives and the hydrolysis of these adducts to 2,3-dihydro-2,3-dihydroxy-AFB_1. From Svenson *et al.* (1974) *Biochem. Biophys. Res. Commun.* **60,** 1036.

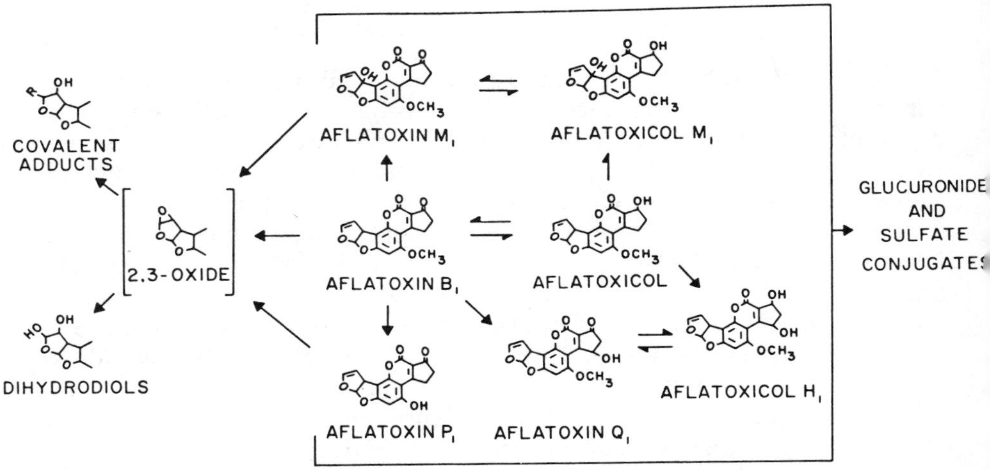

Figure VI-44. Pathways of metabolism of aflatoxin B_1. From Essigmann *et al.* (1982) *Drug Metab. Rev.* **13**(4), 581.

Formation of these primary metabolites does not alter the double bond in the terminal furan ring, leaving it available for a possible secondary oxidative reaction that results in production of 2,3-oxide, a very reactive electrophile. Much effort has been spent to synthesize AFB_1-2,3-oxide. To this end, only a model compound that binds to DNA was made using 3-chloroperbenzoic acid as one of the mildest of the peracids for epoxide synthesis. Reaction of AFB_1 with this peracid gives rise to three different hydroxy esters, but no epoxide. These esters are presumed to arise through initial 2,3-oxide and subsequent nucleophilic attack by the acid. Although the presumed epoxide intermediate was not isolated in the presence of water, the peracid reaction gives rise to the 2,3-dihydrodiol of AFB_1, suggesting an attack by water at the C-2 position with ring opening (Figure VI-43).

c. FORMATION OF DNA ADDUCTS

The determination of the site of AFB_1 binding to DNA *in vitro* was performed by incubation of DNA with liver microsomes from phenobarbital-treated rats and NADPH-generating system. DNA was also isolated from the liver of rats treated with AFB_1 *in vivo.* After hydrolysis of DNA in dilute HCl, the purine bases and AFB_1–base adducts wre resolved by HPLC into approximately 12 discrete chromatographic peaks (Figure VI-45). The main target in DNA is the N^7 atom of guanine. The structure of the principal adduct was deduced from a combination of chemical, NMR, UV, and mass spectral evidence and has been identified in Wogan and Miller's laboratories as 2,3-dihydro-2-(N^7-guanyl)-3-hydroxyaflatoxin B_1 (AFB_1-N^7-Gua). In rat liver 2 hr after dosing, the AFB_1-N^7-Gua adduct makes up approximately 80% of the

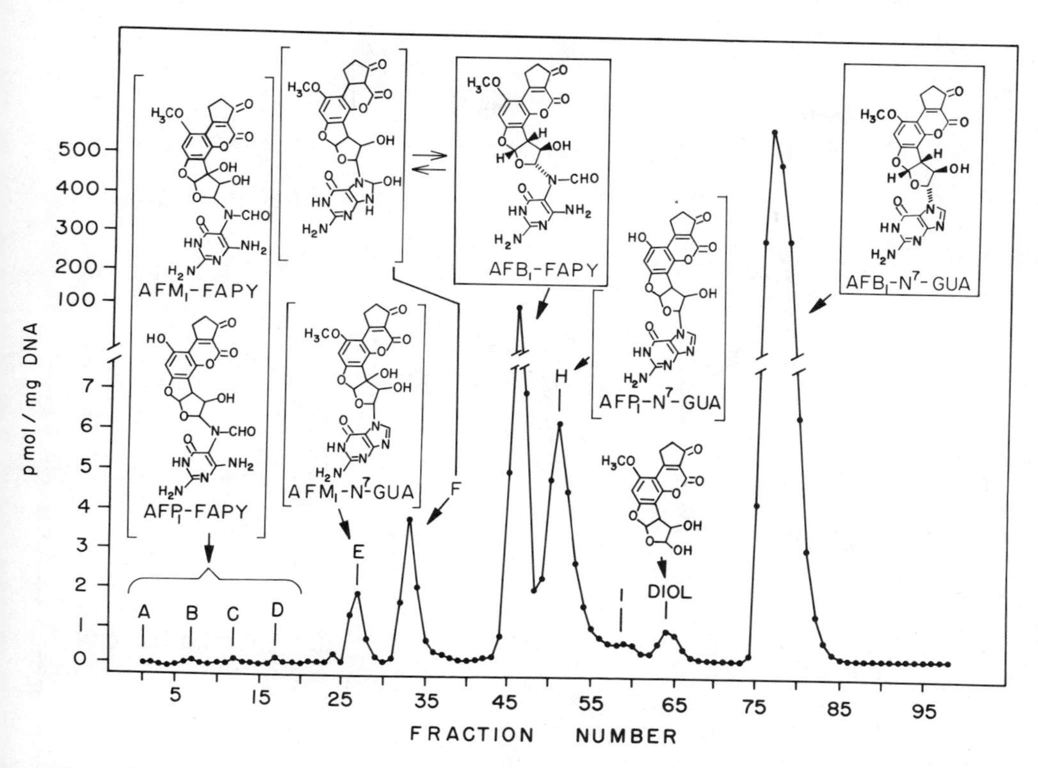

Figure VI-45. DNA adducts formed from AFB_1 in rat liver. HPLC analysis of rat liver DNA hydrolysate 2 hr after administration of [^{3}H]-AFB_1. The structures of AFB_1-N^7-Gua and AFB_1-FAPY have been established unambiguously; the remaining adducts have been tentatively identified. From Essigmann *et al.* (1983) *Cold Spring Harb. Symp. Quant. Biol.* **47,** 327.

DNA adducts that are present. The formation of this adduct in DNA induces a positive charge in the imidazole ring of guanine, leading to alkali-labile sites in DNA. As a consequence of the relatively weak glycosidic bond that is characteristic of all N^7-substituted deoxyribonucleotides (see Chapter IV), AFB_1-N^7-Gua is rapidly removed from DNA (Figure VI-46). In addition, however, there is some enzymatic removal of AFB_1-N^7-Gua from DNA, too.

As a further consequence of the charged imidazole ring of guanine, the C-8 becomes susceptible to attack by hydroxide ion, resulting in imidazole ring opening to an AFB_1–formamidopyrimidine structure (AFB_1-FAPY) (Figure VI-46). This adduct represents approximately 7% of the DNA-bound forms of aflatoxin 2 hr after dosing. It is interesting that the bond linking AFB_1-FAPY to deoxyribose of the DNA backbone becomes chemically stabilized. Although the structures for the remaining DNA-bound derivatives of AFB_1 are tentative, there is a considerable amount of evidence that allowed the reasonable structural proposal shown in Figure VI-45. Peaks A–D might represent the ring-opened derivatives of AFM_1 (AFM_1–FAPY) and AFP_1

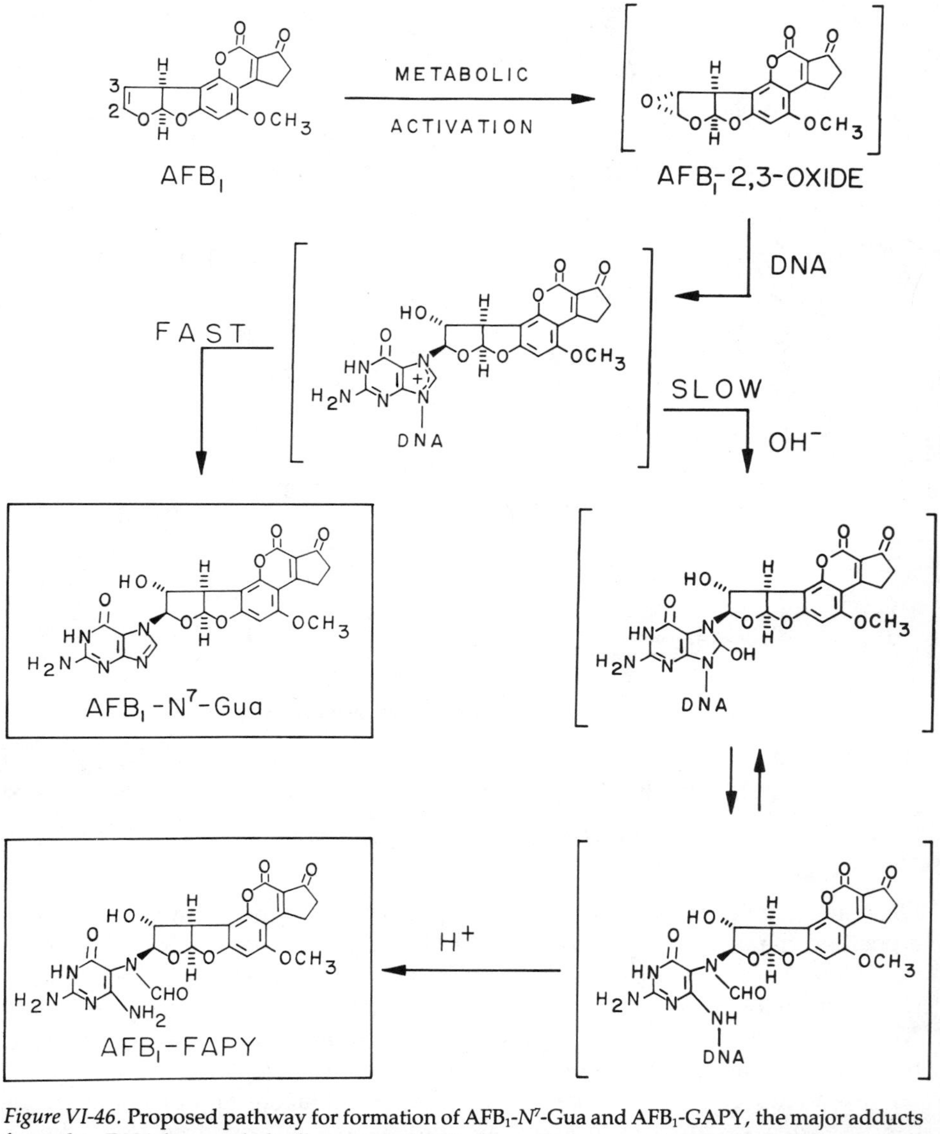

Figure VI-46. Proposed pathway for formation of AFB_1-N^7-Gua and AFB_1-GAPY, the major adducts formed in DNA by metabolically activated AFB_1. From Essignman *et al.* (1983) *Cold Spring Harb. Symp. Quant. Biol.* **47**, **327**.

(AFP_1–FAPY). Preliminary structural data on peak H indicate that it has properties consistent with those of an adduct formed by attack by the N^7 atom of guanine on an activated form of AFP_1 (AFP_1-N^7-Gua). Similarly AFM_1, another well-characterized AFB_1 metabolite, forms an adduct with N^7 of guanine (AFM_1-N^7-Gua in peak E).

All the available structural information strongly supports the generalization that the N^7 atom of guanyl residues is the principal target in DNA for metabolically activated AFB_1 or its metabolites. Furthermore, the main reaction leading to the formation of DNA-bound forms of aflatoxin is epoxidation of the vinyl bond in the terminal furan ring.

There is now strong experimental evidence that a good correlation exists between species specificity and AFB_1-DNA adduct formation. The Swiss mouse, in contrast to the Fischer rat, is highly resistant to the hepatocarcinogenic effects of AFB_1. The most striking difference in AFB_1-DNA adducts formed in mouse and rat liver after injection of AFB_1 is the very low level of the adducts in mouse DNA. The major component (AFB_1-N^7-Gua) in mouse liver had a concentration of only about 1 adduct per 10^6 DNA nucleotides, on the order of 1% of the level of the same adduct in rat liver DNA. The other DNA adducts in mouse are proportionately reduced. These results indicate that mouse liver activates AFB_1 essentially through the same pathways as rat liver, but it has either a very limited capacity for activation or possesses a very efficient system for removal of DNA adducts.

The main pathological effect of AFB_1 in the mouse is observed in the kidney, whereas the liver is insensitive. Results with DNA adduct formation are in good agreement with this observation. The major adduct in mouse kidney attained a concentration of 1 adduct per 6.6×10^6 nucleotides, of comparable magnitude to the level of the same component in rat kidney (1 adduct per 8.8×10^6 nucleotides) but about 100-fold lower than that found in rat liver.

Taken together, experiments with rat liver (target tissue) and kidney (nontarget tissue) support the general concept that tissue specificity for carcinogenicity depends on the total binding level of the activated AFB_1 residues to DNA. A similar conclusion is supported by the data in kidney and liver of mice. Although the spectrum of DNA adducts in both tissues is very similar to that present in rat tissues, the difference in total binding levels is very large compared to the rat.

Thus, while the qualitative pattern of DNA adducts is not revealing with respect to mechanisms responsible for tissue susceptibility, total binding level is strongly correlated with the biological response.

Because there is good correlation between DNA adduct formation and carcinogenicity, the quantitation of the adduct may provide a useful indication of biologically effective dose of AFB_1. For this purpose monoclonal antibodies were produced following immunization of mice with AFB_1 metabolite-modified DNA and also with AFB_1-FAPY-DNA adduct. Using these antibodies, 1 adduct in about 300,000 nucleotides would be detectable, which is the level of binding found in the rat *in vivo*. These monoclonal antibodies should therefore prove useful in detecting lesions in animal and human tissue samples exposed to aflatoxins.

Since administration of AFB_1 to rats also results in the urinary excretion of DNA adducts in a dose-dependent manner, the application of the immunoassay is very useful for the detection of the excreted adducts, too. Comparison of the dose–response curve for adduct excretion with that of adduct formation in rat liver DNA shows considerable qualitative similarity. The quantity of adduct excreted in the urine following AFB_1 administration represents a relatively fixed proportion of the total amount of initially formed DNA adducts. Therefore, this approach may be applicable for the assessment of human exposure not only to aflatoxins but also to other environmental carcinogens where the DNA adducts are chemically well characterized.

2. *Sterigmatocystin (ST)*

This is a carcinogenic mycotoxin produced as a secondary metabolite by *Aspergillus, Penicillium,* and *Bipolaris* species. In rats, ST induces mainly hepatocellular carcinomas. It is about one-tenth as potent as AFB_1. The importance of ST as a human health hazard is unknown because its presence in foods has been detected infrequently and at low concentrations, even though ST-producing fungi are widely distributed.

The metabolism of ST and the covalent binding of its activated form to nucleic acids resemble those of AFB_1 and are illustrated in Figure VI-47. Binding of ST to DNA occurs after incubation of phenobarbital-induced rat liver microsomes with DNA and NADPH-generating systems. Subsequent formic acid hydrolysis of ST-modified DNA and HPLC analysis of the hydrolysate yield one major adduct. Spectral and chemical data identified this adduct as 1,2-dihydro-2,2-(N^7guanyl)-1-hydroxysterigmatocystin. This structure indicates that the ST-1,2-oxide is the metabolite that reacts with DNA. The same major adduct was detected in the liver of rats perfused with medium-containing ST. Since the activation pathway through epoxidation of the vinyl double bond is identical for AFB_1 and ST, the differences in potencies of the two compounds therefore probably relate to differences in the rate of metabolism and/or repair of their reaction products.

3. *Safrole*

Safrole (4-allyl-1,2-methylenedioxybenzene) is a naturally occurring plant constituent. It is a major component of oil of sassafras which is used in the cosmetic and food flavoring industries. Formerly it was used as a flavoring agent in soft drinks and up to 27 μg/liter was present in root beer. Although its use is no longer permitted in the United States and most other countries, its occurrence in certain spices and other plant derivatives leads to ingestion in small amounts by many humans.

Safrole fed to adult rats and mice has weak hepatocarcinogenic activity. When injected into mice during the first week after birth, it shows moderate hepatocarcinogenic activity.

The excretion of a conjugate of 1′-hydroxysafrole in the urine of rats and mice fed safrole established 1′-hydroxysafrole as a proximate carcinogenic

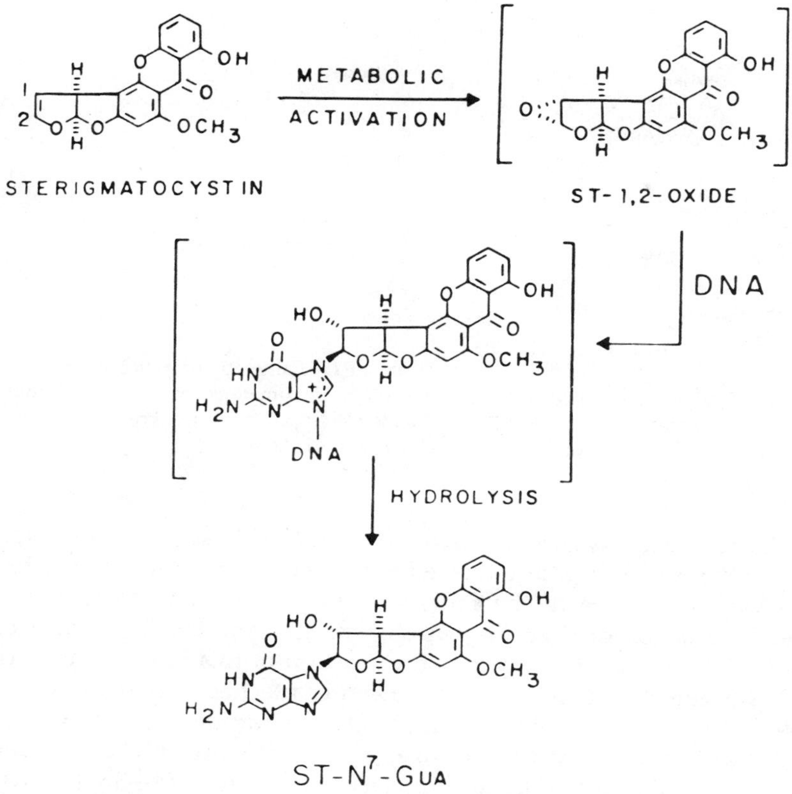

Figure VI-47. Metabolic activation and binding of sterigmatocystin to N^7-G of DNA. From Essigmann *et al.* (1979) *Proc. Natl. Acad. Sci. USA* **76,** 179.

metabolite of safrole. The carcinogenic activity of 1′-hydroxysafrole is greater than that of safrole. A synthetic ester of this metabolite, 1′-acetoxysafrole, reacts with cellular nucleophiles and is carcinogenic at the sites of administration. From the occurrence of sulfotransferase activity for 1′-hydroxysafrole in rat and mouse liver cytoplasm, it appears likely that the sulfuric ester is the principal ester formed *in vivo.* At least two other electrophilic metabolites of 1-hydroxysafrole are also formed *in vivo.* One of these, 1′-hydroxysafrole-2′,3′-oxide, is formed in an NADPH-dependent reaction by liver microsomes. The other possible metabolite, safrole-2,3′-oxide, may also contribute to the carcinogenic activity of safrole. Another electrophilic derivative of safrole, a 1′-oxosafrole, does not appear to be carcinogenic. The metabolic pathways of 1′-hydroxysafrole to electrophilic species are presented in Figure VI-48.

Administration of 1′-[2′,3′-^{3}H]hydroxysafrole to mice resulted in the formation of DNA-bound adducts. Comparison by HPLC of deoxyribonucleoside adducts obtained from the hepatic DNA with those formed by reaction of dG

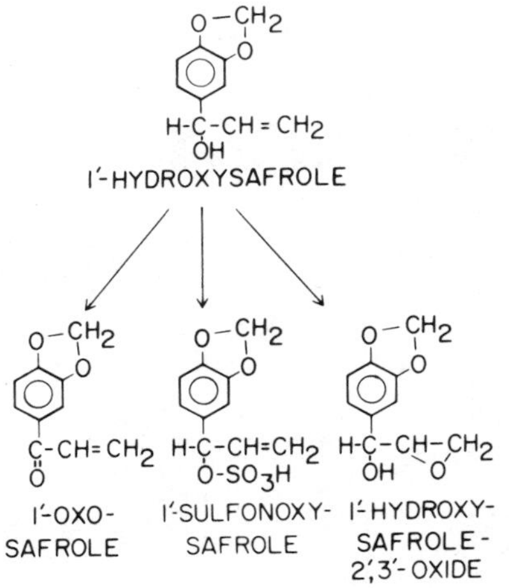

1'-HYDROXYSAFROLE

1'-OXO-SAFROLE

1'-SULFONOXY-SAFROLE

1'-HYDROXY-SAFROLE-2',3'-OXIDE

Figure VI-48. Pathways of metabolism of 1'-hydroxysafrole, a proximate carcinogenic metabolite of safrole, to electrophilic species. From Philips *et al.* (1981) *Cancer Res.* **41**, 2664.

and dA with 1'acetoxysafrole, 1'-hydroxysafrole-2',3'-oxide, and 1'-oxosafrole indicates that the four *in vivo*-formed adducts are derived from an ester of 1'-hydroxysafrole. Three of the four *in vivo* adducts migrated with products formed by reaction of 1'-acetoxysafrole with dG, and the fourth adduct comigrated with the major product of the reaction of this ester with dA (Figure VI-49). The carcinogen moiety was covalently attached to the exocyclic groups of the guanine and adenine residues. One of the dG adducts was further characterized from its NMR spectrum as N^2-(*trans* isosafrol-3'-yl)deoxyguanosine (Figure VI-50a) and the adenine adduct as N^6-(*trans*-isosafrol-3'-yl)deoxyadenosine (Figure VI-50b). The two other dG adducts have not been fully characterized.

4. *Estragole*

Estragole (1-allyl-4-methoxybenzene) is a naturally occurring flavoring agent that is the major constituent of the essential oils of sweet basil, tarragon (estragon), sweet goldenrod, chervil, and Mexican avocado leaves. It is also a minor component of American gum, spirits of turpentine, and oils of fennel, anise, and star anise.

Estragole proved to be hepatocarcinogenic when administeed to preweanling male mice. A major metabolite, 1'-hydroxyestragole, is more carcinogenic than the parent compound in rats and mice.

The metabolism of estragole is very similar to that of safrole. In NADPH-fortified liver microsomal systems, estragole, as well as 1'-hydroxyestragole, is metabolized to its 2',3'-oxide. Further metabolism of 1'-hydroxyestragole to an electrophilic species proceeds by three pathways: (1) by conversion to an ester (probably a sulfuric acid ester catalyzed by cytosolic sulfotransferase);

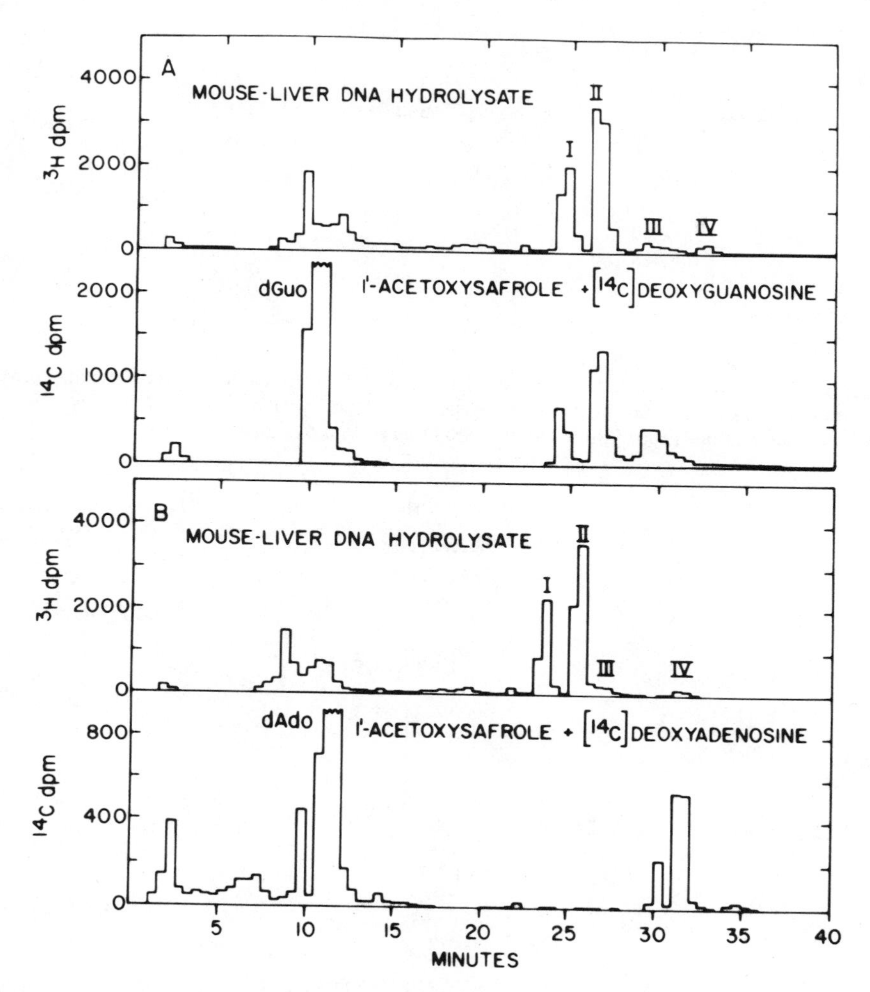

Figure VI-49. HPLC profiles of DNA hydrolysates and marker nucleosides. DNA hydrolysates from the livers of mice given injections of 1′-[^{3}H]hydroxysafrole were cochromatographed with aliquots of the reaction mixture of 1′-acetoxysafrole reacted with [^{14}C]dGuo (A) and [^{14}C]dAdo (B). From Phillips *et al.* (1981) *Cancer Res.* **41,** 2664.

(2) by epoxidation of the 2′,3′ double bond to form 1′-hydroxyestragole-2′,3′-oxide; and (3) by oxidation of the 1′ position to yield 1-oxoestragole (Figure VI-51).

Enzymatic hydrolysates of hepatic DNA isolated from mice given i.p. injections of 1′-[2′,3′-^{3}H]hydroxyestragole contained two major (I and II) and two minor (III and IV) nucleoside adducts (Figure VI-52). Reaction of the model electrophilic species, 1′-acetoxyestragole, with [^{14}C]dG yielded prod-

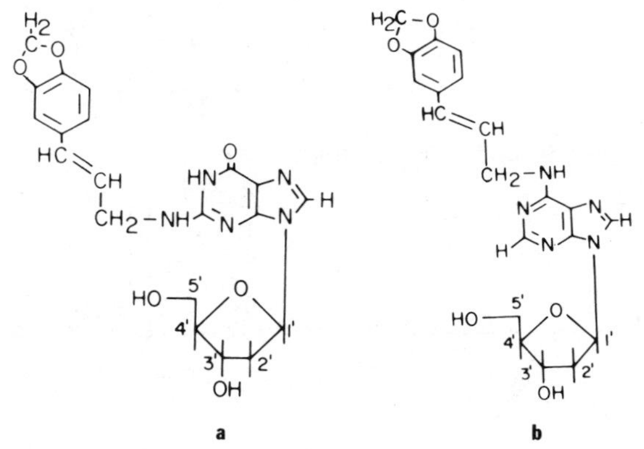

Figure VI-50. Structures of the 1′-hydroxysafrole–deoxyribonucleoside adducts that are formed in mouse liver DNA *in vivo.* (a) N^2-(*trans*-isosafrol-3′-yl)deoxyguanosine; (b) N^6-(*trans*-isosafrol-3′-yl)deosyadenosine. From Phillips *et al.* (1981) *Cancer Res.* **41,** 2664.

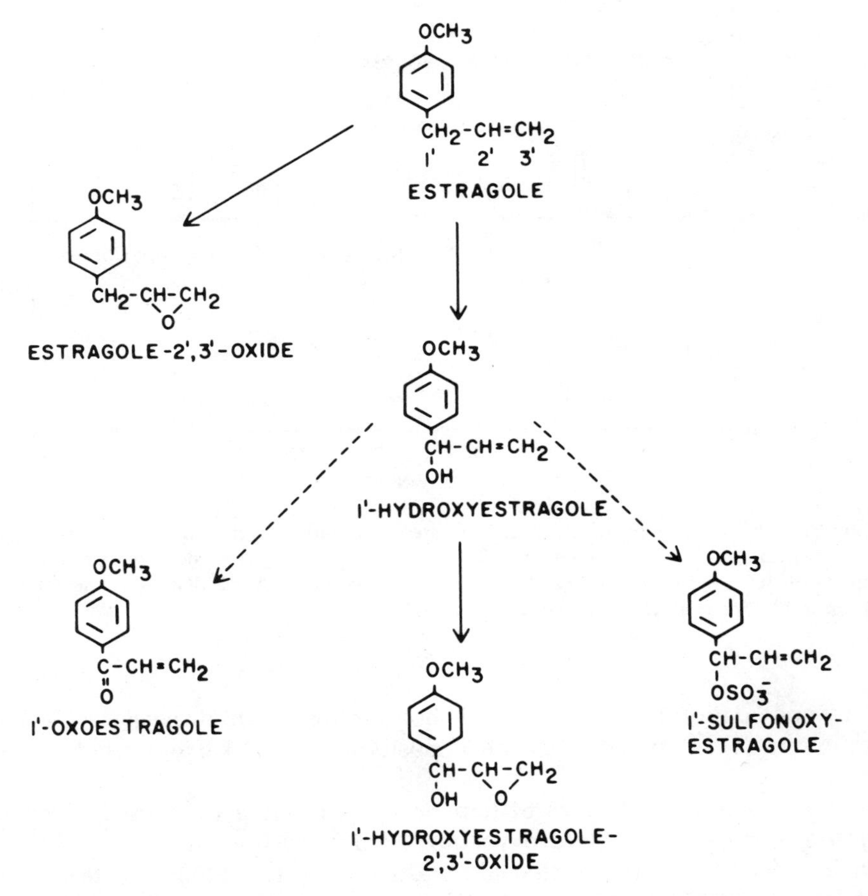

Figure VI-51. The metabolism of estragole to electrophiles. Broken arrows indicate pathways for which there is only indirect evidence. From Phillips *et al.* (1981) *Cancer Res.* **41,** 176.

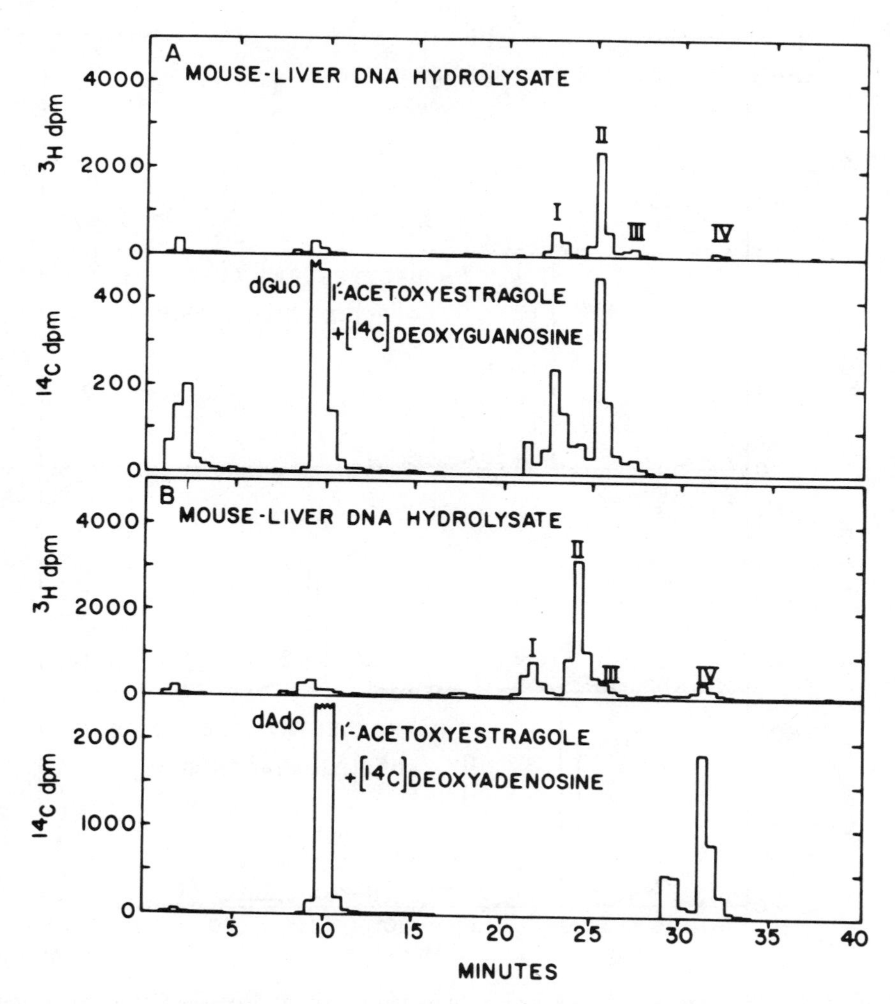

Figure VI-52. HPLC profiles of DNA hydrolysates and marker nucleosides. Cochromatography of mouse liver DNA hydrolysates from mice given injections of 1′-[^{3}H]hydroxyestragole with aliquots of the reaction mixture of 1′-acetoxyestragole reacted with (A) [^{14}C]dGuo and (B) [^{14}C]dAdo. From Phillips *et al.* (1981) *Cancer Res.* **41,** 176.

ucts that coeluted on HPLC with adducts I, II, and III (Figure VI-52A). Adduct IV coeluted with a product of interaction of 1′-acetoxyestragole with [^{14}C]dA (Figure VI-52B). On the other hand, the *in vivo* adducts did not comigrate with any major reaction products formed in reaction of either 1′-hydroxyestragol-2′,3′ oxide or 1′-oxoestragole with dG or dA (Figure VI-53). As with safrole, all adducts obtained from enzymatic digests of liver DNA involved covalent attachment of the carcinogen moiety to the exocyclic group of guanine or adenine. Synthetic samples of adducts II and IV were characterized

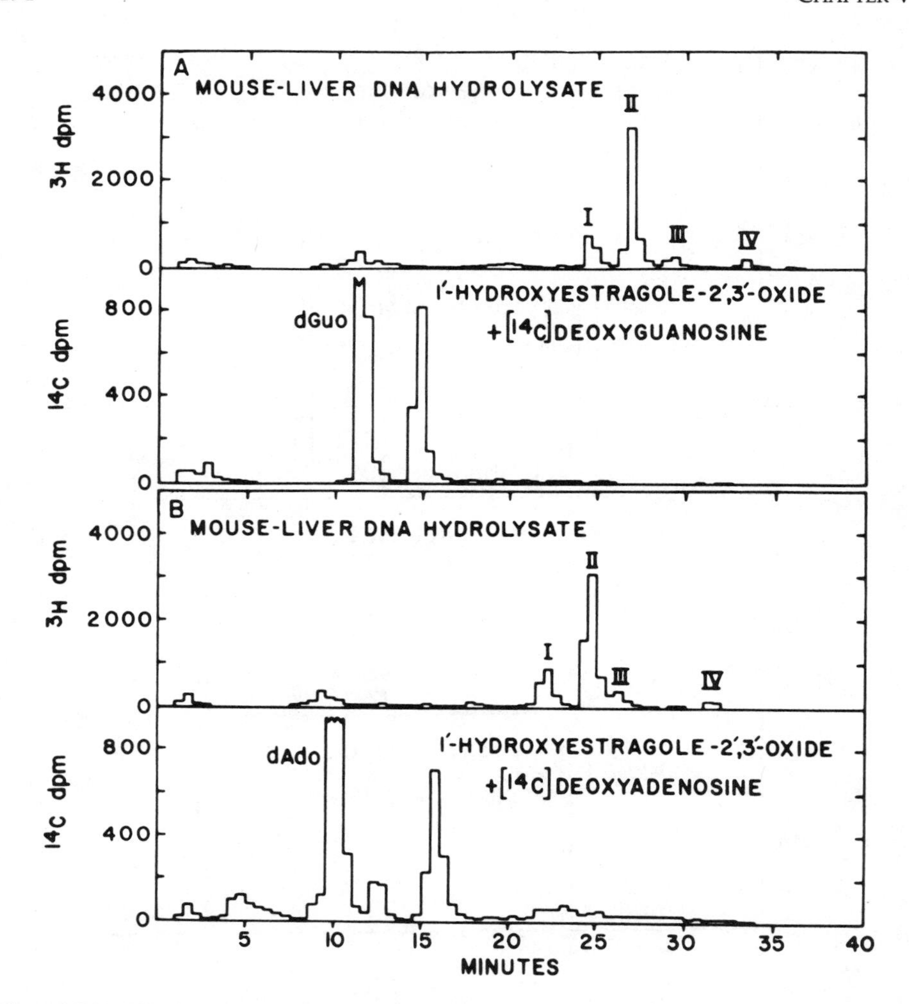

Figure VI-53. HPLC profiles of DNA hydrolysates and marker nucleosides. Cochromatography of mouse liver DNA hydrolysates from mice given injections of 1′-[^{3}H]hydroxyestragole with aliquots of the reaction mixture of 1′-oxoestragole reacted with (A) [^{14}C]dGuo and (B) [^{14}C]dAdo. From Phillips *et al.* (1981) *Cancer Res.* **41,** 176.

from their NMR spectra and assigned the structures N^2-(*trans*-isoestragol-3′-yl)deoxyguanosine and N^6-(*trans*-isoestragol-3′-yl)deoxyadenosine, respectively. Adduct I was assigned the structure N^2-(estragol-1′-yl)deoxyguanosine and adduct III has been tentatively regarded as N^2(*cis*-isoestragol-3′-yl)deoxyguanosine (Figure VI-54).

The binding of estragole metabolites to N^2-guanine is similar to that of other carcinogens, including BP, and some *N*-substituted aromatic compounds such as AAF and MAB. In contrast to those adducts that persist in

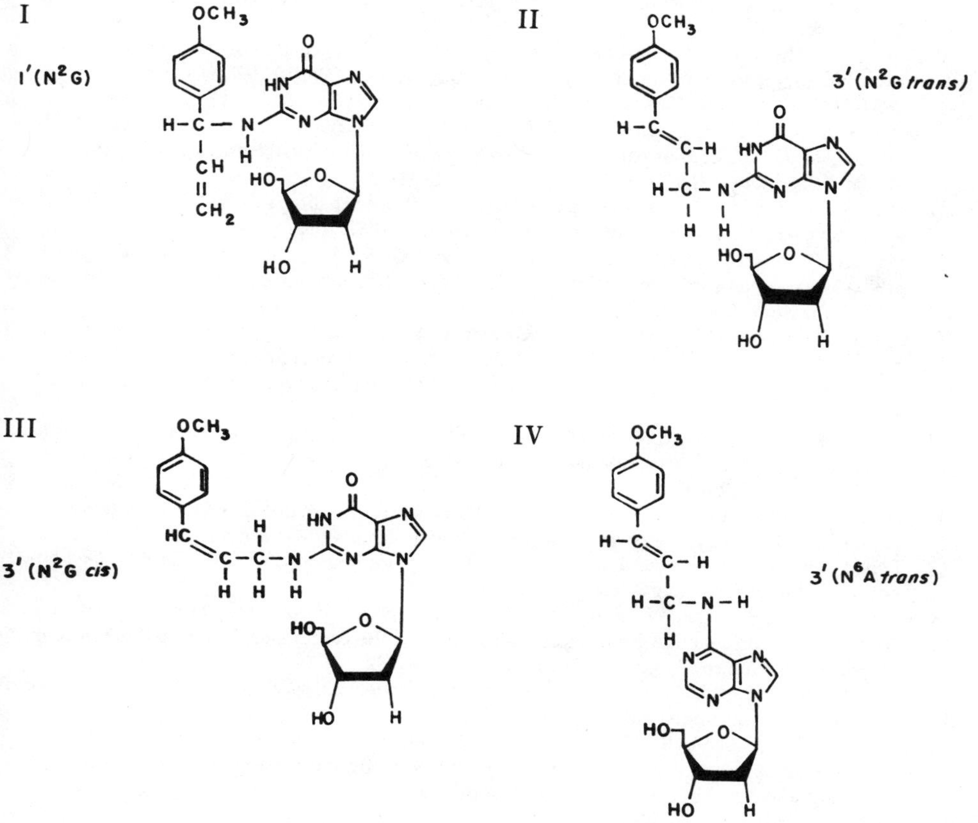

I $1'(N^2G)$

II $3'(N^2G$ *trans*)

III $3'(N^2G$ *cis*)

IV $3'(N^6A$ *trans*)

Figure VI-54. Proposed structures of the 1′-hydroxyestragole–deoxyribonucleoside adducts that are formed in mouse liver DNA *in vivo*. From Phillips *et al.* (1981) *Cancer Res.* **41,** 176.

DNA for at least several weeks following treatment, most of the N^2-guanine and N^6-adenine adducts of 1′-hydroxyestragole are removed from mouse liver DNA quite rapidly. There is also a possibility that persistent but quantitatively minor adducts not yet detected account in part for the binding to DNA in the liver.

SELECTED REFERENCES

Backer, J. M., and Weinstein, I. B. (1982) Interaction benzo(*a*)pyrene and its dihydrodiol-epoxide derivative with nuclear and mitochondrial DNA in C3H10T½ cell cultures. *Cancer Res.* **42,** 2764–2769.

Beland, F. A., Tullis, D. L., Kadlubar, F. F., Straub, K. M., and Evans, F. E. (1981) Identification of the DNA adducts formed *in vitro* from *N*-benzoyloxy-*N*-methyl-4-aminoazobenzene and in rat liver *in vivo* after administration of *N*-methyl-4-adminoazobenzene. In The International

Conference on Carcinogenic and Mutagenic N-Substituted Aryl Compounds, *Natl. Cancer Inst. Monogr.* **58,** 153–161.

Billings, P. C., Uwaifo, A. O., and Heidelberger, C. (1983) Influence of benzoflavone on aflatoxin B_1-induced cytotoxicity, mutation and transformation of C3H/10T1/2 cells. *Cancer Res.* **43,** 2659–2667.

Essigmann, J. M., Green, C. L., Croy, R. G., Fowler, K. W., Büchi, G. H., and Wogan, G. N. (1983) Interactions of aflatoxin B_1 and alkylating agents with DNA: Structural and functional studies. *Cold Spring Harbor Symp. Quant. Biol.* **47,** 327.

Gupta, B. C., Reddy, M. V., and Randerath, K. (1982) ^{32}P-postlabeling analysis of non-radioactive aromatic carcinogen–DNA adducts. *Carcinogenesis* **3,** 1081–1092.

Haugen, A., Groopman, J. D., Hsu, I. C., Goodrich, G. R. Wogan G. N., and Harris, C. C. (1981) Monoclonal antibody to aflatoxin B_1-modified DNA detected by enzyme immunoassay. *Proc. Natl. Acad. Sci. USA* **78,** 4124–4127.

Hemminki, K., Cooper, C. S., Ribeiro, O, Grover, P. L., and Sims, P. (1980) Reactions of "bay-region" and non-"bay-region" diol epoxides of benz(*a*)anthracene with DNA: Evidence indicating that the major products are hydrocarbon–N^2-guanine adducts. *Carcinogenesis* **1,** 177–286.

Hertzog, P. J., Lindsay Smith, J., and Garner, R. C. (1982) Production of monoclonal antibodies to guanine imidazole ring-opened aflatoxin B_1DNA, the persistent DNA adduct *in vivo. Carcinogenesis* **3,** 825–828.

Jeffrey, A. M., Kinoshita, T., Santella, R. M., Grunberger, D., Katz, L., and Weinstein, I. B. (1980) The chemistry of polycyclic aromatic hydrocarbon–DNA adducts. In *Carcinogenesis: Fundamental Mechanisms and Environmental Effects* (B. Pullman, P.O.P. Ts'o, and H. Gelboin, eds.) Academic Press, New York, pp. 505–579.

Kriek, E., Miller J. A., Juhl, W., and Miller, E. C. (1967) 8-(*N*-2-Fluorenylacetamido)-guanosine, an arylamidation reaction product of guanosine and the carcinogen *N*-acetoxy-*N*-2-fluorenylacetamide in neutral solution. *Biochemistry* **6,** 177–182.

Miller, E. C., and Miller, J. A. (1981) Searches for ultimate chemical carcinogens and their reactions with cellular macromolecules. *Cancer* **10,** 2327–2345.

Miller, E. C., Swanson, A. B., Phillips, D. H., Fletcher, T. L., Liem, A. and Miller, J. A. (1983) Structure–activity studies on the carcinogenicities in the mouse and rat of some naturally occurring and synthetic alkenylbenzene derivatives related to safrole and estragole. *Cancer Res.* **43,** 1124–1134.

Moore, P. D., Korreda, M., Wislocki, P. G., Levin, W., Conney, A. H., Yagi, H., and Jerina, D. M. (1977) In vitro reactions of the diastereomeric 9,10-epoxides of (+) and (−) *trans*-7,8-dihydroxy-7,8-dihydrobenzo(*a*)pyrene with polyguanylic acid and evidence for formation of an enantiomer of each diastereomeric 9,10-epoxide from benzo(*a*)pyrene in mouse skin. In *Drug Metabolism Concepts* (D. M. Jerina, ed.) American Chemical Society, Washington, D. C., pp. 127–154.

Moschel, R. C., Baird, W. M., and Dipple, A. (1977) Metabolic activation of the carcinogen 7,12-dimethylbenz(*a*)anthracene for DNA binding. *Biochem. Biophys. Res. Commun.* **76,** 1092–1098.

Osborne, M. R., Jacobs, S., Harvey, R. G., and Brookes, P. (1981) Minor products from the reaction of (+) and (−) benzo(*a*)pyrene-*anti*-diolepoxide with DNA. *Carcinogenesis* **2,** 553–558.

Perera, F. P., Poirier, M. C., Yuspa, S. H., Nakayama, J., Jaretzki, A., Curnen, M. M., Knowles, D. M., and Weinstein, E. B. (1982) A pilot project in molecular cancer epidemiology: Determination of benzo(*a*)pyrene–DNA adducts in animal and human tissues by immunoassays. *Carcinogenesis* **3,** 1405–1410.

Poirier, M. (1981) Antibodies to carcinogen–DNA adducts. *J. Natl. Cancer Inst.* **67,** 515–519.

Poirier, M. C., Santella, R., Weinstein, I. B., Grunberger, D., and Yuspa, S. H. (1980) Quantitation of benzo(*a*)pyrene–deoxyguanosine adducts by radioimmunoassay. *Cancer Res.* **40,** 412–416.

Rinaldy, A., Lockhart, M. L., Deutsch, J. F., Ungers, G. E., and Rosenberg, B. H. (1982) Effects of benzo(*a*)pyrene diol epoxide adducts on DNA synthesis in mammalian cells. *Mutation Res.* **94,** 383–391.

Sawicki, J. T., Moschel, R. C., and Dipple, A. (1983) Involvement of both *syn-* and *anti-*dihydrodiol-epoxides in the binding of 7,12-dimethylbenz(*a*)anthracene to DNA in mouse embryo cell cultures. *Cancer Res.* **43,** 3212–3218.

Stark, A. A. (1980) Mutagenicity and carcinogenicity of mycotoxins: DNA binding as a possible mode of action. *Annu. Rev. Microbiol.* **34,** 235–262.

Sugimura, T. (1982) Mutagens, carcinogens and tumor promoters in our daily food. *Cancer* **49,** 1970–1984.

Weinstein, I. B. (1981) Current concepts and controversies in chemical carcinogenesis. *J. Supramol. Struct. Cell. Biochem.* **17,** 99–120.

Weinstein, I., Jeffrey, A. M., Leffler, S., Pulkrabek, P., Yamasaki, H., and Grunberger, D. (1978) Interactions between polycyclic aromatic hydrocarbons and cellular macromolecules. In *Polycyclic Hydrocarbons and Cancer,* Vol. 2 (H. V. Gelboin and P.O.P. Ts'o, eds.), Academic Press, New York, pp. 4–36.

Westra J. G., Kriek, E., and Hittenhausen, H. (1976) Identification of the persistently bound form of the carcinogen *N*-acetyl-2-minofluorene to rat liver DNA *in vivo. Chem. Biol. Interact.* **15,** 149–164.

Wogan, G. N., Essigmann, J. M., Croy, R. G., Busby, W. F., Jr., Groopman, J. D., and Stark, A. A. (1979) Macromolecular binding of AFB_1 and sterigmatocystin: Relationship of adduct patterns to carcinogenesis and mutagenesis. In *Naturally-Occurring Carcinogens: Mutagens and Modulators of Carcinogenesis* (E. C. Miller, J. A. Miller, I. Hirono, T. Sugimura, and S. Takayama, eds.), University Park Press, Baltimore, pp. 19–33.

VII

Conformation of DNA Modified by Bulky Aromatic Carcinogens

A. Introduction

From the results described previously, it became evident that many, and perhaps all, carcinogens become covalently bound to nucleic acids in the target tissue. It also appears that this interaction is critical to the carcinogenic process. This covalent modification can introduce important changes, not only in the primary structure of nucleic acids, but also in the conformation (three-dimensional structure) of the nucleic acids at the sites of modification. Since distortions in nucleic acid structure are likely to have functional consequences, a detailed description of these conformational changes seems essential if we are to ultimately understand the carcinogenic process at the molecular level.

X-Ray diffraction and a variety of spectral techniques have provided extensive information on the conformations of nucleosides, oligonucleotides, and single- and double-stranded nucleic acids. These studies indicate that two major structural aspects influence the conformation of nucleosides and polynucleotides. These are the relative orientation of the base and sugar residues at the glycosyl bond and the type of pucker in the sugar ring (Figures VII-1, 2). In polynucleotides, the series of single bonds that form the phosphodiester linkages between adjacent nucleotides can also provide sites of rotation during conformational distortions of the polymer.

The major conformational parameter is the relative orientation of the base to the sugar, which is defined in terms of a torsion angle θ CN. The latter refers to the angle formed between the plane of the base and the C1′-O bond of the sugar, viewed along the glycosyl bond (Figure VII-1). There are two ranges of torsional angles within which stable conformations of nucleosides are assumed, the *anti* and the *syn*, and each of these ranges covers more than 90°. X-Ray crystallographic data of normal nucelosides have revealed that the vast majority of these compounds correspond to the *anti* conformation. The NMR and circular dichroism (CD) data are in agreement with the X-ray data. On the other hand, bulky substituents in position 6 on pyrimidines or in

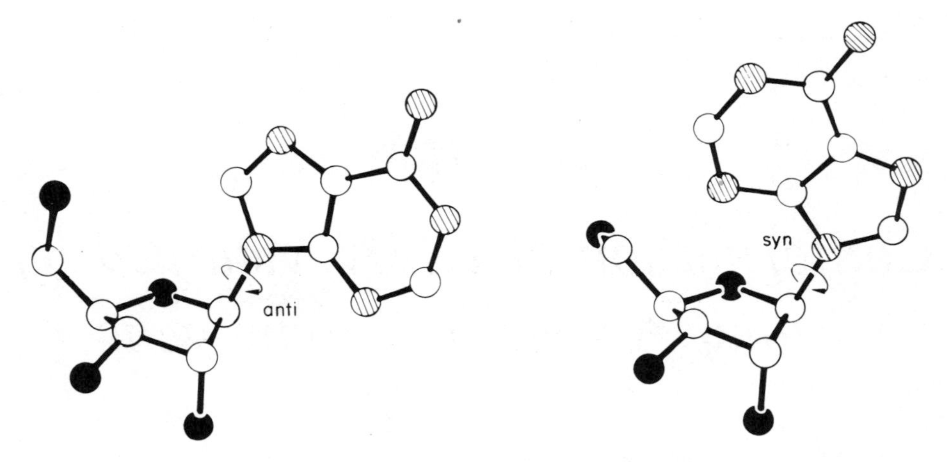

Figure VII-1. Illustration of *anti* and *syn* conformations about the glycosyl bond.

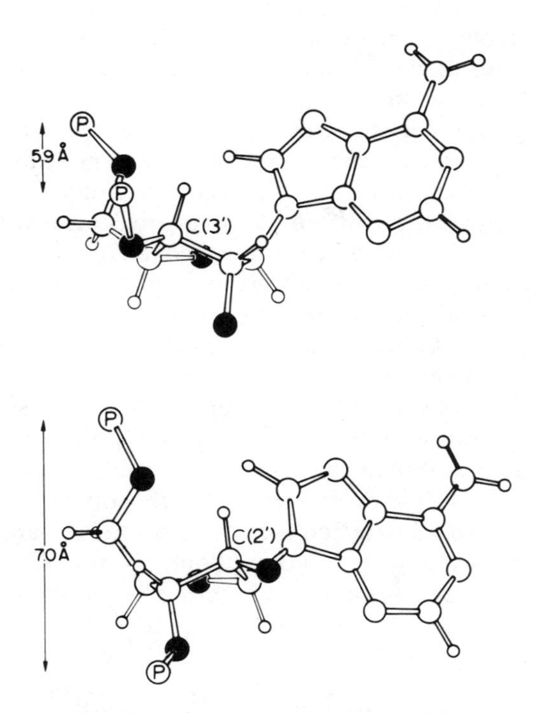

Figure VII-2. The C3′ *endo* and C2′ *endo* pucker forms of the deoxyribose residue.

position 8 on purines force the nucleosides to change their conformation from the usually preferred *anti* to the *syn* position. Well-known examples are the 8-bromo-substituted purine nucleosides and 6-methylpyrimidine nucleosides. The nucleoside conformation in double-stranded DNA helices is usually restricted to *anti* since this conformation is indispensable for ordinary basepairing in structures of the Watson–Crick type.

In nucleosides, the five-membered furanose ring of the sugar residue is puckered so that all of the carbon atoms are displaced by about 0.5 Å from the plane formed by the remaining four atoms (Figure VII-2). In most nucleosides, the sugar residues have either the C2′ *endo* or the C3′ *endo* form, i.e., the respective carbons are located on the same side of the sugar plane as C5′. In a DNA helix with Watson–Crick geometry (B-DNA), all of the nucleoside residues are in the *anti* conformation and the deoxyribose residues have the C2′ *endo* structure.

Recently a novel double-helical left-handed DNA structure was discovered, named Z-DNA (Figure VII-3). This structure was solved at atomic resolution (0.9 Å) from crystals of the hexanucleotide pentaphosphate d(CpGpCpGpCpG). The left-handed structure has also been observed in solution by a salt-induced inversion of the CD spectrum of poly(dG-dC) · poly(dG-dC) and by an additional peak in the ^{31}P NMR spectrum. DNA in Z form is characterized by a left-handed helical sense and a dinucleotide repeat unit resulting in 12 basepairs per helical unit. The internal dG sugars have C3′ *endo* and the dC sugars have C2′ *endo* ring puckering. The dG residues are in *syn* conformation, whereas the dC residues remain in *anti* conformation. As a direct consequence of dG being in the *syn* conformation, the N-7 and C-8 positions, which are so often attacked by carcinogens, are on the outside of the helix (Figure VII-4). This enhanced exposure of the N-7 and C-8 positions of G is particularly relevant with respect to the binding of carcinogens. Although there is as yet no evidence for the existence of Z-DNA *in vivo*, stretches of alternating purine–pyrimidine sequences do occur in certain naturally occurring DNAs. If such a region were modified, this would favor its transition to the Z form. This change in conformation would then induce important functional changes in that region of the genome.

What are the general implications of the above-described conformational aspects of nucleic acids in terms of covalent attachment of various carcinogens? Carcinogens are known to bind to various bases and positions in DNA as has been previously described in some detail (Chapters IV and VI). *A priori*, we might predict that depending on the size of the carcinogen and its site of substitution, there might be appreciable changes in nucleic acid conformation, resulting from alterations in conformation at the glycosyl bond, rotations of the backbone residues, and possible changes in sugar puckering.

The methylation or ethylation of bases by the simple alkylating agents does not appear to cause major conformational distortions of the helix. Some of these modifications can, however, interfere with the Watson–Crick base-

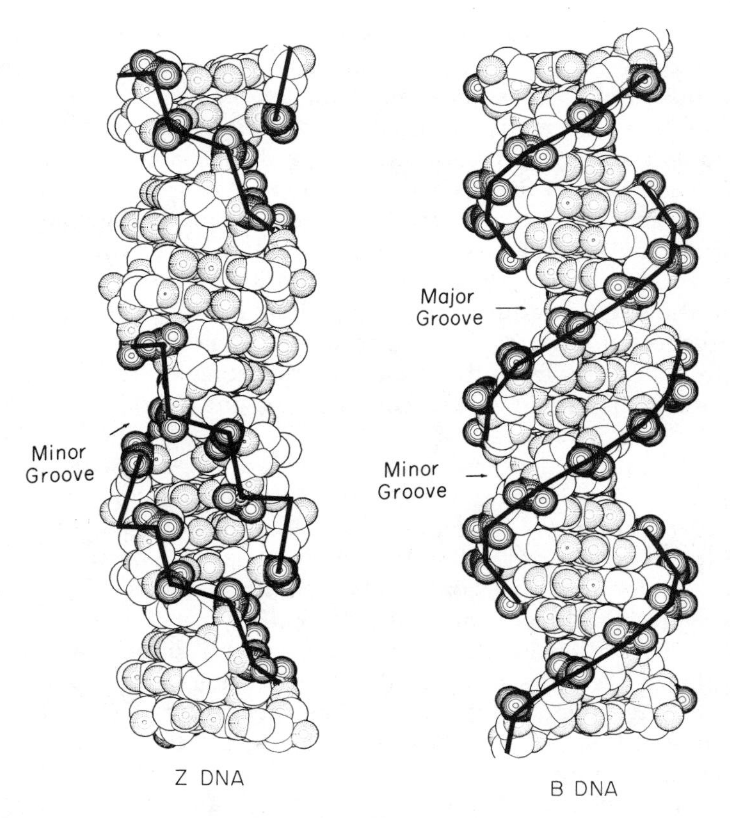

Figure VII-3. Van der Waals' sideviews of Z-DNA and B-DNA. From Wang *et al.* (1979) *Nature (London)* **282,** 680.

pairing and N-3 or N-7 substitutions on the purines, resulting in depurination due to labilization of the glycosyl bond (Chapter III).

The covalent attachment of bulky carcinogens such as *N*-2-acetylaminofluorene (AAF), benzo[*a*]pyrene, naphthylamine, and others does present steric problems and, depending on the sites of substitution, may be associated with major distortions in the native conformation of nucleic acids. We will here confine our discussion to results obtained with these three bulky carcinogens.

B. Methods of Analysis

Fundamental to an understanding of the conformational changes and functional consequences of nucleic acid modification by bulky carcinogens is information on two points: (1) orientation of the covalently bound carcinogen within DNA, and (2) possible alterations in the native helical conformation

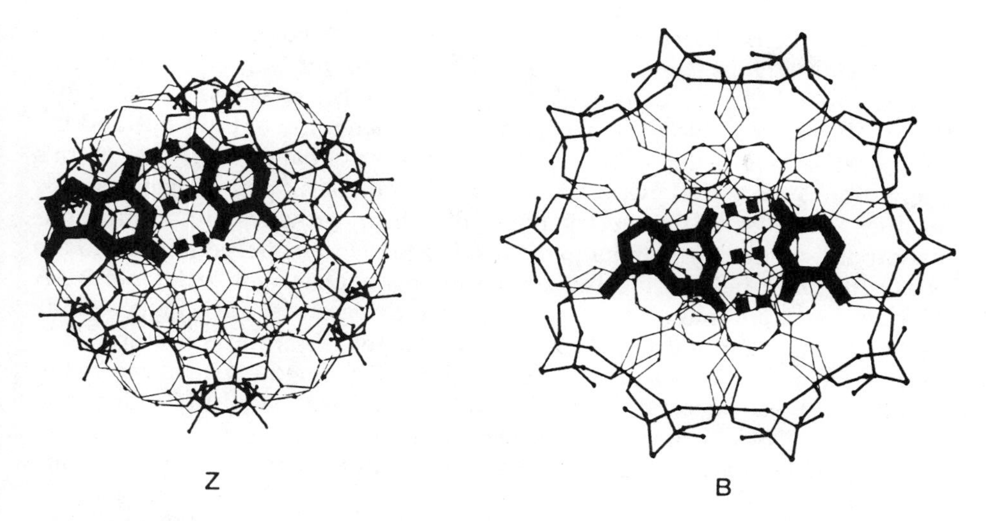

Figure VII-4. End views of the helical forms of Z-DNA and B-DNA. Heavier lines are used for the phosphate–sugar backbone. A guanine-cytosine basepair is shown by shading. They are near the center of the B-DNA but at the periphery of Z-DNA. From Wang *et al.* (1981) *Science* **211,** 172.

of DNA resulting from this modification. By utilizing a variety of spectroscopic, enzymatic, and immunological techniques, one can obtain adequate information on these two points.

Conformational changes associated with covalent modification by chemical carcinogens can be studied either with low-molecular-weight oligonucleotides, which mimic, relatively well, the situation in macromolecules, or they can be studied with polymeric nucleic acids as model compounds. The smallest unit that can be used that has many of the local interactions also present in a polymer is a dinucleoside monophosphate (dimer). As in naturally occurring nucleic acids, it has a phosphate linkage between the 3′ and the 5′ position of the two sugar residues. Furthermore, neighboring bases interact strongly in dimers, as in polymers. CD and NMR spectroscopy are the two most powerful methods used for studying the conformation of dimers in solution, and the results are more readily interpreted than those obtained with higher-molecular-weight polymers. CD can be used to monitor changes that are characteristic of the loss, or presence, of base-stacking interactions. Similarly, NMR spectra can be used to monitor stacking interactions between the bases and other effects of modified residues.

Localization of the carcinogen in the DNA can be studied by UV absorption spectrometry or, if the carcinogen has a characteristic fluorescence spectrum by fluorescence spectroscopy and by its quenching. Application of these techniques will be discussed in connection with conformation of DNA modified by benzo[*a*]pyrene derivatives.

One of the best techniques to ascertain about the orientation of the carcinogen in the DNA is electric linear dichroism. In this technique the electric field is applied to an aqueous solution containing the modified DNA (Figure VII-5). The electric field gives rise to a partial orientation of the DNA and the carcinogen attached to it. The change in the absorbance (ΔA) due to this orientation is measured by polarized light. In the case of pyrene-like aromatic residues, the sign of the electric linear dichroism provides important information on whether the carcinogenic residue tends to be perpendicular to the axis of the DNA helix or is lying along the phosphate–sugar backbone. If ΔA is larger than 0 and has a positive sign, the carcinogen residue is lying outside the helix. On the other hand, if ΔA has a negative sign, the chromophore is perpendicular to the axis of the DNA and is intercalated between the bases. This method was successfully applied with AAF, 1-OH-naphthylamine, and BP-modified DNA. Other physicochemical methods that have been used include thermal denaturation, formaldehyde unwinding, and intrinsic viscosity techniques.

The best evidence for localized regions of denaturation in double-stranded DNA modified by bulky carcinogens comes from using a single-strand-specific S_1 endonuclease from *Aspergillus oryzae*. This enzyme splits only single-stranded regions on the modified molecule. Therefore, by measuring the kinetics of digestion of a modified DNA, information can be obtained concerning the extent to which DNA is in native or a partially denatured form. Additional evidence on the state of DNA can be obtained by using *antinucleoside antibodies*. Since modification occurs mostly on G residues, anticytidine antibodies have mainly been used. These react with cytidine residues only in single-stranded

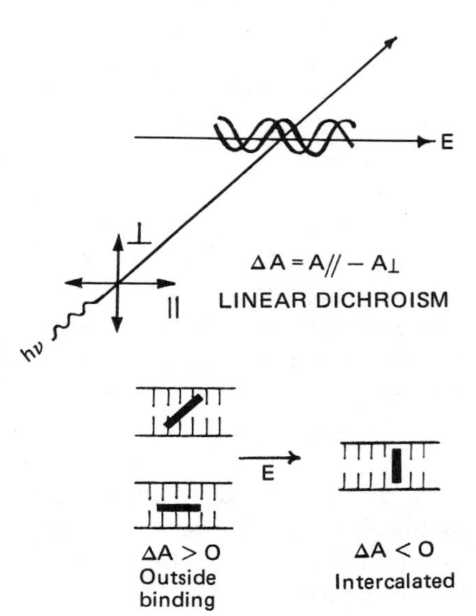

Figure VII-5. Scheme of electric linear dichroism. DNA is oriented in an electric field (E). Absorbance (A) of chromophore is measured using polarized light. Courtesy of N. E. Geacintov.

regions of a polymer and thus they can be used to detect the extent of the possible denaturation in the modified molecules. Both these methods were applied in studies of the conformation of AAF- and BP-modified DNA and will be discussed later.

C. CONFORMATION OF DNA–7,8-DIHYDRODIOL-9,10-OXIDE-BENZO[A]PYRENE (BPDE) ADDUCTS

As was described previously, the major reactive form of benzo[*a*]pyrene involved in DNA binding *in vitro* as well as *in vivo* is the 7β,8α-dihydroxy-9α,10α-epoxy-7,8,9,10-tetrahydrobenzo[*a*]pyrene (BPDE I) metabolite. The covalent binding of BPDE to DNA gives rise to one major product involving a covalent bond between BPDE and the N^2 exocyclic amino group of guanine, but other products have also been detected.

Conformation of the DNA modified by BPDE was investigated by Geacintov and co-workers. They have used absorption, fluorescence, and electric linear dichroism techniques to detect the orientation of covalently bound BPDE residues in DNA and to compare it with intercalated noncovalently bound BP-tetraol (the hydrolysis product of BPDE) residues.

UV absorption spectral changes are different for the intercalation-type binding and for an exterior-type binding of a chromophore. In aqueous solutions the BP chromophore exhibits absorption maxima at 313, 327, and 343–344 nm. Figure VII-6 shows that upon covalent binding of BPDE to DNA there is only a very small spectral shift to the red of the absorption maximum at 343 nm. On the other hand, upon noncovalent binding of free tetraol to

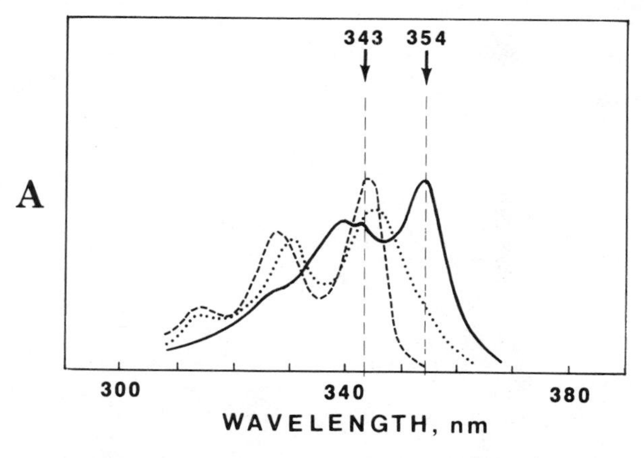

Figure VII-6. Absorption spectra of free BPDE (----), DNA-BPDE covalent adduct (····), and DNA-BPDE noncovalent complex (——). Adapted from N.E. Geacintov (1981) 6th Int. Symp. Polynuclear Aromatic Hydrocarbons (Cooke, Dennis, and Fisher, eds.), Battelle Press, Columbus, Ohio.

DNA by an intercalation mechanism, the absorption maximum is shifted by 10 nm to the red from 343 nm to 353 nm. Thus, it appears that the covalently bound BPDE residues are not intercalated, but lie on the outside of the DNA molecule.

Fluorescence spectral results confirmed the exterior type of binding of BPDE in DNA. The experimental approach utilized in this study is outlined in Figure VII-7. If an external quencher of fluorescence, such as molecular oxygen, that cannot penetrate readily into the double helix is used, it has a large quenching effect on aromatic molecules that are located on the outside of the helix, but it has no quenching effect on the chromophore intercalated between the basepairs. On the other hand, if internal quenchers such as Ag^+ or Hg^{2+}, which bind to DNA by a base-specific interaction mechanism, are utilized, the reverse is true. In the case of a chromophore located on the outside of the helix, the internal quenchers show little or no effect but they exhibit large quenching with chromophores intercalated between the basepairs.

Since Ag^+ and Hg^{2+} influenced the fluorescence of the covalently bound BPDE-DNA adduct only negligibly, in contrast to the strong effect on the fluroescence of the noncovalently intercalated benzo[*a*]pyrene molecules (Figure VII-8), it is possible to conclude that the covalently bound BPDE residue is not located within the DNA, but rather externally to the helix.

The orientation of the BPDE chromophore with respect to that of the DNA bases in covalently bound DNA-BPDE complexes was further investigated by utilizing the electric field-induced linear dichroism technique. Since the ΔA value of the electric linear dichroism spectra has a positive sign, one can assume that the BPDE residue covalently bound to DNA is located outside the helix (Figure VII-9). From the linear dichroism data, Geacintov and co-

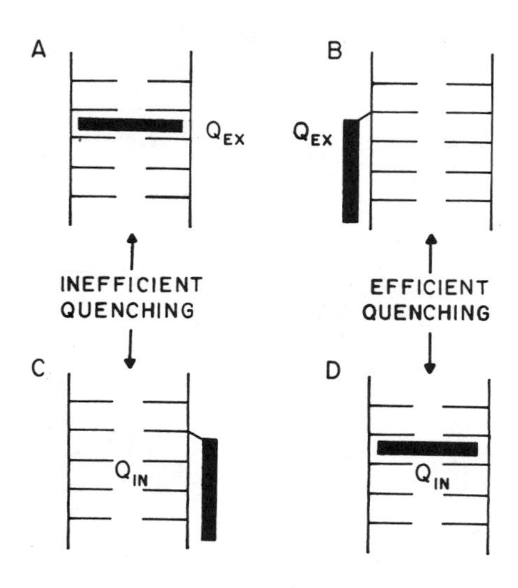

Figure VII-7. Principles of fluorescence quenching method. Q_{IN} internal quencher (Ag^+ or Hg^{2+}); Q_{EX} external quencher (O_2) chromophore. Adapted from Prusik *et al.* (1979) *Photochem. Photobiol.* **29,** 223.

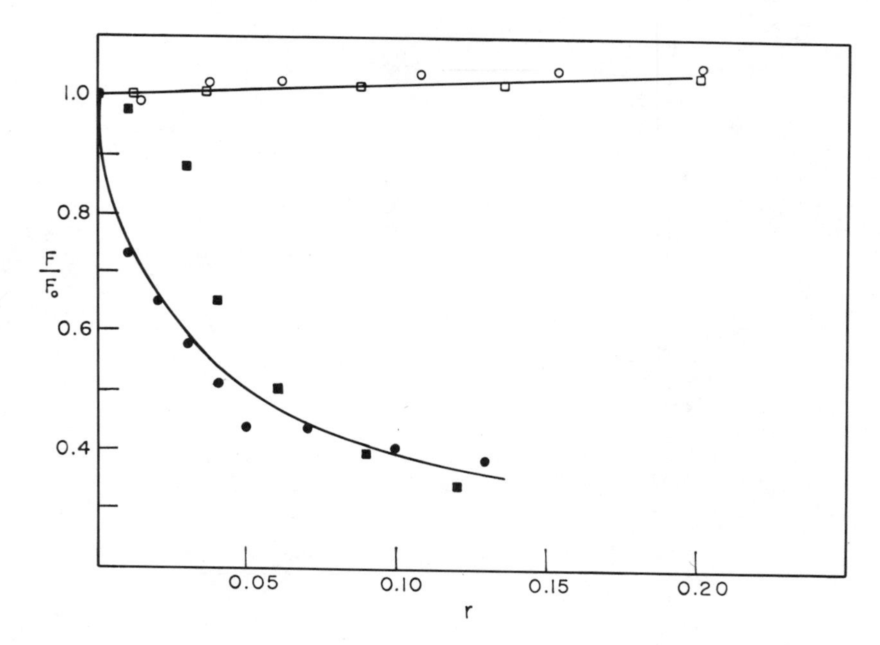

Figure VII-8. Effect of metal ions on fluorescence of BP-DNA complexes. (●) Ag^+–BP-DNA noncovalent complex; (■) Hg^{2+}–BP-DNA noncovalent complex; (○) Ag^+–BPDE-DNA adduct; (□) Hg^{2+}–BPDE-DNA. Adapted from Prusik *et al.* (1979) *Photochem. Photobiol.* **29,** 223.

workers calculated that the maximum angle at which the pyrene chromophore could reside with respect to the long axis of the DNA is 35° or less.

Examination of a model of DNA indicates that the 2-amino group of dG is relatively exposed in the minor groove of the helix. Since the orientation angle of the BPDE chromophore bound to DNA is 35° or less, model building suggests that this chromophore is located in the minor groove of the DNA helix. All spectral data are consistent with this type of orientation of BPDE residue in the modified DNA. A computer-generated stereoscopic display of a model consistent with this orientation is given in Figure VII-10.

The location of the BPDE residue in the minor groove would cause little distortion of the native DNA conformation. This conclusion is supported by the findings that there is only a very slight decrease in the melting temperature (T_m) during heat denaturation and a very small increase in susceptibility of BPDE-modified DNA to the single-strand-specific S_1 nuclease. This is in contrast to the marked conformational distortion and rather large localized regions of denaturation associated with AAF modification of DNA.

Thus, the effect of BPDE modification on the template function of nucleic acids could be explained by the impairment of the basepairing potential of the modified guanine residues rather than by distortion of the conformation of the modified DNA.

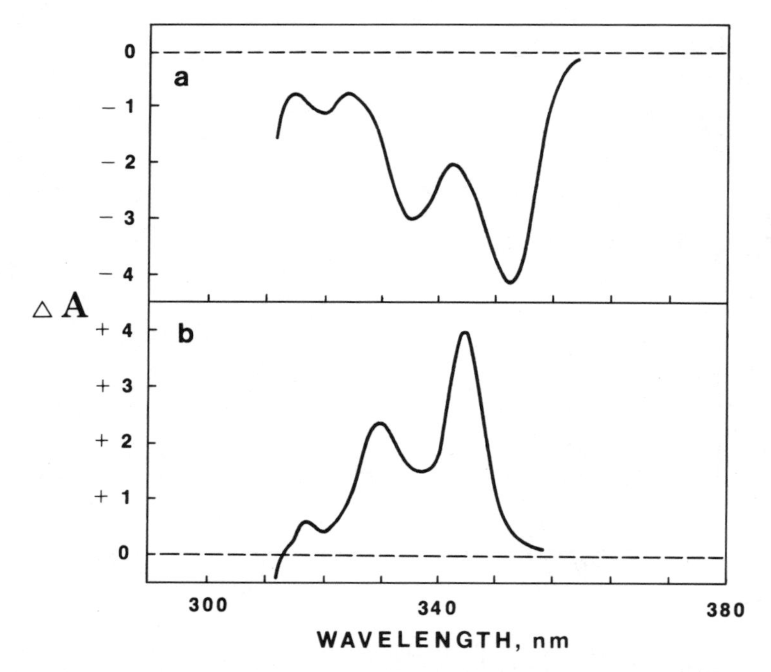

Figure VII-9. Electric linear dichroism spectra (ΔA) of the noncovalent DNA-7,8,9,10-tetrahydro-tetrahydroxybenzo[*a*]pyrene complex (a) and of the covalent DNA-BPDE adduct (b). Adapted from N.E. Geacintov (1981) 6th Int. Symp. Polynuclear Aromatic Hydrocarbons (Cooke, Dennis, and Fisher, eds.), Battelle Press, Columbus, Ohio.

D. Conformation of DNA Modified by *N*-Substituted Aromatic Compounds

1. *Conformation of Acetylaminofluorene- and Aminofluorene-Modified DNA*

It has been described that AAF modifies DNA at the C-8 and N^2 positions of G residues. A third product, the deacetylated C-8 G adduct, is also formed *in vivo*. All three adducts may cause different types of changes in the conformation of the modified DNA molecules.

Binding of AAF at the C-8 position results in a large distortion of the DNA helix referred to as the "base displacement" or the "insertion–denaturation" model. The scheme of this model is presented in Figure VII-11. The first feature of this model is that the attachment of the AAF residue to the 8 position of G is associated with a change in glycosyl conformation from *anti* to *syn*. Evidence for this is restricted to a study of molecular models of AAF-G that indicates severe steric hindrance between AAF and the deoxyribose of the nucleoside, unless the guanine base is rotated about the glycosyl

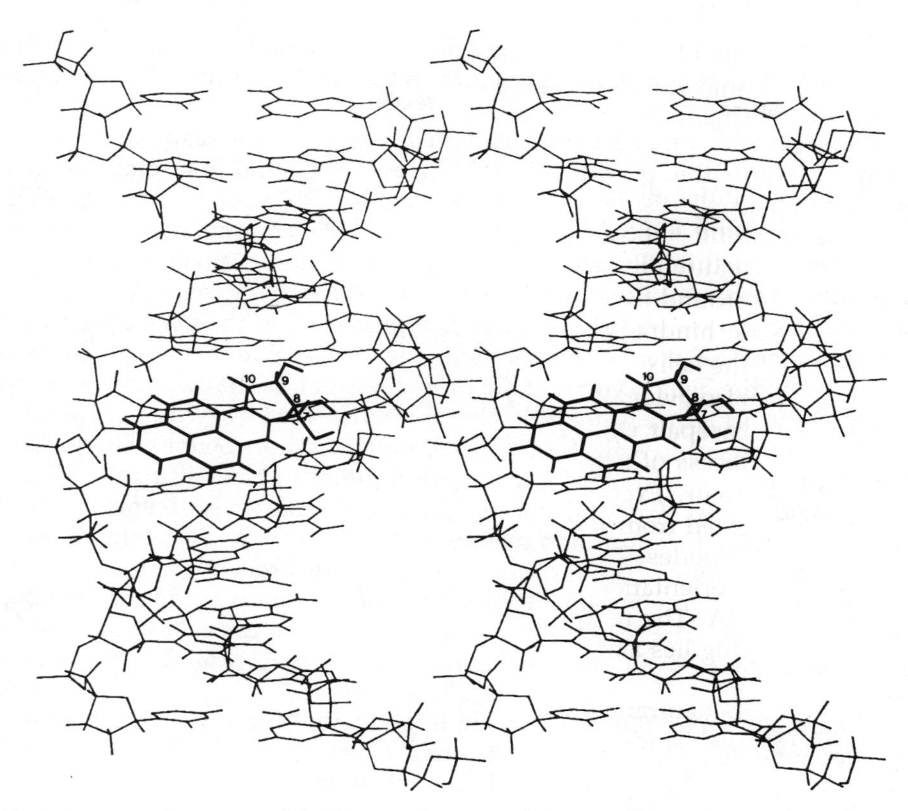

Figure VII-10. A stereoscopic view of the DNA-BPDE adduct. In this model the conformation of the guanine and the coplanarity of the N^2 amino group and the base to which the BPDE is attached are retained. The 7- and 8-hydroxyl groups and the 9- and 10-hydrogens of BPDE are quasiequatorial. The angle between the plane of the pyrene and the axis of the DNA is 28°. (To view this image in stereo, two lenses of about 20-cm focal length should be mounted to about 14 cm above the images and 6.5 cm apart. Viewers of this type are sold by Taylor Merchant Corp., 24 W. 45 St., New York, N.Y. 10036 under the name of Stereoptician 707.) From Jeffrey *et al.* (1980) In *Carcinogenesis: Fundamental Mechanisms and Environmental Effects* (Pullman, Ts'o, and Gelboin, eds.), Reidel, Dordrecht, p. 565.

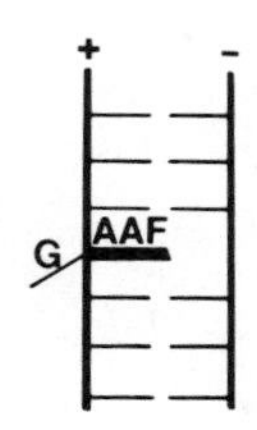

Figure VII-11. Schematic representation of the "base displacement" model of DNA-AAF adduct. AAF cannot basepair and causes base deletion, a "frameshift" mutation.

bond from the *anti* to the *syn* conformation. The second major feature of this model is that there is a stacking interaction between AAF and a base adjacent to the substituted G residue.

These changes are best illustrated on drawings of base-displace minimum energy conformation of dCpdG-AAF (Figure VII-12) and in a computer-generated stereoscopic display of a double-stranded DNA fragment (Figure VII-13). The computer display allows one to readily perform rotation around appropriate bond angles while obtaining a three-dimensional image of the molecular structure on a video screen. In the display, the modified base has been rotated around the glycosyl bond from the *anti* to the *syn* conformation to avoid steric hindrance. In addition, the planar fluorene ring system is inserted into the helix occupying the former position of the displaced guanine residue. It is also evident that the G residue displaced by AAF in the double helix cannot basepair with the C residue on the complementary strand, and during the process of replication or transcription no basepairing at this position could occur. Experimental evidence of the base displacement model has been obtained from proton magnetic resonance and CD spectra of modified oligonucleotides. The technique of electric dichroism was then used to determine the orientation of the covalently bound fluorene ring to the long axis of the DNA. The results clearly indicated that in the case of DNA-AAF the fluorene ring lies almost perpendicular to the helix axis, the angle being 80°.

Thermal denaturation studies are in good agreement with the base displacement model, since AAF modification decreased the T_m of DNA. For each 1% of the modified bases there is an approximate 1.5°C decrease in T_m. This is in contrast with intercalating drugs, which stabilize the native structure of DNA and increase its T_m. Similarly, a marked decrease of the intrinsic viscosity of the modified DNA supports the idea that AAF binding causes localized regions of denaturation.

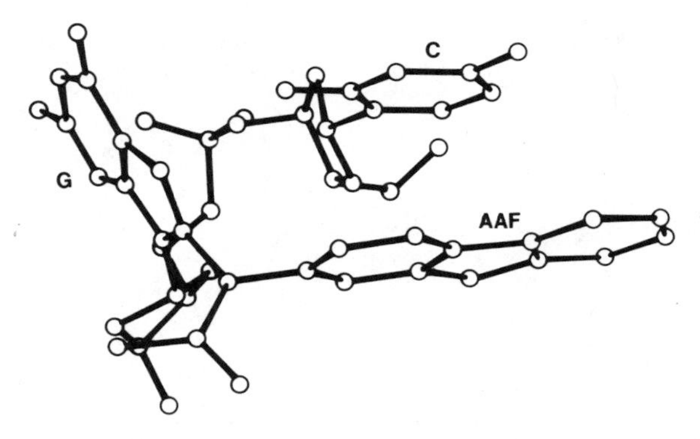

Figure VII-12. Base-displaced minimum energy conformation of dCpdG-AAF. Based on Hingerty and Broyde (1982) *Biochemistry* **21**, 3243.

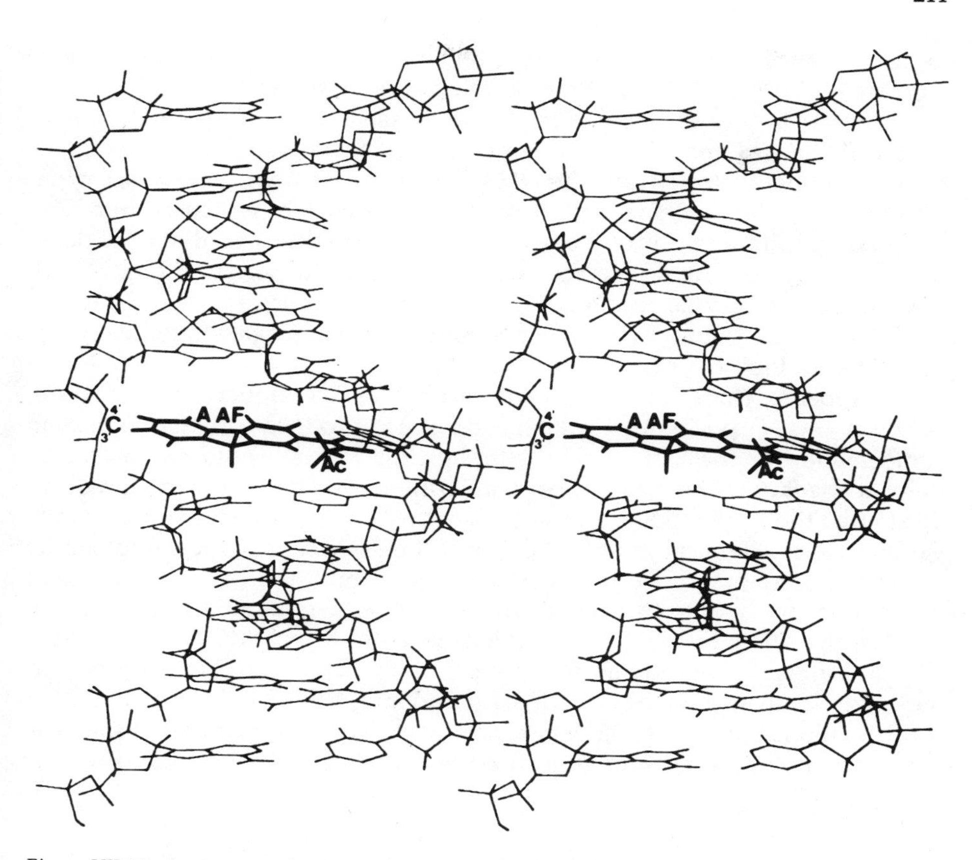

Figure VII-13. A stereoscopic view of the base displacement model of DNA-AAF adduct. The guanine to which the AAF is attached has been rotated out of the helix and the AAF moiety is inserted into the helix and stacked with the bases above and below. Ac designates acetyl group of AAF. The cytosine (marked C) residue on the opposite strand would overlap with the AAF residue; therefore, it has been removed and the 3′ and 4′ carbon atoms of the corresponding deoxyribose in the DNA backbone are indicated. From Jeffrey *et al.* (1980) In *Carcinogenesis: Fundamental Mechanisms and Environmental Effects* (Pullman, Ts'o, and Gelboin, eds.), Reidel, Dordrecht, p. 565.

The best evidence for localized regions of denaturation has come from studies on the susceptibility of AAF-modified DNA to digestion by S_1 nuclease. The estimated number of basepairs destabilized by a single AAF modification is in the range of 5 to 50, depending on the extent of modification of the DNA. Reaction of the modified regions of DNA with anticytidine antibodies supported this conclusion. Since attachment of AAF to G residues requires rotation of the base about the glycosyl bond and there is less hindrance to the rotation of bases in single-stranded than in double-stranded gions, it follows that single-stranded regions of nucleic acids are more susceptible to AAF modificaton than double-stranded ones.

Since there are differences in the steric aspects connected with modification of the C-8 and N^2 positions of G, differences also exist between the conformational distortions in the DNA helix associated with these two types of adducts. Following incubation of the modified DNA with S_1 nuclease, chromatographic analysis of the hydrolyzed fractions showed that S_1 nuclease digested only regions where the adduct is at C-8 but not at N^2 (Figure VII-14). Thus, the enzyme recognized the *N*-deoxyguanosine-8-yl-AAF but not the 3-(deoxyguanosin-N^2-yl)-AAF-modified regions as single-stranded regions on AAF-modified DNA.

It appears, therefore, that in contrast to the C-8 adduct, substitution of AAF on the N^2 position of guanine does not produce a major change in conformation of the DNA helix. Although the precise conformation of the helix at the latter sites has not been determined, model-building studies indicate that the N^2 position is in contrast to the C-8 position of guanine, readily susceptible to chemical modification, and the fluorene residue could simply occupy the minor groove of the DNA helix. Thus, the base displacement model may apply onto to the C-8 and not to the N^2-guanine adduct of AAF.

On the other hand, if AAF binds covalently to poly(dG-dC) · poly(dG-dC) a different conformation emerges. Polymers with alternating purine–pyrimidine sequences in high salt or ethanol have a left-handed Z-DNA conformation. Since in Z-DNA G residues are in *syn* conformation similarly as in AAF-modified DNA, modification of poly(dG-dC) · poly(dG-dC) by AAF

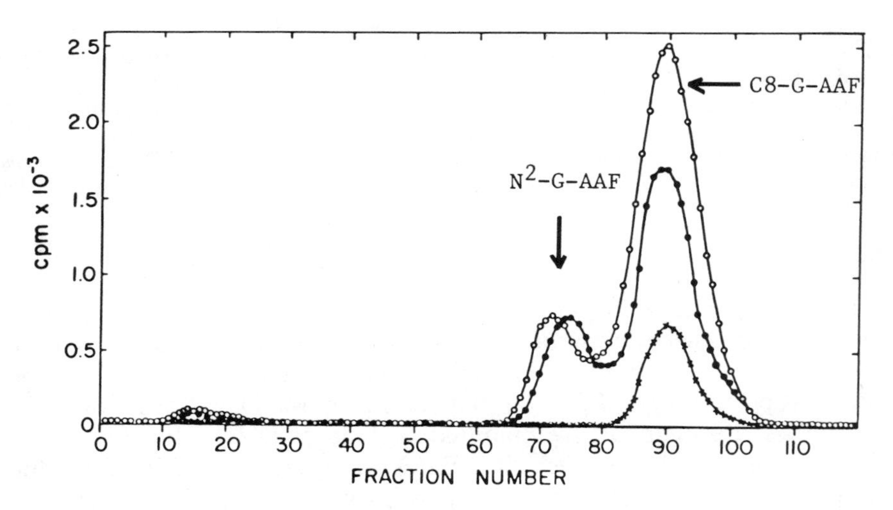

Figure VII-14. Sephadex LH-20 chromatography of nucleoside hydrolysates of total (○), the S_1 nuclease-digested fraction (x), and the S_1 nuclease-resistant fraction (●) of [^{14}C]-AAF-modified DNA. Adapted from Yamasaki *et al.* (1977) *Cancer Res.* **37,** 3756.

also results in a Z-DNA-type conformation. This has been shown by an inversion of CD spectra that is similar to that seen in the unmodified polymer in high-salt or ethanol solutions (Figure VII-15). As a result of Z-DNA-type conformation, the AAF-modified polymer is, in contrast to modified DNA, essentially resistant to S_1 nuclease and to reaction with anticytidine antibodies (Figure VII-16). Potential energy calculations performed on the AAF-modified

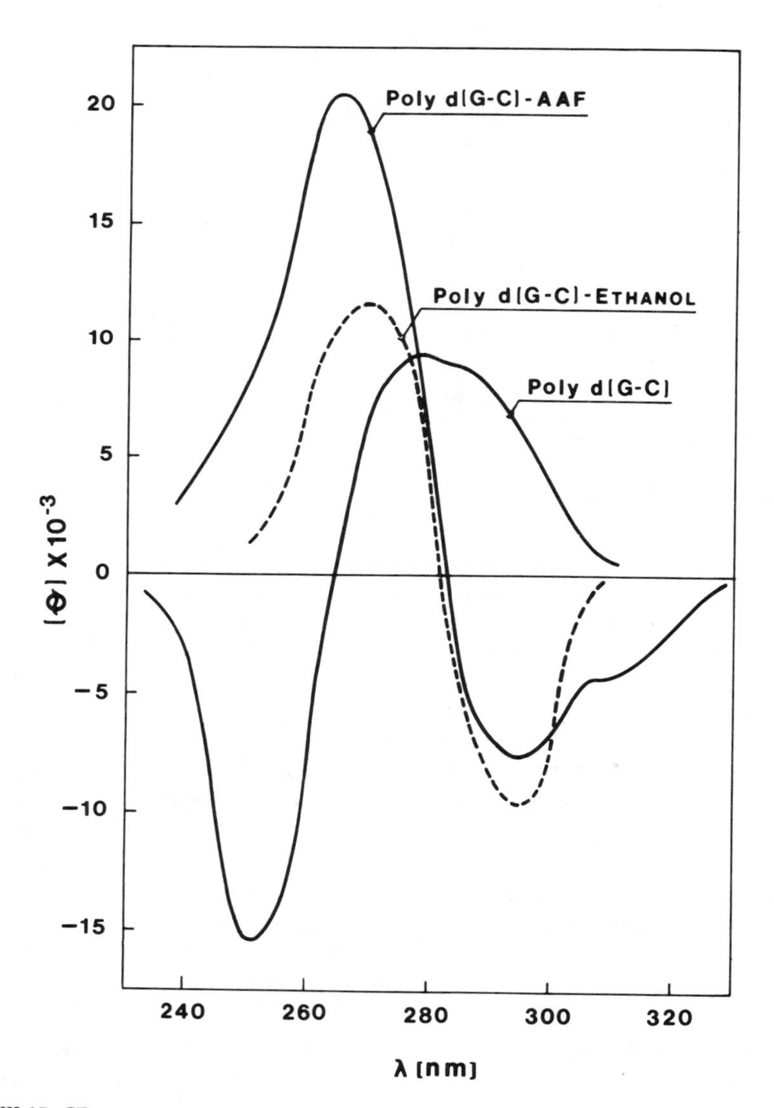

Figure VII-15. CD spectra of (dG-dC) · poly(dG-dC) in 1 mM phosphate buffer, in 60% ethanol, and modified by AAF to an extent of 28% in 1 mM phosphate buffer. From Santella *et al.* (1981) *Proc. Natl. Acad. Sci. USA* **78**, 1451.

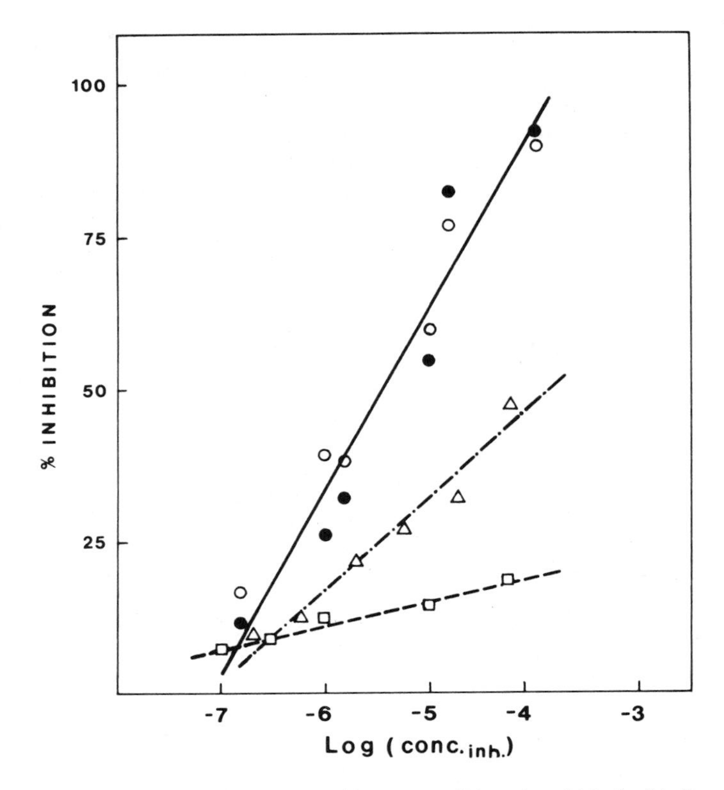

Figure VII-16. Radioimmunoassay at nonequilibrium conditions in which the binding of purified anti-C antibodies to [^{3}H]-DNA was measured in the presence of various concentrations of denatured DNA (●), 11% modified DNA-AAF (○), 21% modified poly(dG-dC) · poly(dG-dC) (□), and 5% modified poly(dG) · poly(dC)-AAF (△). From Santella *et al.* (1981) *Nucleic Acids Res.* **9,** 5459.

dCpdG model system have shown that the Z-DNA conformation is energetically stable. Drawings of the computed dCpdG-AAF conformer in Figure VII-17 show that in contrast to the base displacement model, here G is approximately coplanar with C while the AAF residue is twisted nearly perpendicular to the G residue.

Thus, modification of the C-8 position of G residues results in two different conformations: base displacement and Z-DNA. The specific conformation depends on the presence of AAF-modified dG residues in random or in alternating purine–pyrimidine sequences. This is summarized in Table VII-1.

After *in vivo* administration of *N*-OH-AAF, a major fraction of the DNA-bound carcinogen is *N*-(deoxyguanosine-8-yl)-2-aminofluorene, which has no acetyl group. While modifications with AAF cause large distortions of the DNA as a result of rotation of the G residues from the preferred *anti* to *syn*

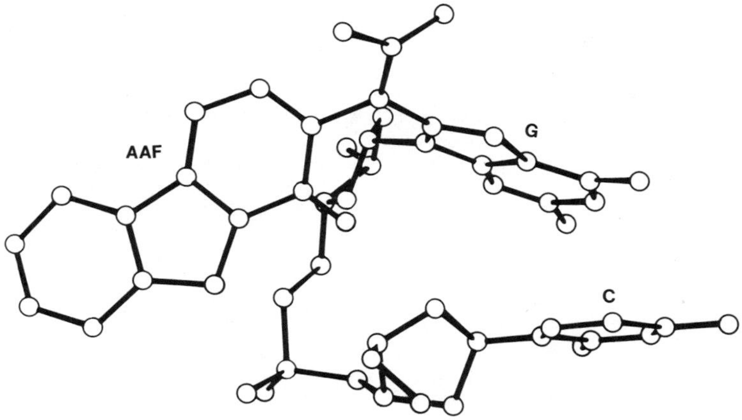

Figure VII-17. Minimum energy conformation of dCpdG-AAF with DNA backbone angles similar to dCpdG segment in Z-DNA. From Santella *et al.* (1981) *Nucleic Acids Res.* **9,** 5459.

conformation, such large conformational changes may not be necessary when the bulky acetyl group is not present. Experimental evidence for this was obtained from NMR and CD spectra of AF-modified oligonucleotides. Similarly, decreased S_1 nuclease digestion of DNA modified by AF and weaker interactions of this DNA with anticytidine antibodies indicate that the local denaturation induced by AF substitution is less than that associated with AAF modification.

Relevant to the *in vivo* significance of the alternative conformations of modified DNA, it is of interest that the C-8-dG-AAF adduct is rapidly removed from rat liver with a half-life of approximately 10 hr. However, the N^2-dG-AFF and C-8-dG-AF adducts remained constant for 14 hr and then were removed at a slow rate (Figure VII-18). See Chapter VIII (Section D) for a more detailed discussion of this topic.

Table VII-1. Comparison of Properties of AAF-Modified DNA and Poly(dG-dC) · Poly(dG-dC)

	Type of AAF-modified polymers	
Properties	DNA	Poly(dG-dC) · poly(dG-dC)
Conformation	Base displacement	Z-DNA
CD spectra	B-DNA type	Inverted B-DNA
Basepairing	Disrupted	Intact
Nuclease S_1 susceptibility	Sensitive	Resistant
Anticytidine antibodies	Reactive	Nonreactive

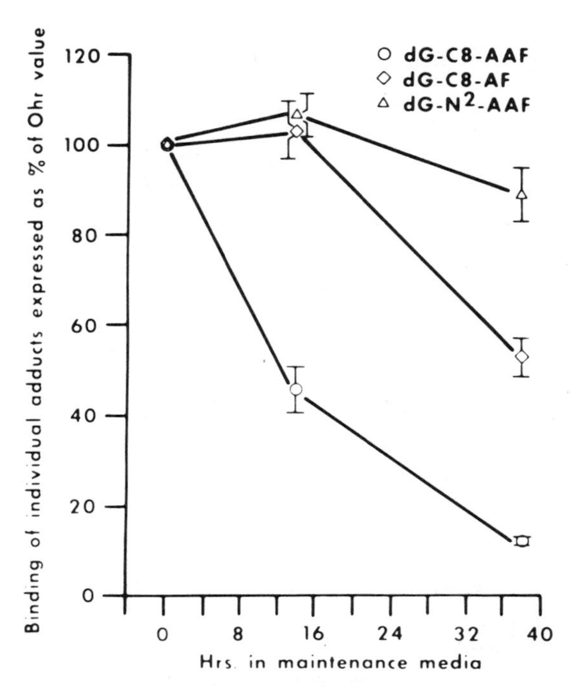

Figure VII-18. Repair of DNA adducts. The binding is expressed as a percentage of the 0-hr pmole/mg DNA value for that adduct. Adapted from Howard *et al.* (1981) *Carcinogenesis* **2,** 97.

2. *Conformation of DNA Modified by N-Hydroxy-1-Naphthylamine*

N-Hydroxy-1-naphthylamine (*N*-OH-1NA) selectively substitutes the O^6 atom of guanosine in DNA. Although this type of modification of DNA is only a minor site of substitution by many carcinogenic *N*-substituted aromatic compounds, modification at this position is frequent with small alkylating agents and represents a mispairing lesion that is believed to play a critical role in mutagenesis or in the initiation stage of carcinogenesis (Chapter II, Section E). The conformational consequences of the O^6 modification by *N*-OH-1NA were studied by fluorescence spectroscopy, electric dichroism, and thermal denaturation. Fluorescence spectroscopic studies showed that the fluorescence decay rate for the naphthyl residue in DNA was similar to that for *N*-OH-1NA in solution. This is in contrast to carcinogens bound to DNA by an intercalation mechanism that displays a decreased fluorescent lifetime. Therefore, it is concluded that *N*-OH-1NA residue bound to DNA is not subject to strong interaction with neighboring DNA bases, and thus is not intercalated. Furthermore, the naphthyl fluorescence was efficiently quenched by oxygen, which means that it is freely accessible to molecular oxygen. On the other hand, Ag^+ ions that bind preferentially to guanosine bases in DNA had only a small effect on the naphthyl fluorescence yield (Figure VII-19). This is again in contrast with the results obtained when Ag^+ ions were added to DNA containing intercalated polycyclic aromatic hydrocarbons. In such a case, fluorescence was quenched by 60 to 80%. Thus, results with O_2 and Ag^+ fluorescence quenching indicate that adducts exist in a relatively open environment in the DNA molecule.

The relative orientation of the napthylamine moiety and the DNA base were determined by Kadlubar and co-workers by electric linear dichroism (Figure VII-20). The negative ΔA values indicate that the short axis of the naphthyl ring tends to be perpendicular to the helical axis.

Construction of space-filling molecular models of DNA containing NA bound to O^6 of G results in a conformation that is in good agreement with the fluorescence and electric dichroism data. In this model, the naphthyl group is situated in the major groove of the DNA and is freely accessible to O_2 and also relative unaffected by Ag^+ fluorescence quenching. Furthermore, the short axis of the naphthyl ring is indeed perpendicular to the helical axis in this model. Based on these data, it is possible to conclude that binding of NA residues to the O^6 of G will not cause a significant perturbation in the DNA *in vivo*.

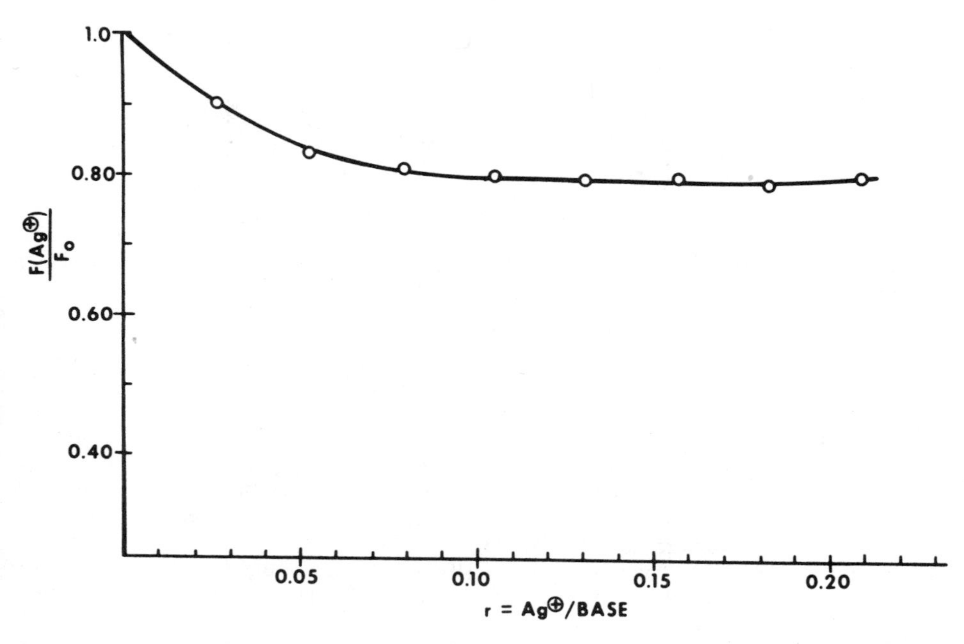

Figure VII-19. Effect of Ag^+ ions on the relative fluorescence of 1-naphthylamine covalently bound to DNA. $F(Ag^+)/F_o$, ratio of the fluorescence intensity with and without added Ag^+; r, moles Ag^+ divided by moles DNA base. From Kadlubar *et al.* (1981) *Cancer Res.* **41,** 2168.

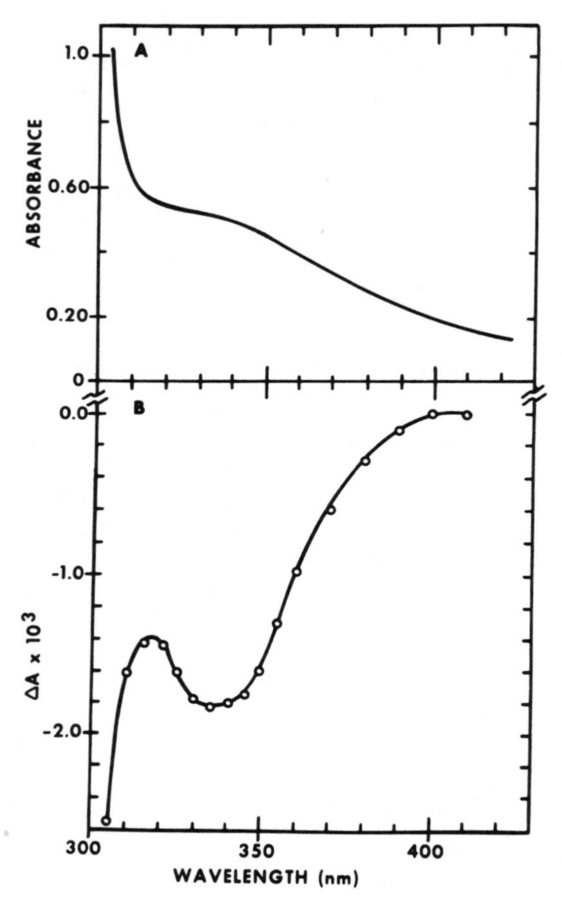

Figure VII-20. Absorption (A) and electric linear dichroism spectra of DNA-1-naphthylamine adduct. From Kadlubar *et al.* (1981) *Cancer Res.* **41,** 2168.

SELECTED REFERENCES

Behe, M., and Felsenfeld, G. (1981) Effects of methylation on a synthetic polynucleotide: The B–Z transition in poly(dG-m^{5}dC) · poly (dG-m^{5}dC). *Proc. Natl. Acad. Sci. USA* **78,**1619–1623.

Drinkwater, N. R., Miller, J. A., Miller, E. C., and Yang, N.-C. (1978) Covalent intercalative binding to DNA in relation to the mutagenicity of hydrocarbon epoxides and *N*-acetoxy—acetylaminofluorene. *Cancer Res.* **38,**3247–3255.

Evans, F. E., Miller, D. W., and Beland, F. A. (1980) Sensitivity of the conformation of deoxyguanosine to binding at the C-8 position by *N*-acetylated and unacetylated 2-aminofluorene. *Carcinogenesis* **1,**955–959.

Fuchs, R. P. P. (1975) In vitro recognition of carcinogen-induced local denaturation sites in native DNA by S_1 endonuclease from *Aspergillus oryzae. Nature (London)* **257,**151–152.

Fuchs, R., and Daune, M. (1972) Physical studies of deoxyribonucleic acid after covalent binding of carcinogen. *Biochemistry* **11,** 2659–2666.

Fuchs, R. P. P., and Daune, M. P. (1974) Dynamic structure of DNA modified with the carcinogen *N*-acetoxy-*N*-2-acetylaminofluorene. *Biochemistry* **13,** 4435–4440.

Geacintov, N. E., Gagliano, A. G., Ivanovic, V., and Weinstein, I. B. (1978) Electric linear dichroism study on the orientation of benzo(a)pyrene 7,8-dihydrodiol 9,10-oxide covalently bound to DNA. *Biochemistry* **17,** 5256–5262.

Geacintov, N. E., Yoshida, H., Ibanez, V., and Harvey, R. G. (1981) Non-covalent intercalative binding of 7,8-dihydroxy-9,10-epoxybenzo(*a*)pyrene to DNA. *Biochem. Biophys. Res. Commun.* **100,** 1569–1577.

Grunberger, D., and Santella, R. M. (1981) Alternative conformations of DNA modified by *N*-2-acetylaminofluorene. *J. Supramol. Structure Cell. Biochem.* **17,** 231–244.

Grunberger, D., and Weinstein, I. B. (1976) The base displacement model. In *Biology of Radiation and Carcinogenesis* (J. M. Yuhas, R. W. Tennant, and J. D. Regan, eds.), Raven Press, New York, pp 175–187.

Grunberger, D., and Weinstein, I. B. (1978) Conformational changes in nucleic acids modified by chemical carcinogens. In *Chemical Carcinogens and DNA,* Part 2 (P. L. Grover, ed.), CRC Press, Boca Raton, Fla., pp. 60–93.

Hamada, H., Petrino, M. G., and Kakunaga, T. (1982) A novel repeated element with Z-DNA-forming potential is widely found in evolutionary diverse eukaryotic genomes. *Proc. Natl. Acad. Sci. USA* **79,** 6465–6469.

Higerty, B., and Broyde, S. (1983) AAF linked to the guanine amino group: a B-Z junction. *Nucleic Acids Res.* **11,** 3241–3254.

Hogan, M. E., Dattagupta, N., and Whitlock, Jr., J. P. (1981) Carcinogen-induced alteration of DNA structure. *J. Biol. Chem.* **256,** 4504–4513.

Nickol, J., Behe, M., and Felsenfeld, G. (1982) Effect of the B–Z transition in poly (dG-m^{5}dC) · poly (dG-m^{5}dC) on nucleosome formation. *Proc. Natl. Acad. Sci. USA* **79,** 1771–1775.

Nordheim, A., Pardue, M. L., Lafer, E. M., Möller, A., Stollar, B. D., and Rich, A. (1981) Antibodies to left-handed Z-DNA bind to interband regions of *Drosophila* polytene chromosomes. *Nature (London)* **294,** 417–422.

Prusik, T., Geacintov, N. E., Tobiasz, C., Ivanovic, V., and Weinstein, I. B. (1979) Fluorescence study of the physico-chemical properties of a benzo(*a*)pyrene 7,8-dihydrodiol-9,10-oxide derivative bound covalently to DNA. *Photochem. Photobiol.* **29,** 223–232.

Rich, A. (1982) Z-DNA: Its chemistry and biology. In *Primary and Tertiary Structure of Nucleic Acids and Cancer Research* (M. Miwa, S. Nishimura, A. Rich, D. G. Söll, and T. Sugimura, eds.), Japan Scientific Societies Press, Tokyo, pp. 153–164.

Sage, E., and Leng, M. (1980) Conformation of poly (dG-dC) · poly (dG-dC) modified by the carcinogens *N*-acetoxy-*N*-acetyl-2-aminofluorene. *Proc. Natl. Acad. Sci. USA* **77,** 4597–4601.

Santella, R. M., Kriek, E., and Grunberger, D. (1980) Circular dichroism and proton magnetic resonance studies of dApdG modified with 2-aminofluorene and 2-acetylaminofluorene. *Carcinogenesis* **1,** 897–902.

Santella, R. M., Grunberger, D., and Weinstein, I. B. (1983) Carcinogens can induce alternate conformations in nucleic acid structure. *Cold Spring Harbor Symp. Quant. Biol.* **47,** 339–346.

Undeman, O., Lycksell, P. O., Gräslund, A., Astlid, T., Ehrenberg, A., Jernström, B., Tjerneld, F., and Norden, B. (1983) Covalent complexes of DNA and two steroisomers of benzo(*a*)pyrene 7,8-dihydrodiol-9,10-epoxide studied by fluorescence and linear dichroism. *Cancer Res.* **43,** 1851–1860.

Wells, R. D., Miglietta, J. J., Klysik, J., Larson, J. E., Stirdivant, S. M., and Zacharias, W. (1982) Spectroscopic studies on acetylaminofluorene-modified $(dT\text{-}dG)_n \cdot (dC\text{-}dA)_n$ suggest a left-handed conformation. *J. Biol. Chem.* **257,** 10166–10171.

Zacharias, W., Larson, J. E., Klysik, J., Stirdivant, S. M., and Wells, R. D. (1982) Conditions which cause the right-handed and left-handed DNA conformational transitions: Evidence for several types of left-handed DNA structures in solution. *J. Biol. Chem.* **257,** 2775–2782.

VIII
Repair in Mammalian Organs and Cells

A. Introduction

Repair is used as a general term for the removal from DNA of a modified base or nucleotide and, in at least one case, for the removal of a modifying group. It is, of course, implied that following removal, in the case of error-free repair, the proper nucleotide is inserted through repair mechanisms (Figure VIII-1). A number of reviews of DNA repair pathways are listed in the Selected References.

Most of the well-known and well-characterized repair enzymes have been isolated from prokaryotes, notably *E. coli*. These include enzymes repairing UV dimers and apurinic/apyrimidinic sites; glycosylases acting on 3-alkyl A, 7-alkyl G, ring-opened 7-alkyl G, deaminated A and C, fragments of ring-opened pyrimidines (urea), and thymine glycol. A mechanism involving transfer of an alkyl group from O^6-alkyl G to an amino acid has also been demonstrated. In all cases, the best or only substrate is double-stranded DNA. There are reports of an enzyme directly inserting purines at apurinic sites in DNA but there is no conclusive evidence for an "insertase." More recently, evidence has been obtained for enzymes dealkylating O^4-methyl T and for glycosylases acting on O^2-methyl T and O^2-methyl C.

In eukaryotes, many nucleic acid derivatives discussed in Chapters III, IV, and VI are found to be removed/repaired *in vivo* at rates varying according to the chemical stability of the derivative, the amount of derivative, the cell or animal species, and the specific organ or cell studied. Much less is known of enzyme mechanisms, but it is likely that there are great similarities with those from prokaryotes.

B. *In Vivo* Removal of Alkyl Derivatives

There is chemical evidence for the disappearance of all ethyl and methyl base or phosphate-alkylated derivatives identified in mammalian cells or an-

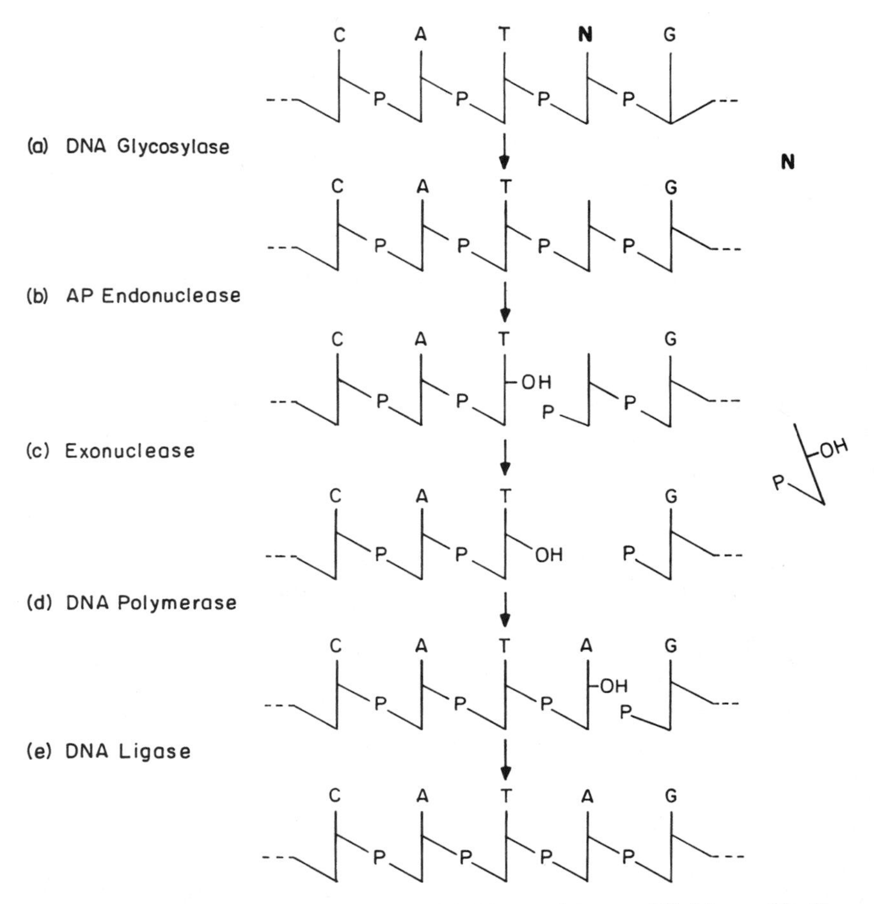

Figure VIII-1. Scheme for base excision repair of DNA containing modified bases (N). The complementary DNA strand is not shown. (a) Glycosyl cleavage of the base–sugar bond, releasing the free base, N, and generating an apurinic/apyrimidinic (AP) site. (b) The AP site is recognized by an endonuclease that catalyzes the formation of a chain break at the 5′ side of the lesion. (c) Excision of the deoxyribosephosphate moiety alone or possibly as part of a small oligonucleotide. (d) Repair replication catalyzed by a DNA polymerase. Ligation catalyzed by a DNA ligase. Adapted from Lindahl *et al.* (1982) In *Molecular and Cellular Mechanisms of Mutagenesis* (Lemontt and Generoso, eds.), Plenum Press, pp. 89–102.

imals treated with alkylating agents. Figures VIII-2 and 3 illustrate that [^{14}C]ethylnitrosourea-treated human fibroblasts or BD IX rat liver have the ability to remove seven derivatives, representing more than 90% of the total ethylation. Phosphotriesters are relatively stable in rat liver or human fibroblasts with estimated $t_{1/2}$ more than 8–12 days. Using a different technique for quantitation, long-term persistence of methyl and ethyl phosphotriesters was found to differ greatly in the liver of C57BL mice given methyl- or ethylnitrosourea. Methyl triesters had a $t_{1/2}$ of about 1 week while about 40% of the

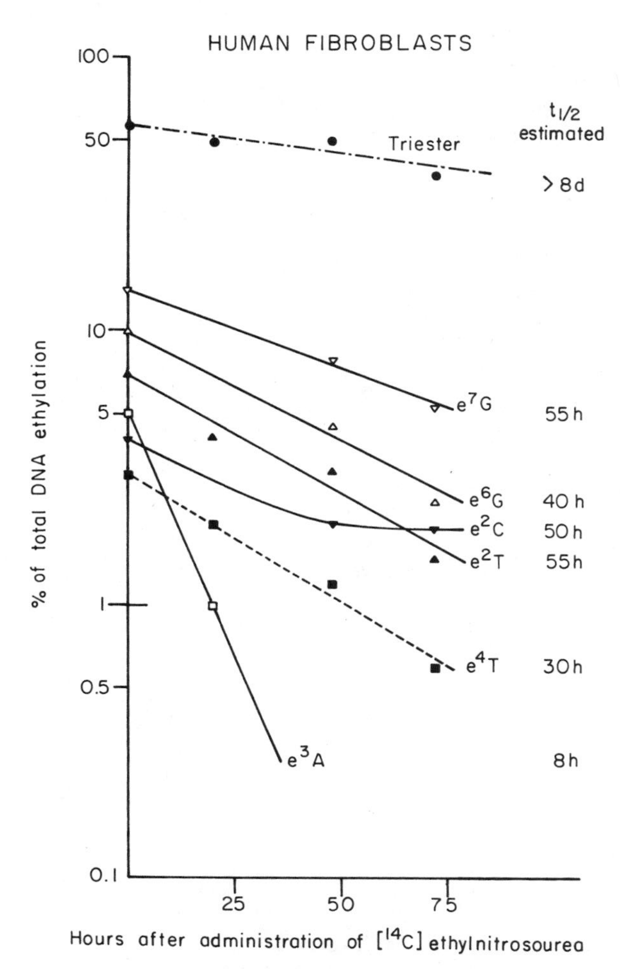

Figure VIII-2. Amounts of [^{14}C]ethyl products from ethylnitrosourea-treated mammalian cells, prelabeled with ^{3}H. The initial proportion of total ethylation is shown on the log scale. All values have been corrected for cell division. The estimated $t_{1/2}$ of derivatives is shown in the figure. Adapted from Bodell *et al.* (1979) *Nucleic Acids Res.* **6,** 2819.

ethyl triesters were still present after 10 weeks (Figure VIII-4). The two derivatives with chemically unstable glycosyl bonds, 7-EtG (e^{7}G) and 3-EtA (e^{3}A), are both lost with half lives greater than the rates of chemical hydrolysis. The *O*-ethyl derivatives, O^6-EtG (e^{6}G), O^2-EtC (e^{2}C), O^4-EtT (e^{4}T), and O^2-EtT (e^{2}T), are also appreciably decreased over the 75-hr period with half-lives of 30–60 hr in both cells and liver. If, however, the decrease in ethylated products is studied in the brain or a pool of other tissues in the same treated BD IX rats, little removal of the *O*-ethyl derivatives is noted (Figures VIII-5, 6). The ca-

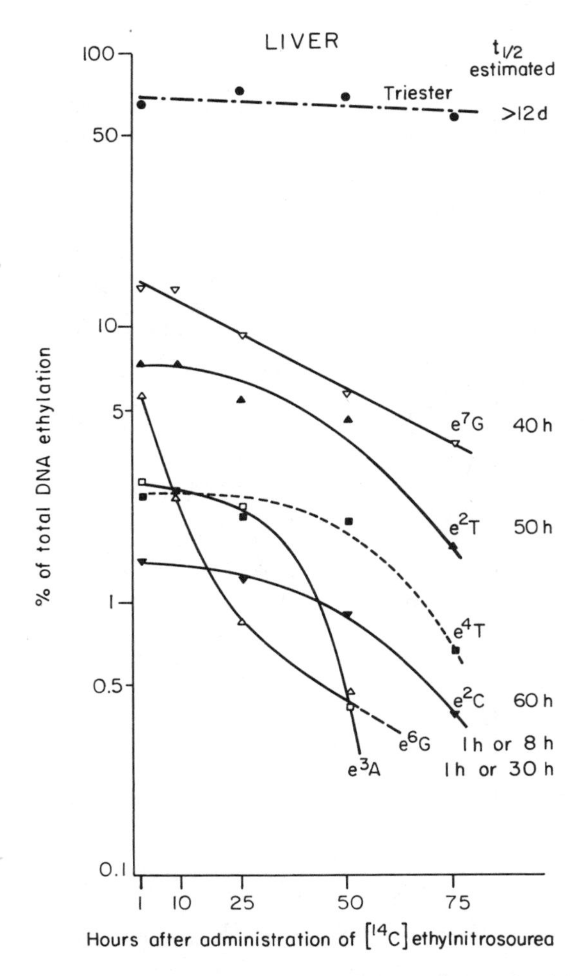

Figure VIII-3. Amounts of ethyl derivatives in liver at 1, 10, 25, 50 and 75 hr after administration of [^{14}C]ethylnitrosourea to 10-day-old BD IX rats. The number of ethyl groups bound in various organs was 1.3–3 per 10^5 DNA-P. The estimated half-lives of each derivative are shown in the figure. Adapted from Singer, Spengler, and Bodell (1981) *Carcinogenesis* **2,** 1069.

pacity to repair DNA damage is highest in liver, which is not unexpected because it has the function of detoxifying and contains a vast store of enzymes.

To further complicate the study of *in vivo* repair, not only does the cell type or organ contain varying amounts of enzyme activity, but the dose and mode of carcinogen treatment affect repair and can saturate constitutive enzymes, or in specific cases, induce additional enzyme activity. Methyl derivatives are probably more readily repaired than ethyl or higher homologs.

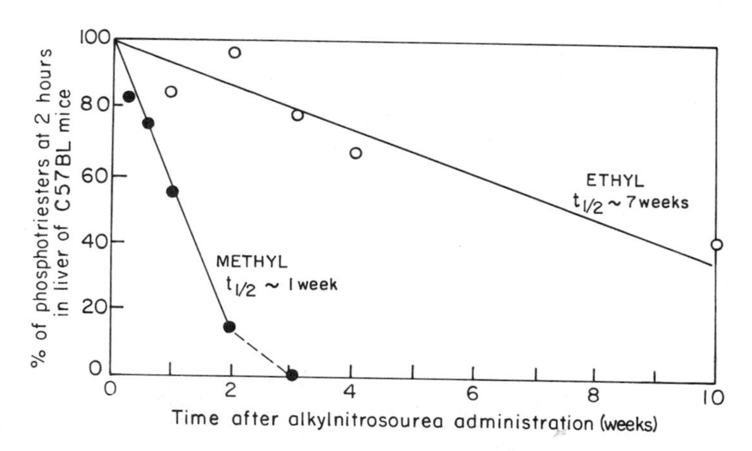

Figure VIII-4. Loss of methyl and ethyl phosphotriesters from liver DNA of C57BL mice given a single injection of methylnitrosourea or ethylnitrosourea. Phosphotriesters were determined from the extent of degradation induced in isolated DNA by alkali. Data from Shooter and Slade (1977) *Chem. Biol. Interact.* **19,** 353.

It was first hypothesized by Goth and Rajewsky that lack of repair of O^6-alkyl G, or its persistence in a specific organ, correlated with organotropic malignancy. This derivative, which was produced to high extents by *N*-nitroso carcinogens, has the potential to mispair with T and would thereby be the initiating mutagenic event. The labile alkyl purines, on the other hand, are produced by poor carcinogens such as dimethylsulfate, and would not persist but rather create apurinic sites, repairable by normal mechanisms (Figure VIII-1). An immense body of data now exists on the production and persistence of O^6-alkyl G under a variety of conditions.

The carcinogens generally used in these studies are methylnitrosourea (MeNU), ethylnitrosourea (EtNU), methylnitrosoguanidine (MNNG), ethylnitrosoguanidine (ENNG), dimethylnitrosamine (DMNA), diethylnitrosamine (DENA), and 1,2-dimethylhydrazine (SDMH). Ethylating agents are much less efficient than methylating agents, but all of these produce O^6-alkyl G in approximate ratios of 0.11 O^6-MeG/7-MeG or 0.7 O^6-EtG/7-EtG. Each of the carcinogens can be made relatively organ-specific by using appropriate experimental animals, as well as by route and time of administration.

O^6-Alkyl G, resulting from a single administration of MeNU, does persist for long periods in rat brain under conditions where it is the target organ (Figure VIII-7). The kinetics of removal are bi- or multiphasic, which is generally true for all enzymatic repair. It can be seen in Figure VIII-5 that O^6-EtG, but also the *O*-ethyl pyrimidines, would persist virtually indefinitely given the kinetics in the figure. When SDMH (also in a single dose) is used under conditions where it is a colon carcinogen, persistence is greater in the

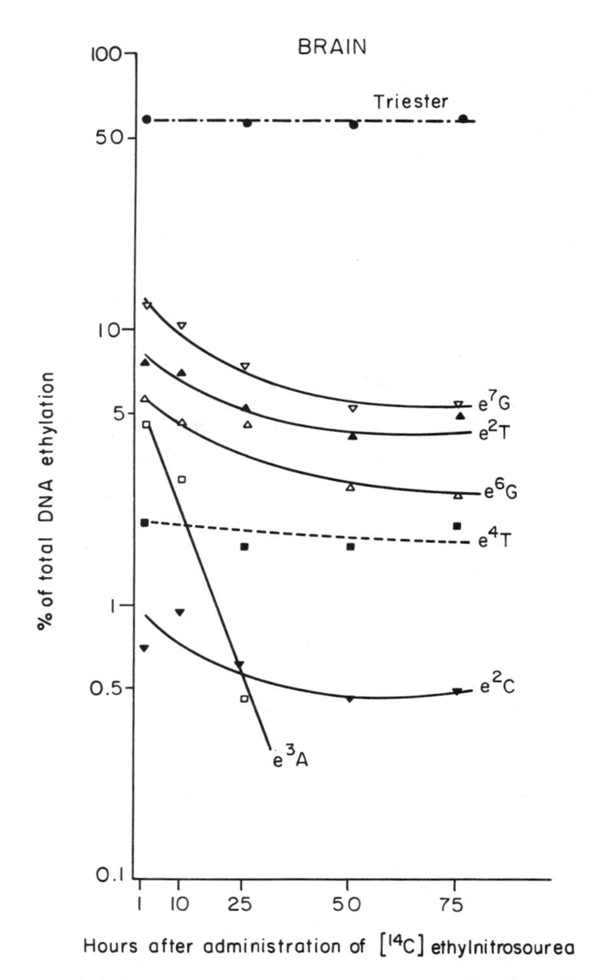

Figure VIII-5. Amounts of ethyl derivatives in brain at 1, 10, 25, 50 and 75 hr after [^{14}C]ethylnitrosourea administration to 10-day-old BD IX rats. Singer, Spengler, and Bodell (1981) *Carcinogenesis* **2,** 1069.

colon than in the ileum or liver (Figure VIII-8). It should be noted, however, that in contrast to MeNU or EtNU, SDMH, which requires metabolic activation, alkylates the liver to a much higher extent than other organs so that at all times (up to 70 to 100 hr) there is more O^6-MeG in liver, the nontarget organ, than in the colon, which eventually becomes malignant. Figure VIII-9 shows that the rate of loss of O^6-MeG and 7-MeG is similar in both liver and colon even though the SDMH dose is fourfold that in Figure VIII-8.

Persistence of O^6-alkyl G has been followed in several species in an endeavor to correlate its presence with malignant transformation. In spite of a number of positive correlations, there also are clear negative correlations. For

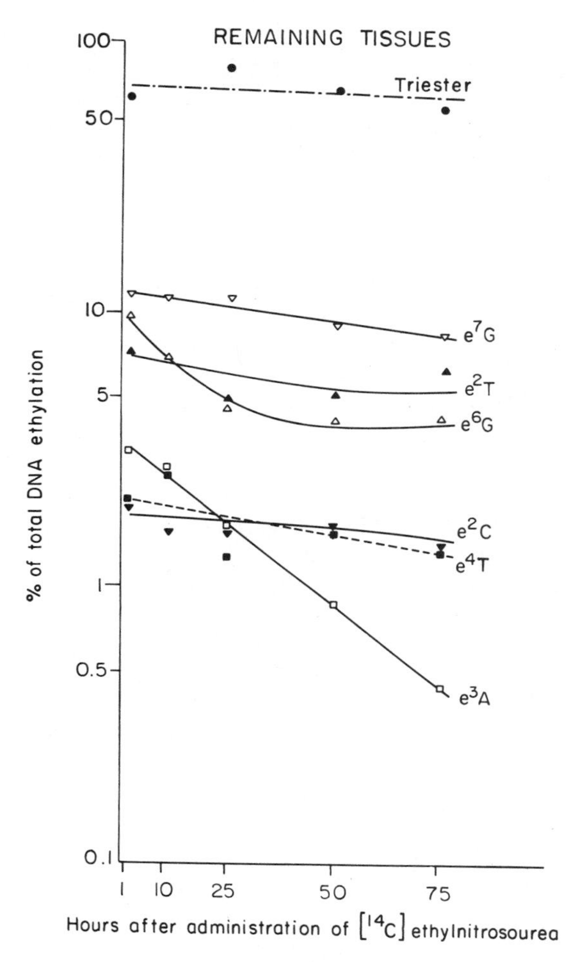

Figure VIII-6. Amounts of ethyl derivatives in a pool of DNA from spleen, lung, thymus, muscle, and intestine at 1, 10, 25, 50 and 75 hr after [^{14}C]ethylnitrosourea administration to 10-day-old BD IX rats. Singer, Spengler, and Bodell (1981) *Carcinogenesis* **2,** 1069.

example, O^6-MeG persists in the brain of gerbils given MeNU but no brain tumors develop, although 40% of the original O^6-MeG is present after 6 months (Table VIII-1). In MeNU- or EtNU-treated mice sensitive to induction of thymic lymphomas, O^6-alkyl persists in brain, lung, and kidney but is rapidly lost from the target organ, the thymus. Syrian hamsters given nitrosoureas (i.p.) mainly develop tumors of the forestomach. However, the persistence of O^6-alkyl G was manyfold greater in the brain and kidney than in the intestine.

All the preceding experiments utilized single-dose administration at a level high enough to induce tumors. Another route to studying repair in organs from whole animals is to vary the amount of carcinogen or to pretreat

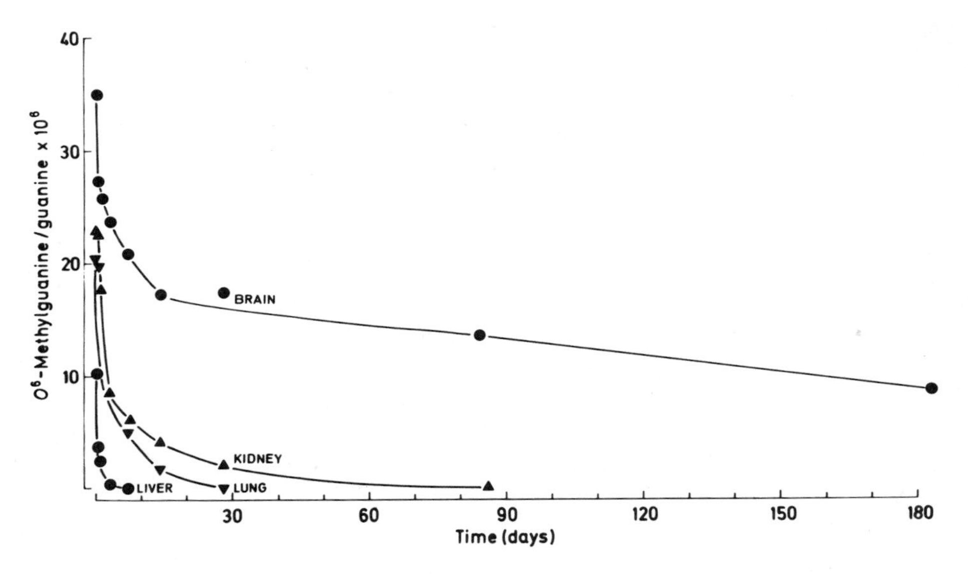

Figure VIII-7. Long-term persistence of O^6-methylguanine in rat brain DNA. Adult BD IX rats received a single i.v. injection of 10 mg/kg [^{3}H]methylnitrosourea and were killed at various time intervals ranging from 4 hr to 184 days. From Kleihues and Bücheler (1977) *Nature (London)* **269,** 625.

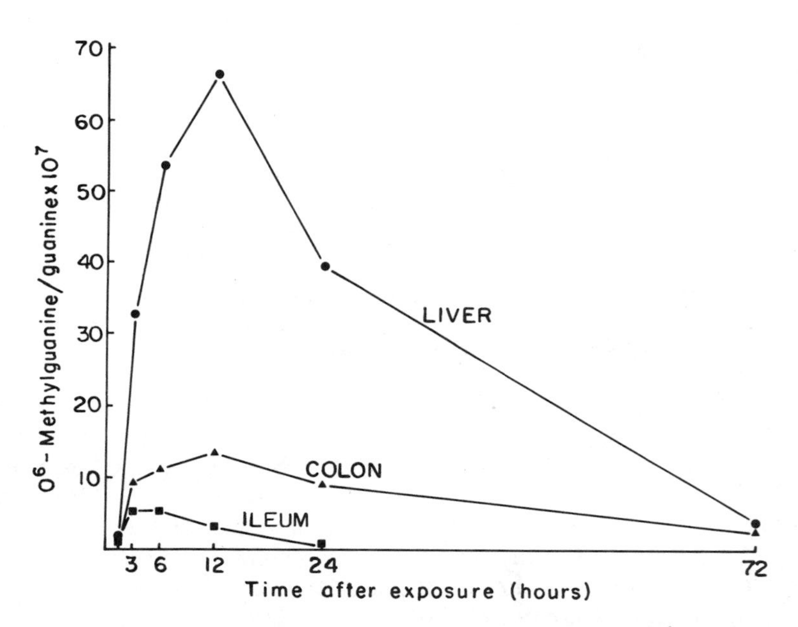

Figure VIII-8. Formation and persistence of O^6-methylguanine in DNA of various rat organs following administration of 1,2-dimethylhydrazine (SDMH). Adult female BD IX rats received a single s.c. injection of 20 mg/kg [^{14}C]-SDMH and were killed at time intervals ranging from 1.5 to 72 hr. From Swenberg *et al.* (1979) *Cancer Res.* **39,** 465.

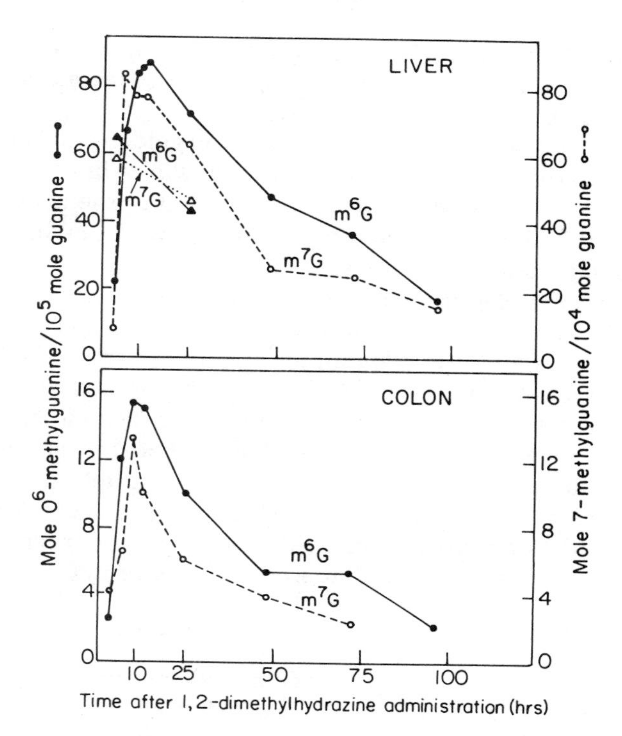

Figure VIII-9. Formation and removal of methyl purines from liver and colon DNA of rats given a single dose of 81.5 mg/kg 1,2-dimethylhydrazine (SDMH). The carcinogen was unlabeled and the derivatives were analyzed using fluorescence spectroscopy. SMDH is a colon carcinogen under the experimental conditions. 7-Methylguanine (m^7G), ○···○; O^6-methylguanine (m^6G), ●—●. Data from Herron and Shank (1981) *Cancer Res.* **41** 3967. Also shown in the top figure are data for two time points using 20 mg/kg [^{14}C]dimethylnitrosamine (DMNA). Data from Pegg and Hui (1978) *Biochem. J.* **173,** 739. m^7G, △···△; m^6G, ▲--▲. 81.5 mg/kg SDMH gives approximately the same initial methylation in liver DNA as 20 mg/kg DMNA. The rate of removal of m^6G and m^7G is similar.

Table VIII-1. Persistence of O^6-Methylguanine in DNA of Various Tissues of the Mongolian Gerbil[a]

Organ	m^6G remaining (mM/M guanine) 1 day	30 days	180 days
Brain	24.6	19.7 (80%)	9.8 (40%)
Lung	20.4	8.2 (40%)	nc[b]
Kidney	41.0	7.8 (19%)	nc
Liver	23.0	nc	nd[c]

[a] Single i.v. injection of 10 mg/kg [^{14}C]methylnitrosourea. From Kleihues *et al.* (1980) *Carcinogenesis* **1,** 111.
[b] nc, not calculable (trace amounts).
[c] nd, not detectable.

the animal with differing amounts of an unlabeled carcinogen followed by a labeled dose. In both of these types of experiments, the amount of labeled alkyl purine is determined over a time period.

The effect of dose of DMNA in rats and Syrian golden hamsters is different (Figure VIII-10). While little removal of m^7G occurs in 24 hr regardless of dose, m^6G is lost rapidly at low doses in both species, but the activity removing m^6G is saturated in hamsters at less than 1 mg/kg, whereas the rat can repair m^6G after administration of 10 mg/kg.

The indication in Figure VIII-10 that O^6-alkyl G repair is greater at low doses of DMNA is also shown in more detail in Figure VIII-11. Here m^6G and m^7G are quantitated 5 min to 180 min after administration of DMNA. It can be seen that the time of highest level of methylation as well as rate of removal of methyl purines is affected by the amount of carcinogen to be metabolized. m^6G is decreased from the expected initial level almost instantly when 5 μg/kg is used. Furthermore, after 3 hr m^6G is almost undetectable, in contrast to its persistence at the larger doses. It should be pointed out that carcinogenesis experiments are performed with levels on the order of 20 mg/kg. When these larger doses are used, there is no loss of m^6G in 4 hr but after

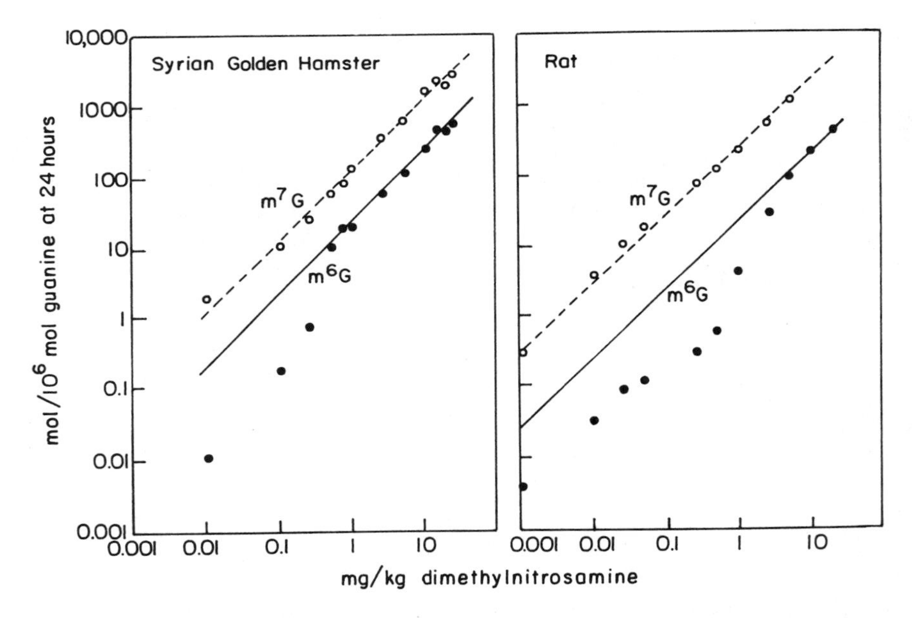

Figure VIII-10. Effect of dose of dimethylnitrosamine on persistence of 7-methylguanine (m^7G) and O^6-methylguanine (m^6G) in hamster liver DNA (left) on rat liver DNA (right) 24 hr after administration of the carcinogen. Data from Stumpf *et al.* (1979) *Cancer Res.* **39**, 50, and Pegg and Hui (1978) *Biochem. J.* **173**, 739. Figure adapted from Montesano, Pegg, and Margison (1980) *J. Toxicol. Environ. Health* **6**, 1001.

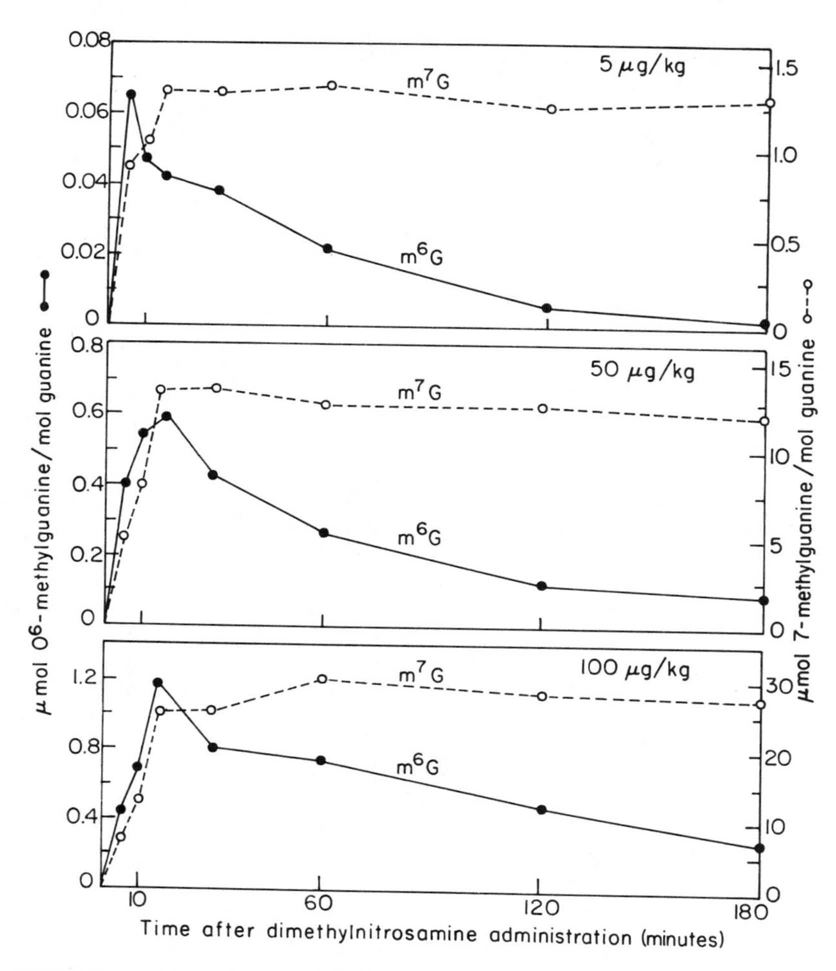

Figure VIII-11. Formation and removal of methyl purines from liver of rats given small doses of dimethylnitrosamine. The scales for O^6-methylguanine (m^6G) and 7-methylguanine (m^7G) have been adjusted to different ratios in the panels and are different from Figure VIII-12. Bottom panel (100 μg/kg) m^6G/m^7G, 0.04; middle panel (50 μg/kg) and top panel (5 μg/kg), 0.05. The calculated ratio of m^6G/m^7G, without chemical or enzymatic removal, is 0.11. With a 5 μg/kg dose (top) the ratio at 5 min for m^6G/m^7G is 0.047 indicating removal of m^6G prior to the first point. Note that the time scale is 5–180 min while in Figure VIII-12 it is 4–24 hr. Data from Pegg and Perry (1981) *Cancer Res.* **41,** 3128.

24 hr about 50% is gone (Figure VIII-12). Lower doses of DMNA clearly indicate a very rapid repair in 4 hr (Figure VIII-12), as would be expected from the early time points for a similar dose (5 μg/kg) shown in the top panel of Figure VIII-11.

There are few data on ethylation, but dose-dependent repair of e^6G is

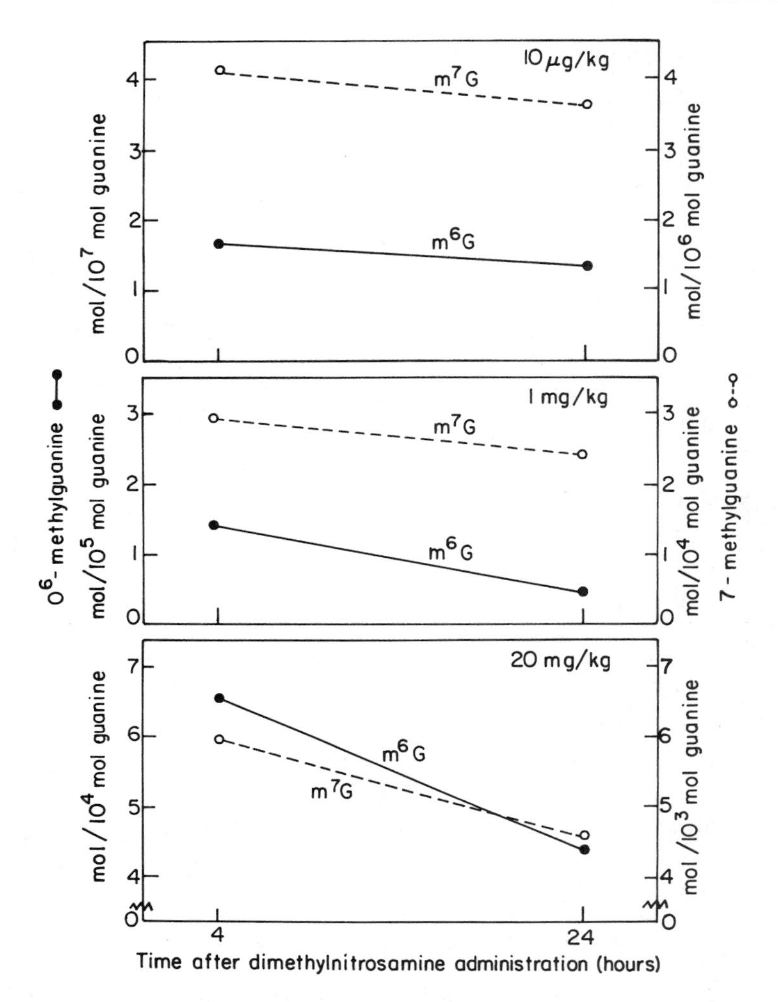

Figure VIII-12. Effect of dose of dimethylnitrosourea on the formation and removal of O^6-methylguanine from rat liver DNA. The amount of carcinogen administered to Sprague–Dawley rats is shown in each panel. The scales for O^6-methylguanine (m^6G) and 7-methylguanine (m^7G) differ by a factor of 10. The calculated ratio of m^6G/m^7G, without chemical or enzymatic hydrolysis occurring, is about 0.11. The ratio at 4 hr at the highest dose (20 mg/kg) is also 0.11. Data from Pegg and Hui (1978) *Biochem. J.* **173,** 739.

illustrated in Figure VIII-13. The amount of e^6G resulting from 0.5 mg DENA/kg (bottom) is approximately the same as that of m^6G formed by 100 μg DMNA/kg in Figure VIII-11 (bottom). Within the limits of the experiment the ethyl derivative decreases at about the same rate as the methyl derivative. Higher doses show substantial repair but there are no data comparable to the effect of dose of DMNA.

Pretreatment with carcinogens can lead to two effects: saturation of repair activities or induction of additional activity. In the case of saturation, a large dose of, e.g., unlabeled MeNU administered to rats followed by a smaller dose of labeled MeNU will have the immediate effect of preventing the repair of m^6G in rat liver (Figure VIII-14, top). As the time between pretreatment and administration of labeled MeNU increases, repair capability returns to that of the control (Figure VIII-14, bottom). The loss of repair ability is also proportional to the level of MeNU used for pretreatment. A number of unrelated carcinogens apparently also affect m^6G repair. However, the most effective are methylating and ethylating agents.

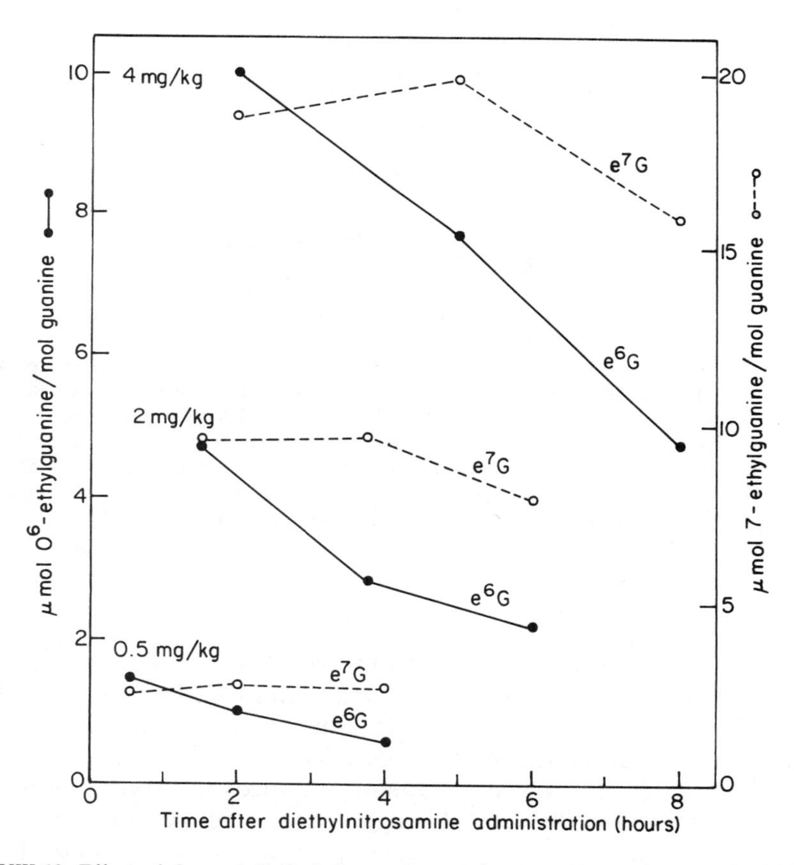

Figure VIII-13. Effect of dose of diethylnitrosamine on the formation and removal of O^6-ethylguanine from rat liver DNA. The amount of carcinogen administered to Sprague–Dawley rats is shown in each section of the figure. The scales for O^6-ethylguanine (e^6G) and 7-ethylguanine (e^7G) differ by a factor of 2. The calculated ratio of e^6G/e^7G, without chemical or enzymatic hydrolysis occurring, is about 0.7. The ratio of e^6G/e^7G at 4 hr at the highest dose (4 mg/kg) is 0.54. Data from Pegg (1980) In *Molecular and Cellular Aspects of Carcinogen Screening Tests* (Montesano, Bartsch, and Tomatis, eds.), IARC Scientific Publication No. 27, pp. 3–22.

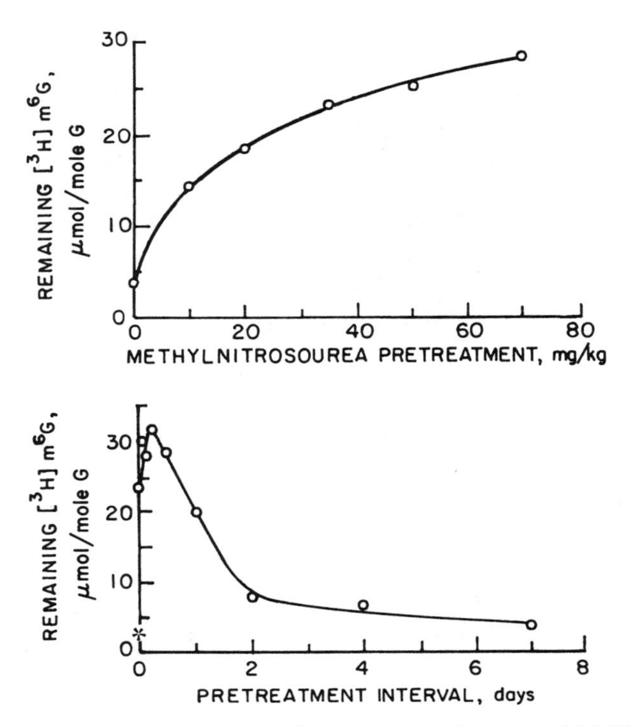

Figure VIII-14. Effect of various pretreatment doses of methylnitrosourea (MeNU) on removal of O^6-methylguanine (m^{6}G) from rat liver DNA *in vivo*. Top: Animals received 10–70 mg/kg MeNU. After 3 hr, they were given 10 mg/kg [^{3}H]-MeNU and killed 6 hr later. The amount of m^{6}G remaining in liver DNA is shown. Bottom: Effect of time after pretreatment with 70 mg/kg unlabeled MeNU on removal of [^{3}H]-m^{6}G 6 hr after administration of 10 mg/kg [^{3}H]-MeNU. The asterisk on the ordinate indicates the amount of [^{3}H]-m^{6}G remaining when no pretreatment is used. Adapted from Kleihues and Margison (1976) *Nature (London)* **259,** 153.

In Raji human lymphoma cells, MNNG or ENNG pretreatment decreases the ability of the cells to repair m^{6}G resulting from a later treatment with [^{14}C]-MNNG. The decreased repair capacity is a function of the amount of MNNG or ENNG. m^{3}A repair is unaffected (Figure VIII-15).

It is postulated that the pretreatment produces O^6-alkyl G, which can exhaust the enzyme in the cell. This hypothesis is supported by the observation that if the time between pretreatment and administration of the labeled carcinogen is increased, repair ability is restored.

In contrast to inhibition of repair after large doses, chronic or repeated exposure to low levels of carcinogen can, in the rat liver, induce enhanced removal of O^6-alkyl G. This effect, illustrated in Figure VIII-16, is not seen for the removal of 7-MeG or 3-MeA. While clear evidence exists for induction of

O^6-alkyl G repair in bacteria and some human cells "adapted" by treatment with very low levels of mutagen/carcinogen, in whole animals only the liver of the rat appears to respond similarly. The O^6-alkyl G repair systems in gerbil, mouse, and hamster have not been found to be inducible. This is more clearly shown by experiments using liver fractions containing repair enzymes to remove derivatives from DNA *in vitro* (Section C).

All of the data on repair in animal organs are derived from mixed cells and often only a single cell type becomes malignant. Thus, if tumorigenesis is to be correlated with the presence and persistence of modified nucleosides, each cell type of organs such as the liver should be separately studied. Recent work from Swenberg's laboratory clearly demonstrates that major differences exist in the ability to repair damaged DNA in different cell populations of the rat liver. Many hepatocarcinogens primarily induce hemangioendotheliomas of the liver, rather than hepatocellular carcinomas. This may be correlated with the much greater ability of hepatocytes to repair O^6-MeG as compared

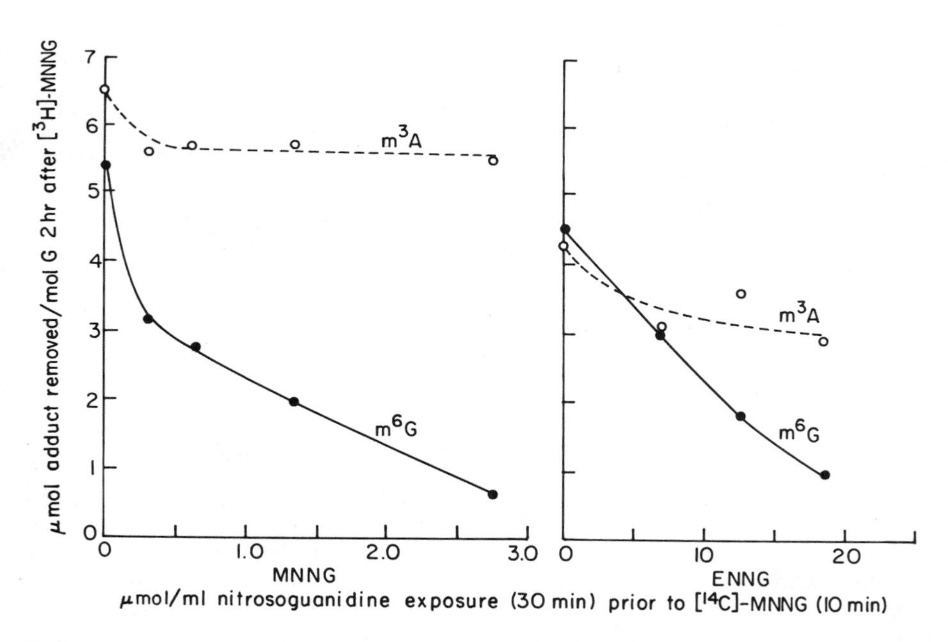

Figure VIII-15. Effect of *N*-methyl-*N*′-nitro-*N*-nitrosoguanidine (MNNG) or *N*-ethyl-*N*′-nitro-*N*-nitrosoguanidine (ENNG) pretreatment on the capacity of the Raji human lymphoma line to remove O^6-methylguanine (m^6G) or 3-methyladenine (m^3A) from its DNA. Cells were pretreated for 30 min with varying amounts of either MNNG or ENNG. After washing and resuspension, the cells were treated for 10 min with either 0.5 μg/ml [^{3}H]-MNNG (MNNG pretreated) or 0.3 μg/ml [^{3}H]-MNNG (ENNG pretreated), reisolated and incubated for 2 hr at 37°C before analysis of the remaining purines. The toxicity of MNNG and ENNG was separately determined and equivalent survival of the Raji mex$^+$ line used was found with a 10-fold greater amount of ENNG compared to MNNG. Data from Sklar, Brady, and Strauss (1981) *Carcinogenesis* **2,** 1293.

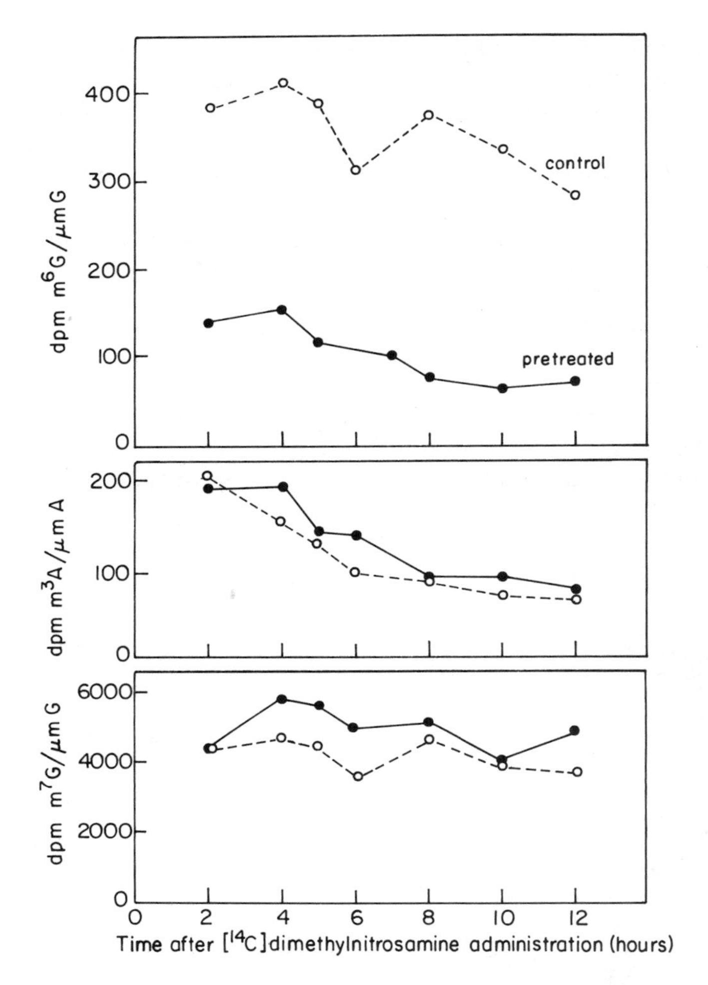

Figure VIII-16. Effect of chronic dimethylnitrosamine administration on repair of methyl purines in DNA from rat liver. Each panel shows the amount of [^{14}C]methyl purine at various times after administration of [^{14}C]dimethylnitrosamine (2 mg/kg) to normal rats (○) or to rats pretreated for a total of 44 days with a total of 88 mg/kg of the unlabeled carcinogen (●). The labeled carcinogen was administered 24 hr after the last unlabeled dose. Top panel, O^6-methylguanine; middle panel, 3-methyladenine; bottom panel, 7-methylguanine. Data from Montesano, Bresil, and Margison (1979) *Cancer Res.* **39,** 1798.

to nonparenchymal cells (NPC) (Table VIII-2). Higher levels of O^6-MeG removal can be induced in hepatocytes but not in NPC by low-level exposure to DENA or SDMH (Figure VIII-17). While this enhanced activity is believed to be a response to cell proliferation resulting from stimulus by the carcinogen, NPC also proliferate to an even greater extent but O^6-MeG removal is not increased. Similarly, proliferation of liver in carcinogen-treated mice, gerbils, and two species of hamster does not enhance O^6-MeG repair.

Table VIII-2. Alkylation of Hepatocyte and Nonparenchymal Cell (NPC) DNA Following Oral Exposure of Rats to 1,2-[^{14}C]Dimethylhydrazine[a]

Cell type[b]	Hours after exposure	Alkylation of DNA per 10^6 guanine: m^6G	m^7G	m^6G/m^7G
Hepatocyte	2	89	1032	0.086
	24	9	753	0.011
NPC	2	65	783	0.083
	24	35	525	0.067

[a] Single administration of 3 mg/kg. Data from Lewis and Swenberg (1980) *Nature (London)* **288,** 185.

[b] Hepatocytes comprise 60–70% of liver cells but > 90% of the mass. NPC consist almost entirely of endothelial and Kupffer cells and account for the remaining 30–40% of cells and contain 10–20% of DNA. Elutriation centrifugation was used for separation. Yields varied from 2–5 × 10^8 NPC and 6–11 × 10^8 hepatocytes per liver and were ≥ 95% pure.

It should be noted that the measurement of persistence of adducts is generally restricted to the alkyl purines, but as shown in Figures VIII–2 and 6, the *O*-alkyl pyrimidines and phosphotriesters are removed at relatively slow rates varying with the organ. Similar studies of the persistence of the alkyl pyrimidines in specific cell types are in progress. Preliminary experiments indicate that O^4-EtdT accumulates in DNA of hepatocytes from rats exposed continuously to diethylnitrosamine.

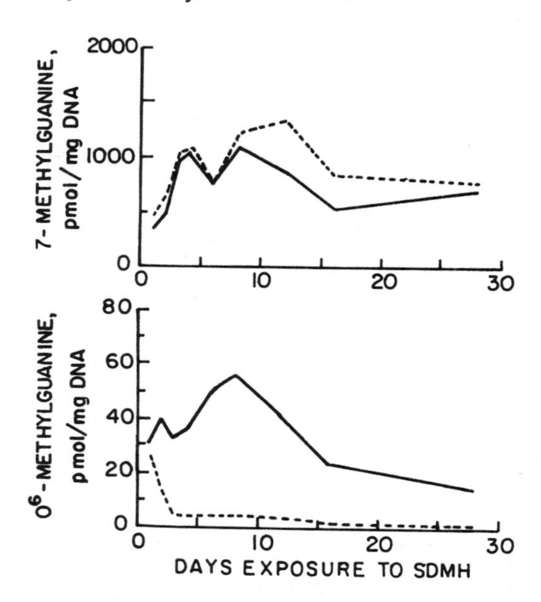

Figure VIII-17. Normalized concentrations of 7-methylguanine (top) and O^6-methylguanine (bottom) remaining in nonparenchymal cells (—) and hepatocytes (····) exposed for intervals up to 28 days to 1,2-dimethylhydrazine (SDMH) via the drinking water. Adapted from Bedell *et al.* (1982) *Cancer Res.* **142,** 3079. See Figure VIII-20 for amounts of O^6-methylguanine methyltransferase activity in cell extracts.

C. Mechanisms of Repair of Alkyl Derivatives in Eukaryotes

In the previous section it was demonstrated that repair/removal of alkyl bases and triesters occurred and was presumably due to enzyme action, as it is in prokaryotes. Isolation and characterization of mammalian repair enzymes has been fraught with difficulties that are primarily due to the trace amounts of such enzymes in eukaryotes and the corresponding problems in identification of the product. Detection of glycosylase activity has been favored by the development of sensitive HPLC systems and the use of substrates of high specific radioactivity. An illustration of separations that are possible is shown in Figure VIII-18. In contrast to *in vivo* repair where the remaining derivative is quantitated, mechanistic studies focus on the identity of the product liberated by enzymatic action.

Glycosylase activity removing 3- and 7-alkyl purines has been found in extracts from human lymphoblasts. The glycosyl bond of both methyl and ethyl purines can be cleaved, but the rate of liberation of methyl purines is

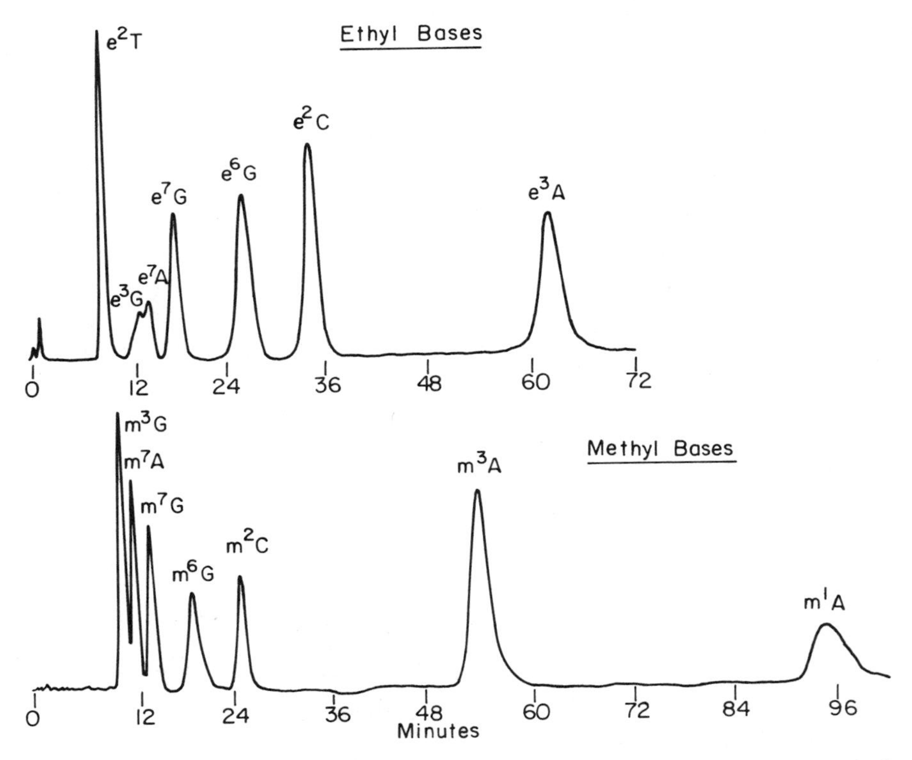

Figure VIII-18. HPLC separation of ethyl bases (top) and methyl bases (bottom) using an Aminex HP-C cation exchange column (NH_4^+ form) with 0.4 M pH 7 ammonium formate as solvent. e^1A, which is not shown, is retained about 120 min. From Singer and Brent (1981) *Proc. Natl. Acad. Sci. USA* **78,** 856.

Table VIII-3. Release by Human Lymphoblasts of Methylated or Ethylated Purines from Alkylated Deoxypolynucleotides[a]

Substrate	Percent depurination					
	m^3A	m^7G	m^3G	e^3A	e^7G	e^3G
Me poly(dA-dT)	100					
Me poly(dG-dC)		7	40			
Me DNA	91	6				
Et poly(dA) · poly(dT)				26		
Et (DNA)				39	5	26

[a] Data from Singer and Brent (1981) *Proc. Natl. Acad. Sci. USA* **78**, 856. Neither 1-alkyladenine nor O^6-alkylguanine were found as products of glycosylase action.

greater than that of ethyl purines (Table VIII-3). Both rat and hamster liver contain enzymes depurinating 3-MeA and 7-MeG, as well as ring-opened 7-MeG (Table VIII-4). Rat liver contains less of the 7-MeG glycosylase than hamster liver, and this is paralleled by the hamster liver's greater efficiency *in vivo* of removing 7-MeG. The available data, both *in vivo* and *in vitro*, suggest that there are at least two glycosylases, but only a 3-MeA glycosylase has been purified from mammalian cells.

Another type of repair enzyme, termed a methyltransferase, transfers methyl or ethyl groups from O^6-alkyl G to an acceptor protein where *S*-alkylcysteine is the product. Although first and thoroughly characterized in *E. coli* by Lindahl's group, this activity is also present in eukaryotes and has been studied in numerous systems, again with the hypothesis that O^6-alkyl G persistence is an important factor in initiation of carcinogenesis.

The first experiments to show that an alkyl transferase existed in eukaryotes were with a crude extract of mouse liver, rat liver, or with a chromatin

Table VIII-4. Release by Cell Extracts of 3-Methyladenine and 7-Methylguanine from Methylated DNA[a]

Extract added	Percent depurination	
	m^3A	m^7G
None	15	5.5
E. coli	86	5.1[b]
Rat liver	42	9.1[b]
Hamster liver	66	12.0[b]

[a] Data from Margison and Pegg, (1981) *Proc. Natl. Acad. Sci. USA* **78**, 861.
[b] Ring-opened m^7G also depurinated from alkali-treated substrate.

fraction, also from rat liver. Only O^6-MeG or O^6-EtG was dealkylated when alkylated DNA or synthetic deoxypolynucleotides containing the labeled derivative were incubated with these crude enzymes. Later it was clearly shown that the alkyl group (both methyl and ethyl) was transferred to a cysteine of an acceptor protein, as was the case for the methyltransferase from *E. coli.* Since guanine was regenerated, this may be called the "perfect" repair system because it is error-free.

Partial purification procedures have been developed so that comparisons of methyltransferase activity can be made for the amount of activity in different animal species and organs and mammalian cells. The effects of varying conditions that may enhance or decrease this repair activity have also been aided since assays may be performed using appropriate substrates *in vitro.*

All mammalian livers studied, including human, contain the methyltransferase. Other organs such as rat and mouse kidney are poor sources compared to liver. Comparison of human and rat liver indicates how much species variation exists. Human liver enzyme, on average, is about 10 times more active in catalyzing demethylation of O^6-MeG than is rat liver (Table VIII-5), which in turn has a higher demethylase activity than mouse or gerbil liver (Table VIII-6). For both human and rat liver, the amount of labeled *S*-methylcysteine found is equivalent to the loss of methyl groups from DNA. Glycosylase activity depurinating 7-MeG or 3-MeA also varies with species (Figure VIII-6). Gerbil liver, which is low in demethylase activity, is high in both glycosylase activities, while mouse liver has little glycosylase activity

Table VIII-5. Removal of O^6-Methylguanine from DNA upon Incubation with Human or Rat Liver Fraction[a]

Protein added (mg)	m^{6}G			
	Human liver		Rat liver	
	pmole	%	pmole	%
0.78	0.71	58	<0.05	
1.56	0.91	75	<0.08	
3.12	1.06	87	0.24	20
6.25	1.11	92	0.55	45
12.50	1.15	95	0.86	71
18.75	1.15	95	0.89	74
25.00	1.16	96	0.95	79

[a] Extracts were incubated with [^{3}H]-MeNU-treated calf thymus DNA for 1 hr. The percentage is the proportion removed relative to the total initially present. Characterization of the product after incubation with protein showed that the methyl groups lost from DNA were quantitatively found as *S*-methylcysteine. Data from Pegg *et al.* (1982) *Proc. Natl. Acad. Sci. USA* **79,** 5162. No difference was found in rate of removal of 7-methylguanine.

Table VIII-6. Capacity of Hepatic Fractions for Removal of Methylated Purines from DNA in Vitro[a]

	Repair activity (fmoles removed/mg/hr)		
Species	m^{6}G	m^{7}G	m^{3}A
Rat	183	6	14
Mouse	129	5	1
Gerbil	40	17	58

[a] Repair activity was measured using as substrate calf thymus DNA methylated with [^{3}H]-methylnitrosourea. Data from Bamborschke *et al.* (1983) *Cancer Res.* **43**, 1306.

toward 3-MeA but more toward 7-MeG. These data lend support to the concept that multiple glycosylases exist.

Methyltransferase activity, also termed AAP (alkyl acceptor protein), has also been assayed in cell extracts from HeLa cells and human lymphocytes. As with human liver samples, there are wide variations in AAP in lymphocytes from humans, and in addition T lymphocytes are more active than B lymphocytes. On the average it is calculated that the number of presumptive acceptor molecules per cell in lymphocytes is between 1.4×10^4 and 1.1×10^5. The same calculation for HeLa cells is about 10^5 acceptor sites per cell. Other estimates of AAP per cell are 2.1×10^4 in Raji human lymphoma cells and $4–8 \times 10^4$ in human fibroblasts or lymphocytes.

There has been a long-standing question regarding whether constitutive repair enzymes such as glycosylases and methyltransferase can be stimulated or induced to produce more enzyme. The answer at the present time appears to be yes and no. First, there are mammalian cells, particularly from human tumors, that are termed mer$^+$ and mer$^-$, or in xeroderma pigmentosum-derived human lymphoblastoid lines they are termed mex$^+$ and mex$^-$. In both terminologies the + indicates ability to remove O^6-MeG rapidly while the − indicates a deficiency in this activity. An *in vitro* assay illustrating the difference between demethylation using extracts from HeLa mer$^+$ and mer$^-$ cell lines is in Figure VIII-19. Both mer$^-$ and mex$^-$ cells apparently cannot be induced or adapted to produce more AAP, while there is evidence that in the mer$^+$ Raji and HeLa cells AAP can be increased about threefold by repeated low levels of alkylation. Glycosylase activities for 7-MeG and 3-MeA in these cells are independent of the m^{6}G removal.

Partial hepatectomy or other liver damage increases the amount of methyltransferase, but not of glycosylases cleaving 3-MeA or 7-MeG (Table VIII-7). When there is apparent increased glycosylase activity it appears to parallel the increase in the amount of DNA polymerase and returns to basal levels, again parallel with DNA synthesis.

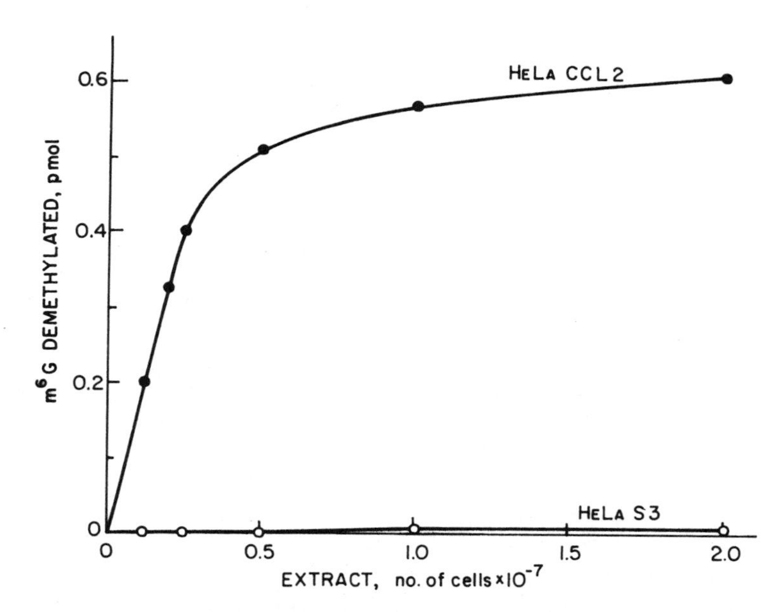

Figure VIII-19. Dependence of demethylation of O^6-methylguanine in poly(dC,dG,[8-^{3}H]m^{6}dG) on cell equivalents of extract from mer$^+$ HeLa CCL2 cells (●) or the mer$^-$strain, HeLa S3 (○), during 4-hr incubations. Adapted from Foote, Pal, and Mitra (1983) *Mutat. Res.* **119,** 221.

3-MeA glycosylase, partially purified from liver of rats fed DENA for 9–15 weeks or acetylaminofluorene for 11–30 weeks (conditions leading to neoplasia), was increased slightly, but this is believed to be due to accompanying cell replication rather than specific induction.

There are also data that induction of increased methyltransferase is a function of liver damage in the rat. This is shown mainly by increases in enzyme activity following hepatic damaging agents (e.g., carbon tetrachlo-

Table VIII-7. Effect of Partial Hepatectomy or Chronic Treatment with Dimethylnitrosamine on the Ability of Rat Liver Extracts to Catalyze the Loss of Methyl Purines from DNA[a]

Treatment prior to administration of [^{14}C]dimethylnitrosamine	Removal activity (fmoles/hr/mg)		
	m^{6}G	m^{3}A	m^{7}G
None	152	94	8.0
Sham operated	158	76	6.2
Chronic dimethylnitrosamine[b]	448	85	7.1
Partial hepatectomy[c]	916	71	7.8

[a] Data from Pegg, Perry, and Bennett (1981) *Biochem. J.* **197,** 195.
[b] 2 mg/kg DMNA administered for 44 days.
[c] 72 hr before measurement.

Table VIII-8. Activity of O^6-Methylguanine Removal System after Treatment with Carcinogens and Other Agents for 21 Days[a]

Treatment		
Agent	mg/kg (21 days)	m⁶G removal (fmoles/mg/30 min)
Dimethylnitrosamine	2	224 ± 23
Diethylnitrosamine	10	245 ± 49
1,2-Dimethylhydrazine	3	247 ± 16
Methylnitrosourea[b]	10	72 ± 14
Streptozotocin[b,c]	30	67 ± 8
Carbon tetrachloride	160	240 ± 55
Saline (0.5 ml/rat/day)		76 ± 8

[a] Data from Pegg and Perry (1981) *Carcinogenesis* **2,** 1195.
[b] Metabolic activation not required.
[c] Streptozotocin, an antibiotic, is a 2-deoxy-D-glucose derivative of methylnitrosourea. It is a potent methylating agent that preferentially methylates liver and kidney DNA. Bennett and Pegg (1981) *Cancer Res.* **41,** 2786.

ride) or carcinogens that are metabolized in the liver (e.g., DMNA), but not following pretreatment with alkylating carcinogens that are directly acting (e.g., MeNU) (Table VIII-8). The increased activity of the partially purified extracts is due to the same enzyme present in normal tissues. This is elegantly demonstrated by the stoichiometry of the loss of methyl groups from O^6-MeG in an *in vitro* substrate and the transfer of labeled methyls to an acceptor protein, with and without induction (Table VIII-9).

While increased cell replication appears to be necessary to increase enzyme activities in rat liver, this does not hold true for C57BL or BALB/C mice,

Table VIII-9. Transfer of Methyl Group of O^6-Methylguanine to Acceptor Protein Catalyzed by a Partially Purified Enzyme Activity from Rat Liver[a]

Treatment of rats prior to isolation of enzyme	m⁶G lost from DNA (10^{-3} × dpm/mg protein)	Radioactivity bound to protein (10^{-3} × dpm/mg protein)
None	13.4	12.2
Partial hepatectomy, 48 hr	68.5	62.3
Carbon tetrachloride, 48 hr	33.5	32.8
1,2-dimethylhydrazine (3 mg/kg/day for 3 weeks)	37.7	33.6
Dimethylnitrosamine (2 mg/kg/day for 3 weeks)	35.4	30.6
Diethylnitrosamine (10 mg/kg/day for 3 weeks)	38.5	37.9

[a] Partially purified enzyme (0.2–2 mg protein) from the liver of rats treated as indicated was incubated with [³H]methylnitrosourea-treated calf thymus DNA. At the end of the incubation half the sample was used to determine the amount of m⁶G left in the DNA and the other half was used to isolate protein containing radioactivity. Data from Pegg and Perry (1981) *Carcinogenesis* **2,** 1195.

gerbil, or hamster. None of the regimes used to stimulate various enzymes in rats were effective in inducing increased O^6-MeG transferase activity.

Returning to the question of repair in individual cell types, it could be shown that the hepatocytes of rat liver responded to exposure to hepatocarcinogens by a two- to threefold enhancement of methyltransferase while the NPC were not inducible (Figure VIII-20). These results obtained with cell-free extracts assayed *in vitro* are virtually identical to the *in vivo* data (Figure VIII-17, bottom), which suggests that neither increased cell replication nor the presence of O^6-alkyl G (resulting from SDMH) was capable of enhancing methyltransferase activity in NPC. This is surprising since the NPC exhibited a 10-fold greater mitogenic response to chronic SDMH exposure than the hepatocytes. The NPC selectively accumulate O^6-MeG 30 times greater than hepatocytes and, moreover, undergo cell replication, which would be likely to lead to a greater possibility of mutation by O^6-MeG mispairing. However, oral doses of SDMH induce hepatomas in 40% of animals under conditions where 90% of the rats have angiosarcomas. This relatively small difference in the site of tumor induction cannot be solely the result of unrepaired O^6-MeG.

Virtually all of this section has, of necessity, been restricted to repair of three alkyl purines that have been studied to the exclusion of other derivatives. While O^6-alkyl G is undoubtedly a biologically significant derivative, the correlations between one derivative and mutation or carcinogenesis appear simplistic. Other promutagenic derivatives such as O^2-alkyl T, O^4-alkyl T, and O^2-alkyl C are also products of alkylation (Chapter IV, Section C5), and their presence or persistence may also contribute to the biological effects of alkylating agents.

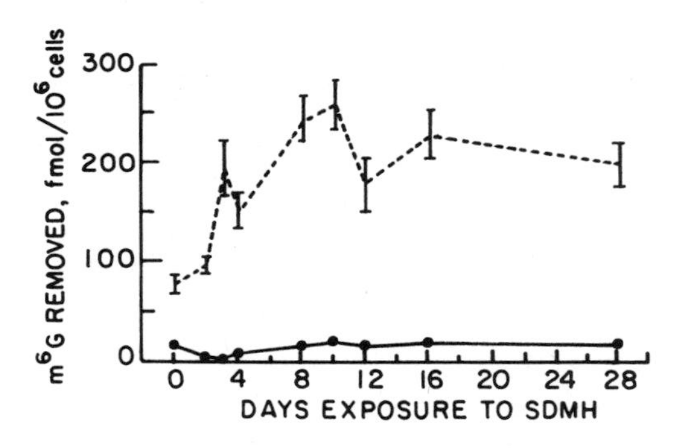

Figure VIII-20. In vitro removal of O^6-methylguanine (m^6G) from [^{3}H-methyl]DNA by extracts from rat hepatocytes (---) and nonparenchymal cells (●) exposed to 1,2-dimethylhydrazine at 30 ppm in the drinking water for up to 28 days. Points represent the mean for cells from three to six rats. Activity is expressed as fmoles of m^6G removed per 30 min per 10^6 cells. Adapted from Swenberg *et al.* (1982) *Proc. Natl. Acad. Sci. USA* **79,** 5499. Figure VIII-17 *(bottom)* gives the amount of m^6G remaining in the two cell types *in situ,* as contrasted to the amount of m^6G removed by cell extracts tested *in vitro* in this figure.

D. *In Vivo* Removal of Aromatic Derivatives

The mechanism of repair of DNA modified by bulky aromatic carcinogens, such as polycyclic aromatic hydrocarbons or aromatic amines, is even less well understood than the repair of DNA damaged by simple alkylating agents. No enzyme has been detected that will specifically repair damage caused by these "bulky" carcinogens. It is thought that a more complex endonucleolytic system that may recognize conformational changes in the DNA backbone (see Chapter VII), rather than specific chemical modification of the DNA bases, is responsible for the removal of the adducts. This endonucleolytic system seems to be similar to the one that initiates the removal of pyrimidine dimers from DNA. The repair of these dimers is associated with the insertion of 35–100 nucleotides for each excised adduct. This type of repair has been called "long patch" in contrast to "short patch" repair initiated by *N*-glycosylases.

In *E. coli* the removal of bulky adducts is initiated by three products of the *uvr* genes, A, B, and C. No corresponding genes or enzymes have been identified in mammalian cells.

Further difficulty encountered in studying removal of bulky adducts is that these types of carcinogens are metabolized by cells to many active intermediates that may bind to different bases of DNA and form many types of adducts (see Chapter VI), which in turn may or may not be substrates for repair enzymes. Some of the DNA adducts are released from cellular DNA due to chemical lability, others remain bound to DNA for relatively longer periods of time.

The relationship of mutagenicity to DNA repair has been studied by comparison of the effects of the chemicals on normal human cells and on repair-deficient human cell lines such as xeroderma pigmentosum (XP). It was found that the cytotoxic effect of some polycyclic aromatic hydrocarbons such as benzo[*a*]pyrene (BP), benz[*a*]anthracene (BA), and dimethylbenz[*a*]anthracene (DMBA) is two or three times higher on XP 12BE cells than on normal human skin fibroblasts. These XP cells perform less than 1% of normal cell repair of UV damage. Similarly, BPDE I and II (see Chapter VI) were much more cytotoxic for XP cells than for normal ones. Many experiments indicate that BPDE adducts are repaired by a cellular excision repair system of normal cells, in contrast to XP cells, which are not able to repair BPDE-induced damage. It was shown that the maximum amount of carcinogen–DNA adduct removal occured 1 hr after the addition of BPDE to normal human fibroblasts (Figure VIII-21). During this first hour about 12% of the adducts had been removed. In another 6 hr, DNA adducts were removed at a rate four times slower than that observed during the first hour. On the other hand, XP cells did not remove the DNA-BPDE adducts at all.

In similar experiments, DNA from normal human fibroblasts harvested immediately, or after 2, 4, or 8 days after treatment with [^{3}H]-BPDE, was hydrolyzed and analyzed by HPLC (Figure VIII-22). The single peak detected

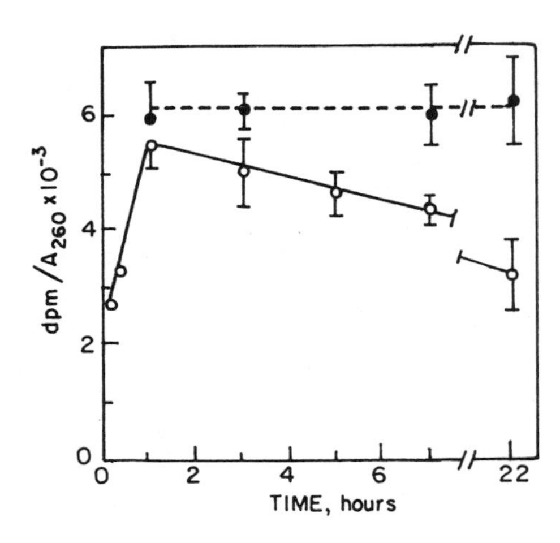

Figure VIII-21. Loss of BP-DNA adducts from normal human fibroblasts (○) and XP fibroblasts (●). Adapted from Koostra (1982) *Carcinogenesis* **3**, 953.

cochromatographed with the BPDE-N^2-G adduct. The kinetics of decrease of the ^{3}H label in this specific peak corresponded to the decrease in radioactivity of the total DNA with time and with the kinetics of recovery of the cells from the potentially cytotoxic and mutagenic effects of BPDE. These results suggest that the N^2-G adduct is responsible for the biological effects, and that excision repair of this lesion by the normal cells is "error-free."

An interesting correlation between repair of BP-induced damage and carcinogenicity was made on human epithelial lung cells, which are considered as target cells for BP. Treatment of these cells with BPDE caused formation of short daughter strands within the first 3 hr that reached control levels within 15 hr. During the 30-hr posttreatment incubation, only 30% of dG-BPDE I and 50% of dG-BPDE II were excised. This low level of removal of BPDE I and II adducts may be related to the transformation of human epithelial lung cells by BP.

In hamster tracheal epithelial cells, within 20 hr after treatment with BP the BP–adenosine adducts are rapidly removed, while the other adducts remain almost unchanged. It is suggested that the differences between the rates of the removal of modified adenine and guanine residues may be due to a not yet understood conformational difference between these two modified residues.

When nondividing human lymphocytes are reacted with 7-methylbromobenz[*a*]anthracene (BMBA), three types of adducts are formed, all on exocyclic amino groups. The major one is the product of the interaction with the N^2 of G and the minor ones with the amino groups of A and C. The adenine adducts are excised more rapidly than those of guanine in mammalian cells. The half-life of the BMBA-G adduct in DNA is almost twice that of the adduct in the BMBA-treated Chinese hamster V79 cells and about fourfold that in the HeLa S3 cells (Figure VIII-23). At low doses of BMBA, all these derivatives are removed. Again no information exists explaining the difference

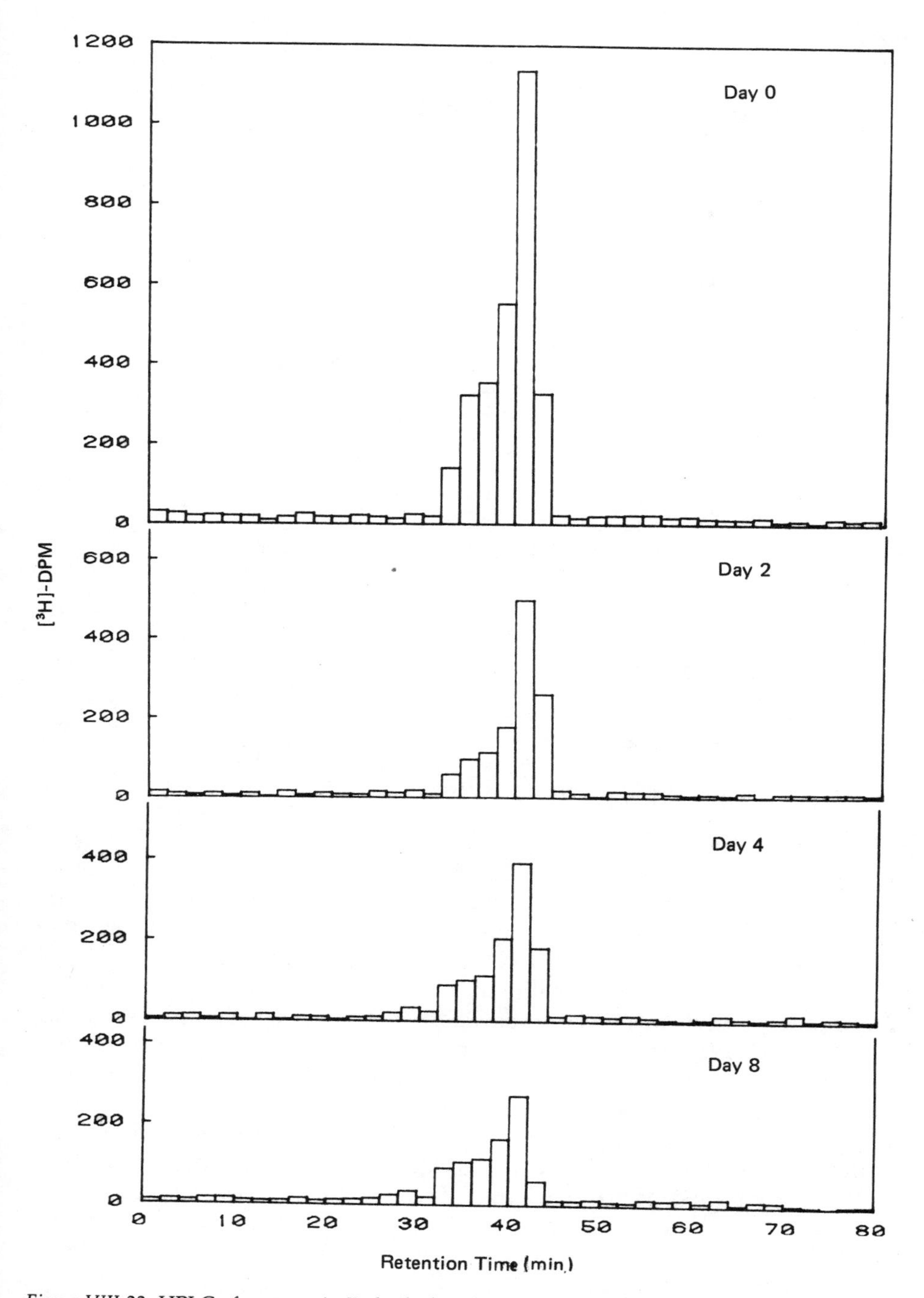

Figure VIII-22. HPLC of enzymatically hydrolyzed DNA adducts remaining in normal fibroblasts treated with [^{3}H]-BPDE and held in confluence for 0, 2, 4, or 8 days. The radioactivity peak cochromatographed with authentic N^2-G-BPDE. From Yang, Maher, and McCormick (1980) *Proc. Natl. Acad. Sci. USA* **77**, 5933.

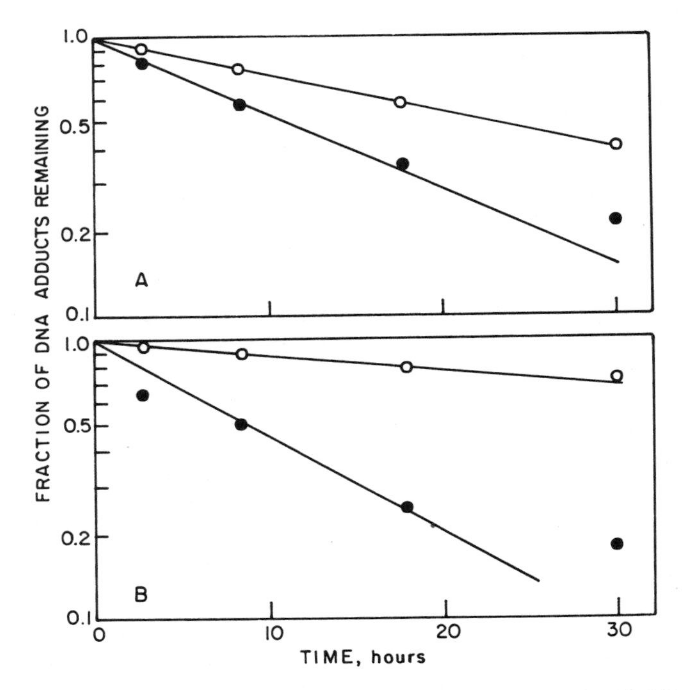

Figure VIII-23. Fraction of initial products present in DNA of cells treated with 0.2 μM 7-BMBA. N^2-G adduct (●); N^6-A adduct (○). (A) Chinese hamster V79 cells; (B) HeLa S3 Cells. Adapted from Dipple and Roberts (1977) *Biochemistry* **16,** 1499.

in conformation between the modified A and G regions. The preferential excision of the A adduct does not hold true for BMBA-treated *E. Coli,* which may indicate different enzymatic mechanisms in prokaryotes and eukaryotes.

Repair of modifications of DNA by *N*-2-acetylaminofluorene (AAF) is complicated by the fact that treatment of the cells, tissues, or whole animals results in formation of three types of G adducts, one of which is deacetylated. The two main adducts are at the C-8 position of G while the minor adduct is at the N^2 position of G (see Chapter VI). After the administration of *N*-hydroxy-AAF to rats, the major adduct (80%), identified as dG(8)-AAF, disappeared from liver DNA with a half-life of about 7 days (Figure VIII-24). On the other hand, the minor adduct (20%), identified as dG(N^2)-AAF, remained in DNA for a period of up to 8 weeks. Primary rat lymphocytes exposed to N-hydroxy-AAF also formed a third derivative, the deacetylated dG(8)-AF adduct. This G-8-AF residue persisted much longer than the G-8-AAF adduct. The differences in the removal of the three adducts can be explained by different conformational changes induced by AAF derivatives at G residues (see Chapter VII). Studies using repair-deficient XP cells indicated that these cells are much more sensitive to the cytotoxic effects of *N*-acetoxy-AAF than the normal fibroblasts, presumably because the XP cells are incapable of repairing the DNA damage induced by AAF.

Nuclear DNA forms a structure with histones, known as the nucleosome. An important question is whether the rate and extent of repair of damage caused by carcinogens can be influenced by such chromatin structure. When normal human fibroblasts are treated with *N*-acetoxy-AAF, the concentration of adducts is found to be higher in the linker DNA than in the nucleosomal core DNA (Figure VIII-25). These adducts are also four times more efficiently removed from linker than from core DNA during the 8 hr immediately after treatment. After 24 hr, the rate of removal slows down. The differential removal of adducts from linker and core DNA may be an indication of the effect of structural accessibility to repair enzymes.

Methylaminobenzene (MAB) resembles AAF in forming derivatives at the C-8 and N^2 of G (Chapter VI). In mice the C-8 adduct is rapidly repaired while the N^2 adduct persists. The repair mechanisms for MAB may be the same as for AAF.

In the case of aflatoxin B_1 (AFB_1), the initial product formed is an adduct on the N-7 of guanine (see Chapter VI). As with all derivatives substituted on the N-7 of G, the resulting quaternary base has a very labile glycosyl bond ($t_{1/2}$, 37°C, pH 7, ~ 12 hr; pH 7.4, ~ 8 hr) (Chapter III, Section B). Secondary reactions include the loss of the AFB moiety, thus regenerating guanine, and also an alkali-catalyzed opening of the imidazole ring, thus stabilizing the glycosyl bond.

Formation and removal of AFB_1-DNA adducts were studied in human

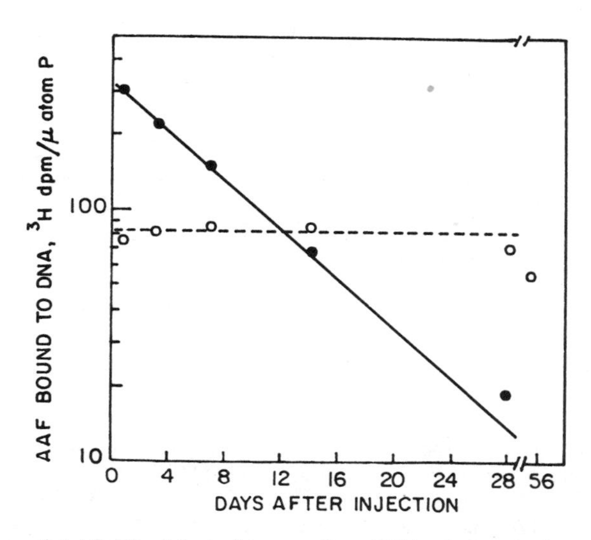

Figure VIII-24. Loss of AAF-dG adducts from rat liver DNA after a single injection of *N*-OH-AAF. C8-dG adduct (●); N^2-dG adduct (○). Adapted from Kriek (1972) *Cancer Res.* **32,** 2042.

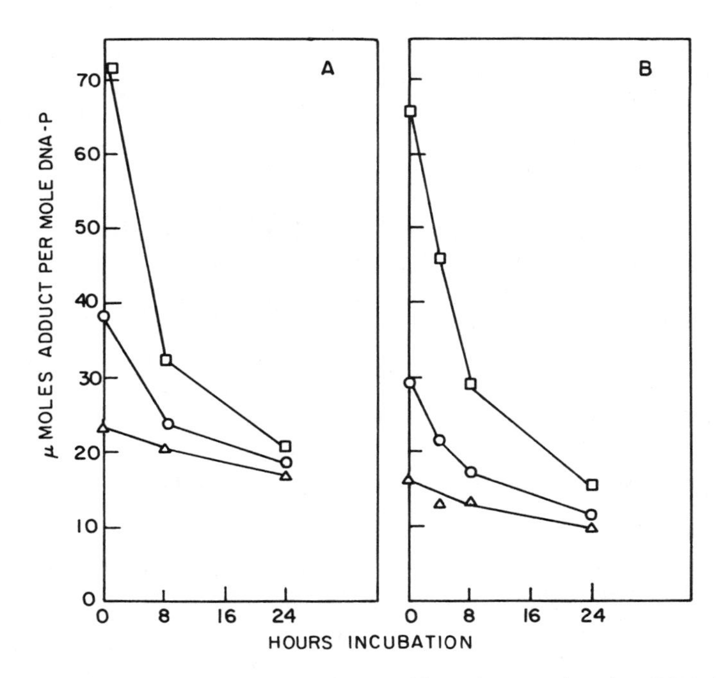

Figure VIII-25. Time course of the removal of AAF adducts from total nuclear DNA (○), purified nucleosomal core DNA (△), and nucleosomal linker DNA (□). Panels A and B: confluent and growing cultures of normal human skin in fibroblasts CRL 1121, respectively. Adapted from Kaneko and Cerruti (1980) *Cancer Res.* **40,** 4313.

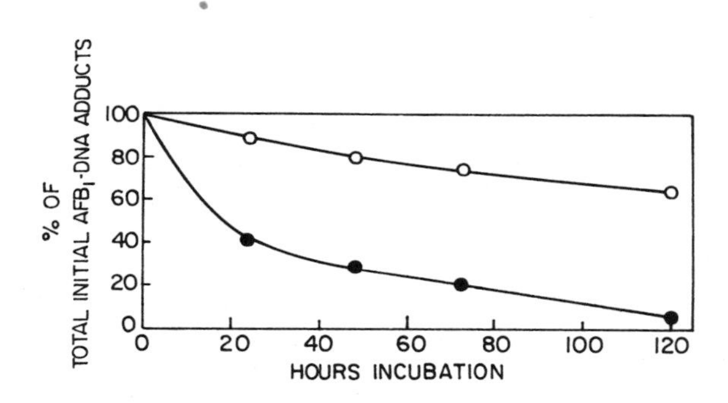

Figure VIII-26. Loss of total AFB_1 adducts from DNA *in situ* in human lung cells A549 following 24-hr treatment with 82 μM [^{3}H]-AFB_1 (●) and from A549 DNA isolated from the same cells and incubated *in vitro* (○). The data are corrected for cell division but do not include the initial rapid loss of the AFB_1-*N*-7G product, which has a $t_{1/2} \sim 8$ hr at pH 7.4, 37°C. Adapted from Wang and Cerutti (1979) *Cancer Res.* **39,** 5165. The product quantitated is likely to be the ring-opened derivative, 8,9-dihydro-8-(2,6-diamino-4-oxo-3,4-dihydropyrimid-5-yl-formamido)-9-hydroxy AFB_1 [Hertzog *et al.* (1982) *Carcinogenesis* **3,** 723]. See Leadon *et al.* (1981) *Cancer Res.* **41,** 5125, for repair in human fibroblasts.

epithelial lung cell line A549 under nontoxic conditions (Figure VIII-26). After 24 hr of exposure to AFB_1, approximately 60% of the total AFB_1 adducts are removed in the following 24 hr of incubation. Within the same time period only 15% of the total adducts are released from AFB_1-DNA under physiological conditions *in vitro*. Therefore, it is concluded that at least a portion of the covalent AFB_1-DNA adducts are enzymatically removed from the DNA in A549 cells, since the rate is only slightly higher than for the chemical hydrolysis of adducts in free DNA.

Lesions persisting during DNA replication are likely to be the ring-opened derivatives of the type known for other 7-alkyl guanines. Although rat liver contains a glycosylase excising ring-opened 7-MeG (Section C), the rate of removal is slow. This is further indicated by the finding that in rats treated with methylating carcinogens a major persistent product is N^5-methyl-N^5-formyl-2,5,6-triamino-4-hydroxypyrimidine (imidazole ring-opened 7-MeG).

SELECTED REFERENCES

Arenez, P., and Sirover, M. A. (1983) Isolation and characterization of monoclonal antibodies directed against the DNA repair enzyme uracil DNA glycosylase from human placenta. *Proc. Natl. Acad. Sci. USA* **80** (in press).

Becker, R. A., Barrows, L. R., and Shank, R. C. (1981) Methylation of liver DNA guanine in hydrazine hepatotoxicity: Dose–response and kinetic characteristics of 7-methylguanine and O^6-methylguanine formation and persistence in rats. *Carcinogenesis* **2,** 1181–1188.

Beland, F. A., Dooley, K. L., and Jackson, C. D. (1982) Persistence of DNA adducts in rat liver and kidney after multiple doses of the carcinogen *N*-hydroxy-2-acetylaminofluorene. Cancer Res. **42,** 1348–1354.

Bodell, W. J., Singer, B., Thomas, G. H., and Cleaver, J. E. (1979) Evidence for removal at different rates of *O*-ethyl pyrimidines and ethylphosphotriesters in two human fibroblast cell lines. *Nucleic Acids Res.* **6,** 2819–2829.

Cerutti, P. A., Wang, V. T., and Amstad, P. (1980) Reactions of aflatoxin B_1 damaged DNA *in vitro* and *in situ* in mammalian cells. In *Carcinogenesis: Fundamental Mechanisms and Environmental Effects* (B. Pullman, P. O. P. Ts'o, and H. Gelboin, eds.), Reidel, Dordrecht, pp. 465–477.

Cleaver, J. E. (1983) Xeroderma pigmentosum. In *The Metabolic Basis of Inherited Disease* (J. B. Stanbury, J. B. Wyngaarden, D. S. Fredrickson, J. L. Goldstein, and M. S. Brown, eds.), 5th ed., McGraw–Hill, New York, pp. 1227–1250.

Craddock, V. M., and Henderson, A. R. (1982) The activity of 3-methyladenine DNA glycosylase in animal tissues in relation to carcinogenesis. *Carcinogenesis* **3,** 747–750.

Defais, M. J., Hanawalt, P. C., and Sarasin, A. R. (1983) Viral probes for DNA repair. *Adv. Radiation Biol.* **10,** 1–37.

Essigmann, J. M., Croy, R. G., Bennett, R. A., and Wogan, G. N. (1982) Metabolic activation of aflatoxin B_1: Patterns of DNA adduct formation, removal and excretion in relation to carcinogenesis. *Drug Metab. Rev.* **13,** 581–602.

Glickman, B. W. (1982) Methylation-instructed mismatch correction as a postreplication error avoidance mechanism in *Escherichia coli.* In *Molecular and Cellular Mechanisms of Mutagenesis* (J. F. Lemontt and W. M. Generoso, eds.), Plenum Press, New York, pp. 65–87.

Goth, R., and Rajewsky, M. F. (1974) Molecular and cellular mechanisms associated with pulse-carcinogenesis in the rat nervous system by ethylnitrosourea: Ethylation of nucleic acids and elimination rates of ethylated bases from the DNA of different tissues. *Z. Krebsforsch.* **82,** 37–64.

Hanawalt, P. C., Cooper, P. K., Ganesan, A. K., and Smith, C. A. (1979) DNA repair in bacteria and mammalian cells. *Annu. Rev. Biochem.* **48,** 783–836.

Karren, P., Helmgren, T., and Lindahl, T. (1982) Induction of a DNA glycosylase for *N*-methylated purines is part of the adaptive response to alkylating agents. *Nature* **296,** 770–773.

Kleihues, P., Doerjer, G., Swenberg, J. A., Hauenstein, E., Bücheler, J., and Cooper, H. K. (1979) DNA repair as regulatory factor in the organotropy of alkylating carcinogens. *Arch. Toxicol. Suppl.* **2,** 253–261.

Kleihues, P., Doerjer, G., Ehret, M., and Guzman, J. (1980) Reaction of benzo(*a*)pyrene and 7,12-dimethylbenz(*a*)anthracene with DNA of various rat tissues *in vivo.* In *Quantitative Aspects of Risk Assessment in Chemical Carcinogenesis, Arch. Toxicol. Suppl.* **3,** 237–246.

Kleihues, P., Hodgson, R. M., Veit, C., Schweinsberg, F., and Wiessler, M. (1983) DNA modification and repair *in vivo:* Towards a biochemical basis of organ-specific carcinogenesis by methylating agents. In *Organ and Species Specificity in Chemical Carcinogenesis* (R. Langenback and S. Nesnow, eds.) Plenum Press, New York, pp. 509–530.

Laval, J., and Laval, F. (1980) Enzymology of DNA repair. In *Molecular and Cellular Aspects of Carcinogen Screening Tests* (R. Montesano, H. Bartsch, and L. Tomatis, eds.), International Agency for Research on Cancer, Lyon, pp. 55–75.

LeBlanc, J.-P., and Laval, J. (1982) Comparison at the molecular level of uracil-DNA glycosylases from different origins. *Biochimie* **64,** 735–738.

Likhachev, A. J., Ivanov, M. N., Brésil, H., Planche-Martel, G., Montesano, R., and Margison, G. P. (1983) Carcinogenicity of single doses of *N*-nitroso-*N*-methylurea and *N*-nitroso-*N*-ethylurea in Syrian golden hamsters and the persistence of alkylated purines in the DNA of various tissues. *Cancer Res.* **43,** 829–833.

Lindahl, T. (1979) DNA glycosylases, endonucleases for apurinic/apyrimidinic sites, and base excision-repair. *Prog. Nucleic Acid Res. Mol. Biol.* **22,** 135–192.

Lindahl, T. (1982) DNA repair enzymes. *Annu. Rev. Biochem.* **51,** 61–87.

Lindahl, T., Rydberg, B., Hjelmgren, T., Olsson, M., and Jacobson, A. (1982) Cellular defense mechanisms against alkylation of DNA. In *Molecular and Cellular Mechanisms of Mutagenesis* (J. F. Lemontt and W. M. Generoso, eds.), Plenum Press, New York, pp. 89–102.

Linn, S. (1982) Nucleases involved in DNA repair. In *The Nucleases* (S. Linn and R. J. Roberts, eds.), Cold Spring Harbor Laboratory, New York, pp. 59–83.

Linn, S., Mosbaugh, D. W., LaBelle, M., Krauss, S. W., and Lackey, D. (1982) Enzymatic studies of DNA repair. In *Primary and Tertiary Structure of Nucleic Acids and Cancer Research* (M. Miwa, S. Nishimura, A. Rich, D. G. Söll, and T. Sugimura, eds.), Japan Scientific Societies Press, Tokyo, pp. 165–179.

Maru, G. B., Margison, G. P., Chu, Y.-H., and O'Connor, P. J. (1982) Effects of carcinogens and partial hepatectomy upon the hepatic O^6-methylguanine repair system in mice. *Carcinogenesis* **3,** 1247–1254.

Montesano, R. (1981) Alkylation of DNA and tissue specificity in nitrosamine carcinogenesis. *J. Supramol. Struct. Cell. Biochem.* **17,** 259-273.

Montesano, R., Pegg, A. E., and Margison, G. P. (1980) Alkylation of DNA and carcinogenicity of *N*-nitroso compounds. *J. Toxicol. Environ. Health* **6,** 1001–1008.

Müller, R., and Rajewsky, M. F. (1983) Elimination of O^6-ethylguanine from the DNA of brain, liver, and other rat tissues exposed to ethylnitrosourea at different stages of prenatal development. *Cancer Res.* **43,** 2897–2904.

O'Connor, P. J., Capps, M. J., and Craig, A. W. (1973) Comparative studies of the hepatocarcinogen *N,N*-dimethylnitrosamine *in vivo:* Reaction sites in rat liver DNA and the significance of their relative stabilities. *Br. J. Cancer* **27,** 153–166.

O'Connor, P. J., Chu, Y.-H., Cooper, D. P., Maru, G. B., Smith, R. A., and Margison, G. P. (1982) Species differences in the inducibility of hepatic O^6-alkylguanine repair in rodents. *Biochimie* **64,** 769–773.

Pegg, A. E. (1982) Guest editorial: Formation and removal of methylated nucleosides in nucleic acids of mammalian cells. *J. Cancer Res. Clin. Oncol.* **103,** 313.

Pegg, A. E., and Bennett, R. A. (1982) Mammalian DNA repair enzymes. In *Enzymes of Nucleic Acid Synthesis and Processing* (S. T. Jacob, ed.), CRC Press, Boca Raton, Fl., pp. 179–205.

Poirier, M., Dubin, M. A., and Yuspa, S. H. (1979) Formation and removal of specific acetylaminofluorene–DNA adducts in mouse and human cells measured by radioimmune-assay. *Cancer Res.* **39,** 1377–1381.

Rabes, H. M., Wilhelm, R., Kerler, R., and Rode, G. (1982) Dose and cell cycle-dependent O^6-methylguanine elimination from DNA in regenerating rat liver after [^{14}C]dimethylnitrosamine injection. *Cancer Res.* **42,** 3814–3821.

Roberts, J. J. (1978) The repair of DNA modified by cytotoxic, mutagenic, and carcinogenic chemicals. *Adv. Radiat. Biol.* **7,** 211–436.

Sarasin, A., Rossignol, J.-M., and Philippe, M. (1983) DNA polymerases and DNA repair in eukaryotic cells. In *New Approaches in Eukaryotic DNA Replication* (A. M. de Recondo, ed.), Plenum Press, New York, pp. 313–332.

Sedgwick, B., and Lindahl, T. (1982) A common mechanism for repair of O^6-methylguanine and O^6-ethylguanine in DNA. *J. Mol. Biol* **154,** 169–175.

Singer, B. (1979) Guest editorial: *N*-Nitroso alkylating agents: Formation and persistence of alkyl derivatives in mammalian nucleic acids as contributing factors in carcinogenesis. *J. Natl. Cancer Inst.* **62,** 1329–1339.

Singer, B., and Brent, T. P. (1981) Human lymphoblasts contain DNA glycosylase activity excising *N*-3 and *N*-7 methyl and ethyl purines but not O^6-alkylguanines or 1-alkyladenines. *Proc. Natl. Acad. Sci. USA* **78,** 856–860.

Singer, B., Spengler, S., and Bodell, W. J. (1981) Tissue-dependent enzyme-mediated repair or removal of *O*-ethyl pyrimidines and ethyl purines in carcinogen-treated rats. *Carcinogenesis* **2,** 1069–1073.

Sirover, M. A., and Gupta, P. K. (1983) Regulation of DNA repair in human cells. In *Human Carcinogenesis* (C. C. Harris and H. N. Autrup, eds.), Academic Press, New York, 225–280.

Sklar, R., and Strauss, B. (1981) Removal of O^6-methylguanine from DNA of normal and xeroderma pigmentosum-derived lymphoblastoid lines. *Nature(London) 289,* 417–420.

Smith, G. J., Grisham, J. W., and Kaufman, D. G. (1981) Cycle-dependent removal of certain methylated bases from DNA of 10T $^1/_2$ cells treated with *N*-methyl-*N'*-nitro-*N*-nitrosoguanidine. *Cancer Res.* **41,** 1373–1378.

Strauss, B., Altamirano, M., Bose, K., Sklar, R., and Tatsumi, K. (1979) Carcinogen-induced damage to DNA. In *Carcinogens: Identification and Mechanisms of Action* (A. C. Griffin and C. R. Shaw, eds.), Raven Press, New York, pp. 229–250.

Swenberg, J. A., Dyroff, M. C., Bedell, M. A., Popp, J. A., Huh, N., Kirstein, U., and Rajewsky, M. F. (1984) O^4-Ethyldeoxythymidine, but not O^6-ethyldeoxyguanosine, accumulates in DNA of hepatocytes from rats exposed continuously to diethylnitrosamine. *Proc. Natl. Acad. Sci. USA* (in press).

Teebor, G. W., and Frenkel, K. (1983) The initiation of DNA excision repair. *Adv. Cancer Res.* **38,** 23–59.

Verly, W. G. (1980) Commentary: Prereplicative error-free DNA repair. *Biochem. Pharmacol.* **29** 977–982.

Yang, L. L., Maher, V. M., and McCormick, J. J. (1980) Error-free excision of the cytotoxic, mutagenic N^2-deoxyguanosine DNA adduct formed in human fibroblasts by $(\pm)$-7β,8α-dihydroxy-9α,10α-epoxy-7,8,9,10-tetrahydrobenzo[*a*]pyrene. *Proc. Natl. Acad. Sci. USA* **77,** 5933–5937.

IX
Site-Directed Mutagenesis

In the preceding chapters dealing with modifications of nucleic acids it is assumed that, in general, these chemical reactions are randomly located along the genome. Exceptions occur when protein–nucleic interactions or nucleic acid secondary structure play a role. When studying mutation *in vivo*, changes in phenotype or in the sequence of newly synthesized DNAs, RNAs, or protein have been used to detect the mutants. These methods can answer only the question of whether a particular reagent, usually modifying multiple nucleic acid sites, is a mutagen; if a cellular function is changed, sequence data may pinpoint the specific nucleic acid regions responsible.

Advances in nucleic acid chemistry during the past decade have made it possible to construct point mutants with a desired modification at a predetermined site. The requirements are to possess extensive sequence information of biologically active nucleic acids (e.g., SV40, φX174, M13, Qβ), as well as techniques for sequencing progeny nucleic acids, and for some systems the availability of specific polymerases, DNases, and ligases for nicking and/or for construction of recombinants.

Although the methodology can be enormously complicated and time-consuming, it is possible to describe the main experimental approaches simply. The genetic background, specific methodology, and critical evaluation of these procedures can be found in the references given in the figures and at the end of the chapter.

The first method used was by Weissmann's group and can be called "forcing." This means that synthesis on a template is terminated at a particular base by omitting the complementary NTP, followed by synthesis using only the desired mutating NTP. Alternatively, single base elongation of a primer can be forced by having only such a single NTP present. In such forcing experiments, it is advantageous to use polymerases lacking endonucleolytic activity. Further elongation to complete the molecule can be achieved with DNA polymerase I. Examples shown in diagrammatic form are: Incorporation of N^4-hydroxy C in place of U in Qβ RNA (Figure IX-l) and similarly of N^4-hydroxy dC in the globin gene (Figure IX-2). This can lead to mutation since

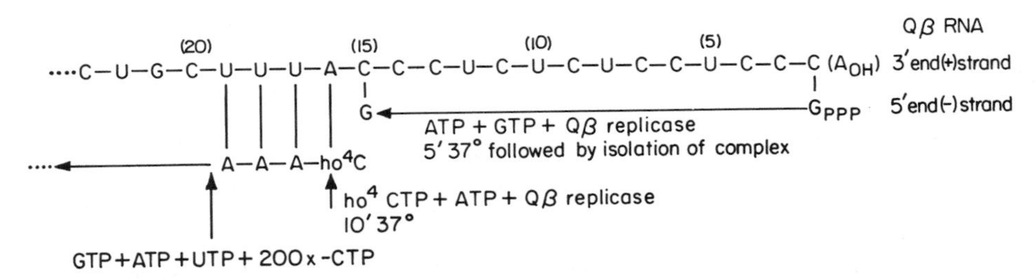

Figure IX-1. Synthesis of bacteriophage Qβ minus strands containing N^4-hydroxy CMP in position 15. The completed virus RNA was used as a template for plus strands 33% of which had a G → A transition in the 16th position from the 3′ end (which corresponds to position 15 of the minus strands). The modified RNA could be replicated by Qβ replicase. Data from Flavell, Sabo, Bandle, and Weissmann (1974) *J. Mol. Biol.* **89,** 255.

the N^4-hydroxy derivatives can be in either the imino (U-like) or the amino (C-like) form as a result of their changed tautomeric equilibrium (Chapter III).

Another use of this principle is incorporation of a mismatched base in ϕX174 *am* 18 DNA, which can be demonstrated to cause a point mutation. Revertant DNAs obtained upon transfection were sequenced in the *am* 18 region and both transitions and transversions are produced at the designated

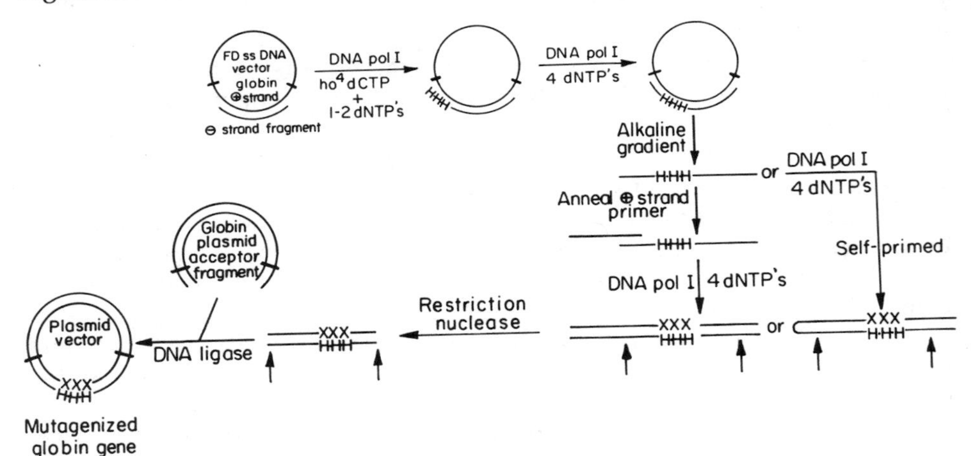

Figure IX-2. General scheme of site-directed mutagenesis of the globin gene. Globin plus strand DNA cloned in a single-stranded fd DNA served as template in this example. The primer is a fragment of globin DNA with an appropriate 3′ end. After incorporating N^4-hydroxy dCMP residues (H) and further elongation, the extended primer was isolated and made double-stranded by DNA polymerase I (Pol I) either in a self-primed reaction or using a suitable plus-strand primer. The mutagenized segment was cleaved out by restriction nucleases (arrowheads) and ligated into an acceptor fragment supplying the missing parts of the globin gene as well as the vector (pBR322) sequences. X denotes sites at which point mutations may arrive *in vitro* due to ambiguous basepairing with N^4-hydroxy C, which can be in either the amino or the imino form (see Chapter IV, Section B). Adapted from Weber *et al.* (1981) In *Developmental Biology Using Purified Genes* (D. D. Brown, ed.), ICN-UCLA Symposia on Molecular and Cellular Biology, Vol. XXIII, Academic Press, New York, pp. 367–385.

site. The protocol (Figure IX-3) in this type of experiment uses an error-prone polymerase, such as that from avian myeloblastosis virus, on the basis that polymerases with an editing function would excise the mismatch.

A modification of this approach is shown in Figure IX-4. The analogs to be incorporated are simply substituted at the predetermined site for the normal dNTP since the modified triphosphates can basepair with the selected complementary base. Mutations can result because of their possible ambiguity in basepairing (Chapter III).

The next type of experimental approach, pioneered by Shortle and Nathans, utilizes single-stranded gaps caused by nicking SV40 DNA with restriction enzymes. After the nick is converted to a short gap by the 5′ to 3′ exonuclease activity of DNA polymerase I, exposed cytosine residues in the

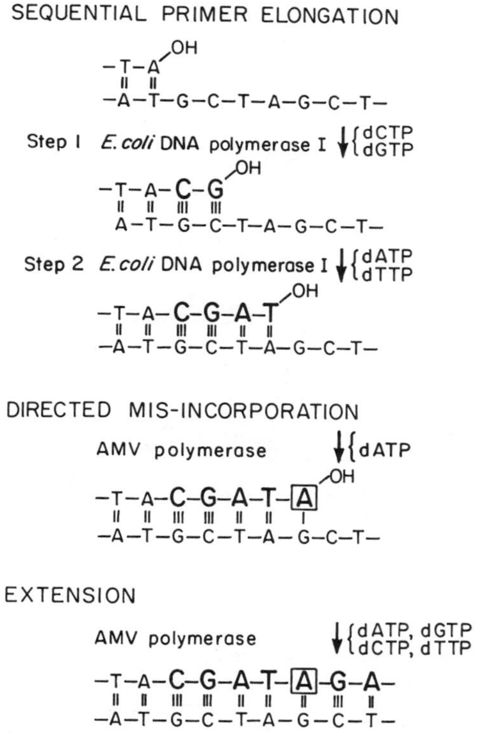

SEQUENTIAL PRIMER ELONGATION

```
      OH
 -T-A
  ‖ ‖
 -A-T-G-C-T-A-G-C-T-

Step 1  E. coli DNA polymerase I  ↓{dCTP, dGTP}
          OH
 -T-A-C-G
  ‖ ‖ ⦀ ⦀
  A-T-G-C-T-A-G-C-T-

Step 2  E. coli DNA polymerase I  ↓{dATP, dTTP}
              OH
 -T-A-C-G-A-T
  ‖ ‖ ⦀ ⦀ ‖ ‖
 -A-T-G-C-T-A-G-C-T-
```

DIRECTED MIS-INCORPORATION

```
 AMV polymerase  ↓{dATP}
                  OH
 -T-A-C-G-A-T-[A]
  ‖ ‖ ⦀ ⦀ ‖ ‖  |
 -A-T-G-C-T-A-G-C-T-
```

EXTENSION

```
 AMV polymerase  ↓{dATP, dGTP, dCTP, dTTP}

 -T-A-C-G-A-T-[A]-G-A-
  ‖ ‖ ⦀ ⦀ ‖ ‖  ‖  ⦀ ‖
 -A-T-G-C-T-A-G-C-T-
```

Figure IX-3. Model for single base substitution by directed misincorporation based on infidelity of DNA synthesis. (Top) The primer is sequentially elongated using *E. coli* DNA polymerase I with proofreading activity and a limited number of dNTPs in order to reach the preselected template position. This is illustrated in steps 1 and 2. (Middle) When the 3′-OH terminus of the primer is adjacent to the template position selected for mutagenesis, the error-prone polymerase from avian myeloblastosis virus (AMV) is used to misinsert the single noncomplementary nucleotide provided. (Bottom) Following the misinsertion, all four dNTPs are added to complete synthesis. Adapted from Zakour and Loeb (1982) *Nature (London)* **295,** 708.

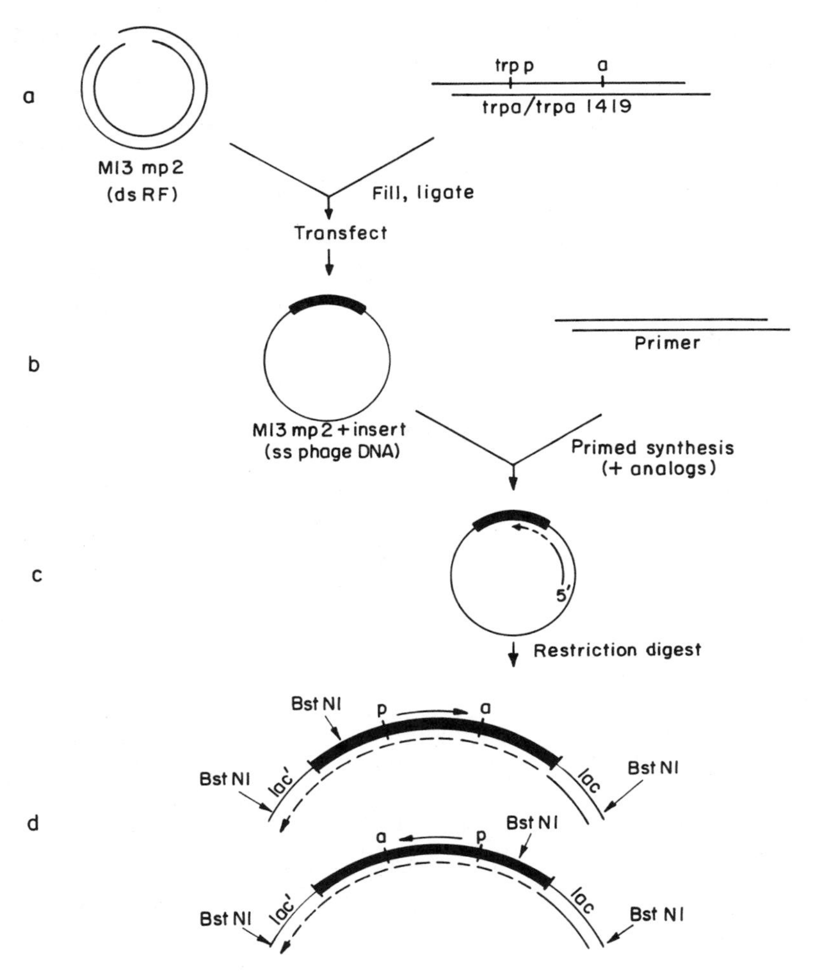

Figure IX-4. Scheme of template preparation. Shown here is the method by which deoxyribonucleotide base analogs were incorporated into the normal and mutant *trp* attenuator DNA. Analogs used included 2,6-diaminopurine, 5-iodo and 5-bromo deoxynucleoside triphosphates. The double-stranded vector M13mp2, which has been cut at the single *Eco*RI site, is mixed with either the *Hpa*II 560-basepair *trp a* fragment or the *Hinf* 510-basepair *trp a* 1419 fragment (a). Each of these fragments contains the *trp* promoter and attenuator regions. The overhanging ends of the vector and fragment are filled and then ligated. 71-18 cells are transfected with the mix and phage containing inserts are identified by using plaque hybridization to RNA and DNA probes. The single-stranded DNA isolated from these phage is used in a primed synthesis reaction with a 90-basepair *lac* primer (b). A specific analog is substituted for its homologous normal triphosphate and incorporated into the strand that is being synthesized (c). A *Bst*NI digest of the resulting double-stranded DNA then produces the template for transcription reactions. As there are two possible orientations for each cloned insert, two different *Bst*NI restriction fragments containing the *trp* promoter and attenuator will be obtained for both *trp a* and *trp a* 1419 (d). From Farnham and Platt (1982) *Proc. Natl. Acad. Sci. USA* **79,** 998.

remaining strand are deaminated by sodium bisulfite (Chapter IV, Section B2), and the gap filled by action of DNA polymerase I (Figure IX-5). Those molecules having deaminated cytosines then acquire an AU basepair where originally there was a GC pair. If this change is located within an original restriction site it is possible to select the mutated DNA for replication *in vivo* since only these molecules remain intact upon appropriate nuclease treatment.

Rather than chemically modifying the exposed region, it is also possible to synthesize an oligonucleotide of the same size as the gap, containing one noncomplementary base, and ligate it into the gap (Figure IX-6). The disadvantage here is that repair of derivatives such as O^6-MeG may occur when the mutated viral DNA is transfected into cells (Chapter VIII, Sections B, C). If repair does not occur, this is a powerful tool to study individual carcinogen products as mutagens.

A combination of creating single-stranded gap and using forcing conditions can also yield specific mutations. This hybrid technique has been termed "misrepair" and includes omission of one dNTP during repair of the gap or misincorporations of an α-thiophosphate nucleotide, which DNA polymerase I is not able to excise (Chapter III, Section D). Both point mutations and deletions result when dTTP (αS) was used as a mismatched base, instead of dTTP.

The last major method can be very versatile. Unique oligonucleotides, complementary to any one portion of a single-stranded genome, can be synthesized with a specific mismatch, then annealed to the (+) or (−) strand of a viral DNA and finally used as a primer to complete the duplex molecules (Figure IX-7). Although this method has already yielded information on gene

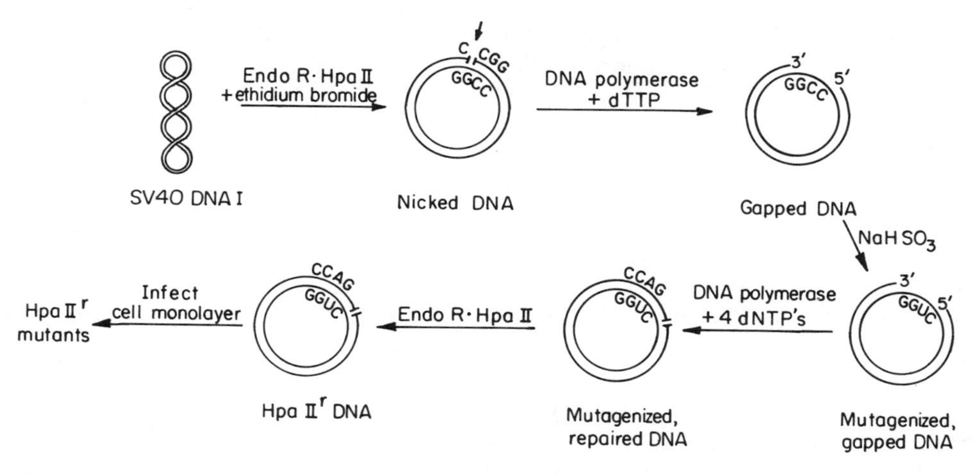

Figure IX-5. Outline of a local mutagenesis method, starting with endo R · *Hpa*II, which recognizes the single 5′ CCGG sequence in SV40 DNA. Bisulfite is used for mutation of a cytosine exposed in the gapped 3′ GGCC site. The resulting uridine then pairs with adensoine when the gap is repaired. From Shortle and Nathans (1978) *Proc. Natl. Acad. Sci. USA* **75,** 2170.

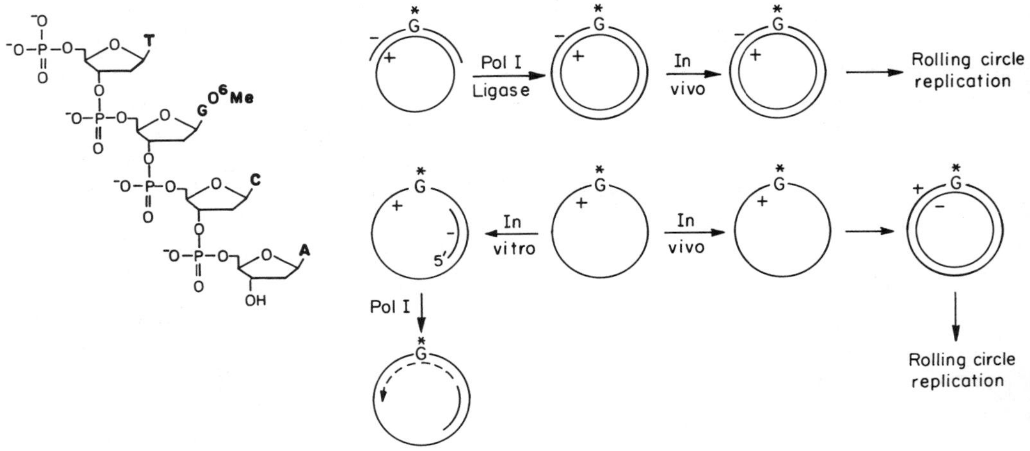

Figure IX-6. Scheme for building O^6-methylguanine into either single-stranded or double-stranded DNA of φX174. (Left) An oligonucleotide is constructed complementary to a specific region of the viral strand of φX174 DNA. The modified base, m^6G, is synthesized into a central site that ordinarily would be occupied by guanine. (Top) Treatment of this DNA with polymerase and ligase produces a duplex molecule that is transfected into *E. coli* spheroplasts. (Bottom) Alternatively, the (+) strand containing the modified G is either copied using a complementary primer *in vitro* or transfected into *E. coli* spheroplasts and replicated. The asterisk denotes the O^6-oligonucleotide shown on the left. The scheme is by courtesy of J. M. Essigmann.

expression, there exist possible complicating problems that can arise as applications increase. The oligonucleotide must interact only with the desired site. The efficiency of mutation depends of the thermal stability of the H-bonded oligomer. When this contains a mismatch, it must apparently be at least a 12-mer. If a carcinogen-modified base were to be introduced, it would

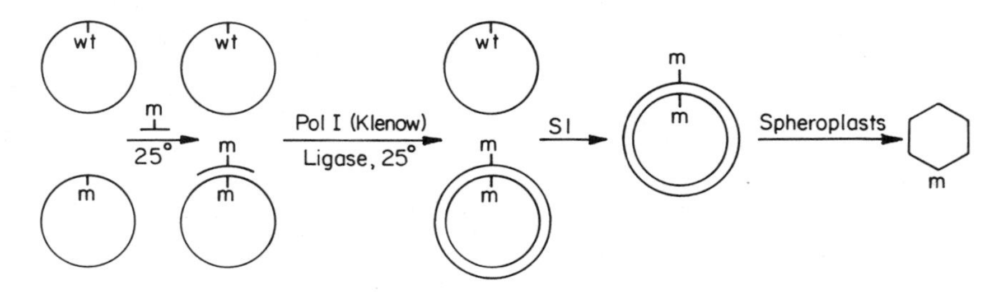

Figure IX-7. Scheme for *in vitro* selection of mutant φX174 DNA using a synthetic oligoribonucleotide (m). At 25°C, the mutant or mismatched oligoribonucleotide is a more efficient primer of *E. coli* DNA polymerase I, large fragment (Pol I, Klenow), on homologous (wild-type) DNA template (wt). Thus, the mutant DNA is selectively converted, after ligation, to closed circular duplex DNA that is resistant to S_1 endonuclease. Transfection of *E. coli* spheroplasts then yields a phage population specifically enriched in the mutant. From Smith and Gillam (1981) In *Genetic Engineering,* Vol. 3 (Setlow and Hollaender, eds.), Plenum Press, pp. 1–32.

be likely to disrupt hybridization more than a normal mismatched base, and thus longer oligomers would be needed. However, as the length of the oligomer increases, secondary structure may play a role for both the primer and template and prevent complete annealing. Mutation efficiency is clearly increased by priming at 0°C, where the primer will be bound most stably. Another problem with this method is to prevent editing out of the mismatched primer by 3′-exonuclease degradation prior to priming DNA synthesis. This may be accomplished by using the Klenow fragment of DNA polymerase I. Smaller primers have been used but these are likely to be restricted to studying mutation on selected genes having an easily detected function (Figure IX-8).

Region-directed mutagenesis has also developed, primarily using the methodology described in this Chapter. In some cases, information can be more easily obtained by using a larger target than a single base. The rationale

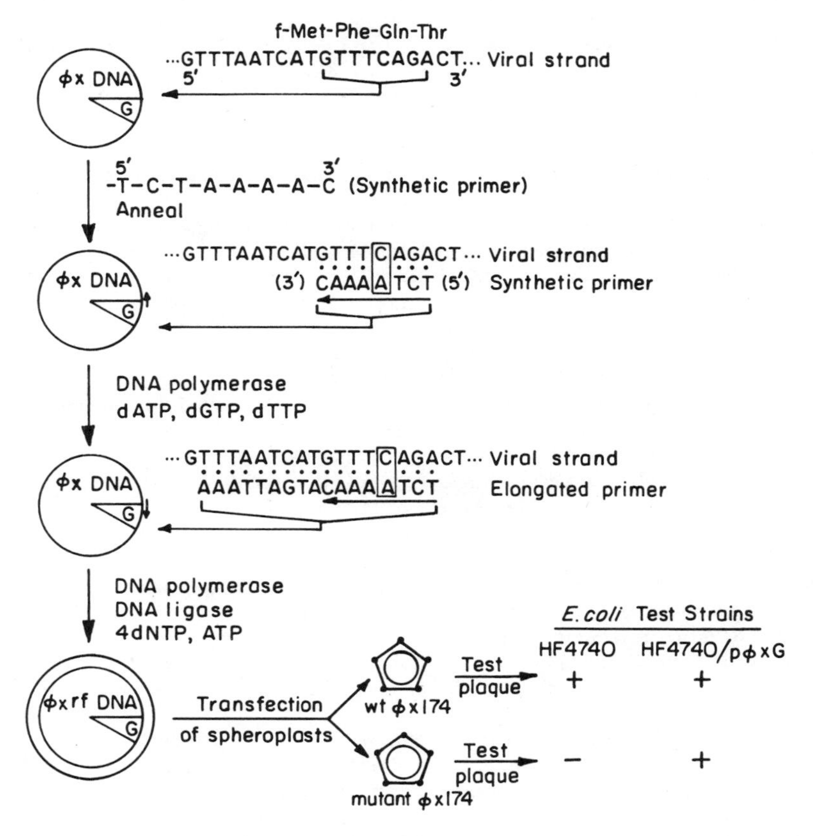

Figure IX-8. Strategy for introduction of a site-specific G → A base change at position 2401 (gene G) of φX174 RF DNA and the detection of mutant phage. The primer is underlined in the figure; the arrow indicates the direction of elongation. The mismatch at the preselected mutation site in gene G is boxed. *E. coli* DNA polymerase lacking the 5′ → 3′ endonuclease activity was used throughout. From Bhanot, Khan, and Chambers (1979) *J. Biol. Chem.* **254,** 12684.

for mutating a segment of viral DNA is basically the greater efficiency of obtaining mutants without necessarily using a restriction enzyme site or synthesizing an oligonucleotide. These methods rely on introducing into the viral DNA a plasmid or restriction fragment with specific characteristics. These molecules usually contain a single-stranded region that can be mutated by bisulfite deamination of C → U.

SELECTED REFERENCES

Caruthers, M. H. (1982) The application of nucleic acid chemistry to studies on the functional organization of gene control regions. In *Primary and Tertiary Strucutre of Nucleic Acids and Cancer Research* (M. Miwa, S. Nishimura, A. Rich, D. Söll, and T. Sugimura, eds.), Japan Scientific Societies Press, Tokyo, pp. 295–306.

Chambers, R. W. (1982) Site-specific mutagenesis: A new approach for studying the molecular mechanisms of mutation by carcinogens. In *Molecular and Cellular Mechanisms of Mutagenesis* (J. F. Lemontt and W. M. Generoso, eds.), Plenum Press, New York, pp. 121–145.

Everett, R. D., and Chambon, P. (1982) A rapid and efficient method for region- and strand-specific mutagenesis of cloned DNA. *EMBO J.* **1,** *433–437.*

Flavell, R. A., Sabo, D. L., Bandle, ER. F., and Weissmann, C. (1974) Site-directed mutagenesis: Generation of an extracistronic mutation in bacteriophage QβRNA. *J. Mol. Biol.***89,** 255–272.

Kalderon, D., Oostra, B. A., Ely, B. K., and Smith, A. E. (1982) Deletion loop mutagenesis: a novel method for the construction of point mutations using deletion mutants. *Nucleic Acids Res.* **10,** 5161–5171.

Livneh, Z. (1983) Directed mutagenesis method for analysis of mutagen specificity: Application of ultraviolet-induced mutagenesis. *Proc. Natl. Acad. Sci. USA* **80,** 237–241.

Montell, C., Fisher, E. F., Caruthers, M. H., and Berk, A. J. (1982) Resolving the functions of overlapping viral genes by site-specific mutagenesis at a mRNA splice site. *Nature (London)* **295,** 380–384.

Peden, K. W. C., and Nathans, D. (1982) Local mutagenesis within deletion loops of DNA heteroduplexes. *Proc. Natl. Acad. Sci. USA* **79,** 7214–7217.

Saffran, W. A., Goldenberg, M., and Cantor, C. R. (1982) Site-directed psoralen crosslinking of DNA. *Proc. Natl. Acad. Sci. USA* **79,** 4594–4598.

Shortle, D., and Botstein, D. (1982) Single-stranded gaps as localized targets for *in vitro* mutagenesis. In *Molecular and Cellular Mechanisms of Mutagenesis* (J. F. Lemontt and W. M. Generoso, eds.), Plenum Press, New York, pp. 147–156.

Shortle D., DiMaio, D., and Nathans, D. (1981) Directed mutagenesis. *Annu. Rev. Genet.* **15,** 265–294.

Shortle, D., Grisafi, P., Benkovic, S. J., and Botstein, D. (1982) Gap misrepair mutagenesis: Efficient site-directed induction of transition, transversion and frameshift mutations *in vitro. Proc. Natl. Acad. Sci. USA* **79,** 1588–1592.

Smith, M., and Gillam, S. (1981) Constructed mutants using synthetic oligodeoxyribonucleotides as site-specific mutagens. In *Genetic Engineering,* Vol. 3 (J. K. Setlow and A. Hollaender, eds.), Plenum Press, New York, pp. 1–32.

Weber, H., Dierks, P., Meyer, F., Van Ooyen, A., Dobkin, C., Abrescia, P., Kappeler, M., Meyhack, B., Zeltner, A., Mullen, E. E., and Weissmann, C. (1981) Modification of the rabbit chromosomal β-globin gene by restructuring and site-directed mutagenesis. In *Developmental Biology Using Purified genes* (D. D. Brown, ed.), ICN-UCLA Symposia on Molecular and Cellular Biology, Vol. XXIII, Academic Press, New York, pp. 367–385.

Weiher, H., and Schaller, H. (1981) Segment-specific mutagenesis: Extensive mutagenesis of a *lac* promotor/operator element. *Proc. Natl. Acad. Sci. USA* **79,** 1408–1412.

Zoller, M. J., and Smith, M. (1983) Oligonucleotide-directed mutagenesis of DNA fragments cloned into M13 vectors. In *Methods of Enzymology, Recombinant DNA,* Part B (R. Wu, L. Grossman, and K. Moldave, eds.), Academic Press, New York, pp. 468–500.

APPENDIX A

UV Spectral Characteristics and Acidic Dissociation Constants of Modified Bases, Nucleosides, and Nucleotides

Table A. UV Spectral Characteristics and Dissociations for Constants of Modified Bases, Nucleosides, and Nucleotides[a]

Compound	Acidic		Basic			
	λ_{max} (nm)	λ_{min} (nm)	λ_{max} (nm)	λ_{min} (nm)	pK_a	References[g]
***ADENINE** (Fig. B-1)						
Monosubstituted						
*1-Methyl- (Fig. B-2)	259	228	270	234	7.0–7.2	12, 56, 68, 153, 192, 196, 229, 230
*1-Ethyl- (Fig. B-3)	260	233	271	242	6.9, 7.0	128, 153, 201
1-Isopropyl-	259		269			118
1-(2-Hydroxyethyl)-	262	235	269	239		222
1-(2-Fluoroethyl)-	260	222	270	242		222
1-(2-Hydroxypropyl)-	259		271		7.2	117
1-(2-Hydroxyethylthioethyl)-	262		271		7.2	192
1-Benzyl-	260	235	271	242	6.5, 7.0	10, 124
1-Methoxy-					6.66 & 11.45	99
1-*N*-Oxide	258		233, 275		2.6	205, 209
2-Methyl-	267	238	270 (280)	239	≈5.1	56, 68
2-Methoxy-	239, 274	227, 250	276 (242)	253	3.6 & 10.2	99
*3-Methyl- (Fig. B-4)	274	235	273	244	5.3, 6.0, 6.1, 6.1	12, 40, 115, 117, 153, 192, 196
3-Ethyl-	274	240	273	247	6.5	116, 153, 201, 240
3-Isopropyl-	274		273			118
3-(2-Hydroxypropyl)-	274		273		6.0	117
3-(2-Hydroxyethylthioethyl)-	275		274		≈6	192
3-(2-Diethylaminoethyl)-	275	236	274	245		162
3-5′-(5-Deoxy-D-ribofuranosyl)-	274		274			76
3-Benzyl-	275		272		5.1	144, 145
*N^6-Methyl- (Fig. B-5)	267	231	273 (280)	238	4.2, 4.2	12, 40, 56, 68, 229, 230
N^6-Ethyl-	268	231	274 (281)	241		41, 201

N^6-Butyl-	270		275			41
N^6-Formyl-			280	244		187
N^6-Acetyl-	281		280			27
N^6-(2-Hydroxyethyl)-	272	233	273	236	3.7	234
N^6-(2-Carboxyethyl)-	268	236	274 (281)	240		184
N^6-(2-Diethylaminoethyl)-	275	233	274	239		162
N^6-Isopentenyl-	273	235	275	240		169
N^6-Glycyl-	275		279			27
N^6-(L-Threonyl)-	275 (245, 283)	230	275	243		182
N^6-(2-Aminoethyl)-	275		274			27
*7-Methyl- [Fig. B-6(A)]	273	237	270 (280)	230	≃3.5, 3.6	40, 145, 161, 166, 201, 207
7-Ethyl-	272	239	270 (280)	234		116, 201
7-Isopropyl-	272		272			118
8-Methoxy-	270		271			56
9-Methyl-	261	230	262	228	2.2, 2.9	29, 40, 117, 153, 196, 207, 243
9-Ethyl-	259	230	262	228	4.1	117, 142, 153, 240
9-Isopropyl-	260		262			118
9-(2-Hydroxypropyl)-	259		261			117
9-(2-Diethylaminoethyl)-	258	227	261	229		162
9-Benzyl-	259		261			145
Disubstituted						
1,N^6-Dimethyl-	261	230	273	245		16, 201, 230
1,N^6-Ethano-	262	234	274	245		222
1,7-Dimethyl-	270					217
1-Butyl,7-methyl-	268					217
1,9-Dimethyl-	260	235	260 (265)	235	9.08	16, 89, 117
1,9-Diethyl-	261		260			54
1,9-Di-(2-hydroxypropyl)-	260		260			117

Table A (continued)

Compound	Acidic		Basic			
	λ_{max} (nm)	λ_{min} (nm)	λ_{max} (nm)	λ_{min} (nm)	pK_a	References[g]
***ADENINE** *(continued)*						
1,9-Di-(2-diethylaminoethyl)-	257	233	261	232		162
1-Ethyl,9-methyl-	261		261		9.16	89
1-Propyl,9-methyl-	261		261		9.15	89
1,9-Dibenzyl-	267		271			141
1-Methoxy,9-methyl-					8.55	99
3,*N*[6]-Dimethyl-	281		287			94
3,*N*[6]-Di-(2-diethylaminoethyl)-	282	243	282	249		162
3,7-Dimethyl-	270–276	246	225, 280	221, 247	11	12, 16, 80, 117, 243
3,7-Di-(2-hydroxypropyl)-	278		281			117
3,7-Di-(2-diethylaminoethyl)-	276	237	279	245		162
3,7-Dibenzyl-	278		281		9.6	145
3,9-Dimethyl-	270, 265					52, 80
3,9-Diethyl-	271					52
3,9-Dimethyliso-	279					80
N[6],*N*[6]-Dimethyl-	276	236	282	245		41, 229
N[6],*N*[6]-Diethyl-	278		282			41
N[6]-Methyl,*N*[6]-nitroso-	216, 252, 295		225, 310		2.0	190
N[6],7-Dimethyl-	279		275			161, 217
**N*[6]-Methyl-7-ethyl- (Fig. B-7)	277 (285)	241	276	244		201
N[6]-Propyl-7-methyl-	281		277			161
N[6]-Butyl-7-methyl-	279					217
N,9-Dimethyl-	265		268		4.02	65, 89, 168, 243
N[6],9-Diethyl-	265		269			54
N[6],9-Di-(2-diethylaminoethyl)-	266	229	270	233		162
N[6]-Ethyl,9-methyl-	265		268		4.08	89, 168
N[6]-Propyl,9-methyl-	266		270		4.14	89

N^6-Butyl,9-ethyl-	266		269			89
N^6-Methoxy,9-methyl-	269		284			51
N^6-Ethoxy,9-methyl-	269		284			51
Trisubstituted						
1,N^6,N^6-Trimethyl-	221, 293	246	232, 301	262		224
2-Methyl-1,N^6-etheno-	275		224, 272, 280, 297 (232)			178
3,N^6,N^6-Trimethyl-	290	243	294	250		87, 224
3,N^6,N^6-Triethyl-	293		296			87
3-Methyl-N^6,N^6-diethyl-	294		297			87
3-Ethyl-N^6,N^6-dimethyl-	291		295			87
3-Benzyl-N^6,N^6-dimethyl-	292		296			87
3,N^6,7-Tribenzyl-	289		unstable		9.4	145
N^6,N^6,7-Trimethyl-	233, 293	250	291	246		224
N^6,N^6,7-Triethyl-	229, 300		298			87
N^6,N^6,9-Trimethyl-	269	234	277	237		87, 168, 224
N^6,N^6,9-Triethyl-	271		279			87
N^6,N^6-Dimethyl,9-ethyl-	270		277			87, 142
N^6,N^6-Dimethyl,9-benzyl-	269		278			87
N^6,N^6-Diethyl,9-methyl-	271		279			87
9-Methyl,1,N^6-etheno-	221, 276	246	227, 270, 278	248, 275		106
***GUANINE** (Fig. B-8)						
Monosubstituted						
1-Methyl-	250 (270)	228	227 (262)	242	3.1, 3.13	40, 49, 66, 68, 155, 166, 186, 203, 229
*1-Ethyl- (Fig. B-9)	251 (274)	229	278 (260)	243		109, 197
1-Isopropyl-	253					118
1-(2-Diethylaminoethyl)-	255, 275		257, 268			162
N^2-Methyl-[b]	251, 279	228	245–255, 278	238	3.3	42, 58, 68, 166, 186, 203, 229

Table A (continued)

Compound	Acidic		Basic			
	λ_{max} (nm)	λ_{min} (nm)	λ_{max} (nm)	λ_{min} (nm)	pK_a	References[g]
GUANINE *(continued)*						
*N^2-Ethyl- (Fig. B-10)	253 (280)	229, 272	≃245, 279			42, 188, 198
N^2-Isopropyl-	252		277			118
N^2-Formyl-	258	228				188
N^2-Acetyl-	260	224	268	245		188
N^2-(α-Ethoxypropionyl)-	260	225	267	232		187, 188
N^2-[2-(Ethylthio)ethyl]-	251, 280		274			174
N^2-(α-Propionamino)-	248, 282	227, 271	278 (250)		1.4	2
N^2-Benzoyl-	267 (247)	222	235, 276, 316	248, 300		188
3-Methyl-	263 (244)	227	273	246		40, 119, 186, 223
*3-Ethyl- (Fig. B-11)	262 (244)	233	273	248		198
3-Isopropyl-	263		273			118
3-Benzyl-	263 (243)	274			4.00	141
3-5′-(5′-Deoxy-D-ribofuranosyl)-	263		273			76
3-Methyliso-	285		243, 285			173
O^6-Methyl-	286		246, 284			3, 122
*O^6-Ethyl- (Fig. B-12)	286	253	284 (246)	259		3, 197
O^6-(2-Hydroxyethyl)-	287	257	284	260		131, 220
O^6-Propyl-	286		246, 283			3
O^6-Isopropyl-	285 (230)		283 (245)			118, 123
O^6-Butyl-	286	252	246, 285			3, 152
O^6-Isobutenyl-	286 (232)		283 (245)			123
7-Methyl-	249 (272)	226	281 (240)	255	3.5	13, 49, 66, 68, 73, 155, 166
*7-Ethyl-(Fig. B-13)	249 (274)	228, 268	280	258	3.[illegible]	13, 153, 197
7-Isopropyl-	249 (274)		278 (240)			118, 123

7-Butyl-	249 (273)	228, 266				152
7-Benzyl-	250		281		3.2	10
7-(2-Carboxyethyl)-	250 (270)	230	280	257		167
7-(2-Oxoethyl)-	211 (273)	229	280	258		181
7-(2-Hydroxyethyl)-	250	229	281	261		13, 130
7-(2-Fluoroethyl)-	252 (275)	228	281	257		130
7-(2-Chloroethyl)-	250 (275)	230	280	257		131, 219
7-(2-Hydroxypropyl)-	250, 272	229	280	257		117
7-(2-Aminoethyl)-	250 (270)	230	280	257		61
7-(2-Diethylaminoethyl)-	253, 270		295 (250)			162
7-[2-(Ethylthio)ethyl]-	250, 276		240, 280			174
7-(β-Hydroxyethylthioethyl)-	250		281			11
7-(Deoxygalactit-1-yl)-	250, 273	230	280	257		85
7-(1-Deoxyanhydrogalactit-1-yl)-	250, 273	230	283	256		85
7-*N*-Oxide or 7-hydroxy-	245, 267	251	227, 283	243		37
8-Propyl-	249, 276		276			123
8-Isobutyl-	249, 278		276			123
8-(3-Methylbutyl)-	249, 277		275			123
9-Methyl-	251, 276		268 (258)		2.9, 2.83	155, 186
*9-Ethyl- (Fig. B-14)	252, 277	230	253, 268	238		197
9-Isopropyl-	253, 276		256, 258			118, 123
Disubstituted						
1,N^2-Etheno-	221, 268, 295	234, 262, 318				179
1,7-Dimethyl-	252 (272)	230	284 (251)	262	3.40	155, 166, 186, 197
*1,7-Diethyl- (Fig. B-15)	252 (275)	232	285 (250)	263		194
1,9-Dimethyl-	254 (277)	229			3.28	155, 186, 207
N^2,3-Etheno-	215, 256		229, 270			179, 246
N^2,N^2-Dimethyl-[b]	255 (289)	229	277–283			42, 58, 68, 203
N^2-Methyl,N^2-nitroso-	256, 293		245, 310 (265)		2.2	190
3,O^6-Dimethyl-	284 (235)		287			223

Table A (continued)

Compound	Acidic		Basic		pK_a	References[g]
	λ_{max} (nm)	λ_{min} (nm)	λ_{max} (nm)	λ_{min} (nm)		
GUANINE (*continued*)						
3-Methyl,*O*[6]-ethyl-	284 (235)		287			223
3,7-Dimethyl-				*c*		80, 243
3,9-Dimethyl-	248		247, 265			80, 88
7,8-Dimethyl-	249, 277		280 (235)		4.4	157
7,9-Dimethyl-	254 (278)	229		*c*		94, 186
7-Methyl,9-ethyl-	254, 281			*c*	7.3	115
7,9-Di-(2-hydroxyethyl)-	254, 281	229, 271		*c*		13
7,9-Di-(2-diethylamino ethyl)-	257, 278			*c*		162
8-Methyl-1-(3-methyl-2-butenyl)-	252 (273)		277 (262)			75
8-Methyl-3-(3-methyl-2-butenyl)-	267 (245)		277			75
8-Methyl-7-(3-methyl-2-butenyl)-	247, 276		278			75
8,9-Dimethyl-	252, 277 (289)		280 (252)		4.11	157
8-Methyl-9-(3-methyl-2-butenyl)-	252, 277		269 (255)			75
Trisubstituted						
1,7,9-Trimethyl-	254, 280			*c*		186
O[6]-Methyl-*N*[2],3-etheno-	215, 264–269		255, 275			179
O[6],3,7-Trimethyl-	267			*c*		179
O[6]-Benzyl-*N*[2],3-etheno-	216, 266		225, 276			179

***CYTOSINE** (Fig. B-16)

Monosubstituted						
1-Methyl-	213, 283	241	274	250	4.55–4.61, 5.4	48, 93, 101, 207, 211, 225, 243
1-Ethyl-	254	245	274	250		197
O^2-Methyl-	260	241	270	246	5.41	210
*O^2-Ethyl- (Fig. B-17)	228, 260	213, 244, 240	223, 270	220, 243		196
3-Methyl-	273		294	251	7.38–7.49	14, 68, 207, 211, 225
*3-Ethyl- (Fig. B-18)	275	241	296	257		212
3-*N*-Oxide			221, 289			209
N^4-Methyl-	278	240	286 (230)	256	4.55	68, 214, 226, 232
N^4-Ethyl-	277	244	284	253	4.58	212, 232
N^4-Hydroxy-	216, 276	240			2.8	24, 91, 180
N^4-Methoxy-	216, 279	241				180
N^4-Butyloxy-	216, 285	249	220, 242, 300	228, 275		180
N^4-(2-Hydroxyethyl)-	280		283			225
N^4-(3-Hydroxypropyl)-	280					225
N^4-Acetyl-	305 (226)		227, 290			20, 38
N^4-(Methylcarbamoyl)-	296		290			38
N^4-Carbethoxy-	298		294			38
5-Methyl-	211, 284	242	288	254	4.6, 4.8	193, 209
5-Methylenesulfonate-	285	243			4.3	72
5-Hydroxy-	298		312			100
6-Methyl-					5.13, 4.8	149, 209
Disubstituted						
1,3-Dimethyl-	281	243	272	247	9.29–9.4	14, 101, 211, 225, 243
1,N^4-Dimethyl-	218, 285	244	274 (235)	250	4.38–4.47	101, 211, 214
1-Methyl-N^4-hydroxy-	220, 280	246			2.9	24
1-Methyl-N^4-methoxy-	221, 288	246				24
1-Methyl-N^4-amino-	213, 280	241	285		4.85	28

Table A (continued)

Compound	Acidic		Basic		pK_a	References[g]
	λ_{max} (nm)	λ_{min} (nm)	λ_{max} (nm)	λ_{min} (nm)		
CYTOSINE *(continued)*						
1-Methyl-*N*[4]-(2-hydroxy-ethyl)-	288		275			225
1,5-Dimethyl-	291	244			4.76	47
5-Methyl-*N*[4]-hydroxy-	218, 280	247			2.8	91
3,*N*[4]-Etheno-	282 (245–257)	221				98
N[4],*N*[4]-Dimethyl-	283	242	290 (235)	250	4.15, 4.25	214, 232
N[4]-Methyl,*N*[4]-acetyl-	213, 245, 309					226
N[4]-Methyl,*N*[4]-(2-hydroxy-ethyl)-			285			225
Trisubstituted						
1,3,*N*[4]-Trimethyl-	212, 287	248	280	247	9.65	101, 162, 211
1-Methyl-3,*N*[4]-etheno-	213, 290	225	214, 272	228		106
1,3-Dimethyl-*N*[4]-hydroxy-	216, 290	250			2.7	24
1,3-Dimethyl-*N*[4]-methoxy-	218, 290	250				24
1,*N*[4],*N*[4]-Trimethyl-	220, 288	248	283	242	4.2	214
1,*N*[4]-Dimethyl-*N*[4]-hydroxy-	222, 290	250			3.8	24
5-HYDROXYMETHYLCYTOSINE	279		284			229
3-Methyl-	278	242	296	256	7.1	115, 202
N[4]-Hydroxy-	222, 282	245			2.2	91
***XANTHINE** (Fig. B-19)						
Monosubstituted						
*1-Methyl-[d] (Fig. B-20)	260–265 (235)	239	283 (245)	257	1.3	126, 156, 197
3-Methyl-[d]	266–270		275 (232)		0.8	126, 156,

*7-Methyl-[d] (Fig. B-21)	267	233	289 (237)	255	0.8	66, 126, 197
7-*N*-Oxide or 7-hydroxy-	272	244	224, 297	267		37
9-Methyl-[d]	260		245, 278		2.0	126, 156
Disubstituted						
1,3-Dimethyl-	266		275		0.7	126, 156
1,7-Dimethyl-	≃260		233, 289		0.5	126, 156
1,9-Dimethyl-	262		248, 277		2.5	126, 156
3,7-Dimethyl-	265		234, 274		0.3	126, 156
3,9-Dimethyl-	265		270 (240)		1.0	126, 156
7,9-Dimethyl-	239, 262			*c*		154
8,9-Dimethyl-	238, 265		245, 278			154
Trisubstituted						
1,3,7-Trimethyl-	266				0.5	126
1,3,9-Trimethyl-	266				0.6	126
1,7,9-Trimethyl-	232, 262			*c*		154
***URACIL** (Fig. B-22)						
Monosubstituted						
1-Methyl-	208, 268	241	265	242	≃1.8	18, 97, 150, 193, 238, 241, 243
1-Ethyl-	268					241
O^2-Methyl-			263		0.7	237
*O^2-Ethyl- (Fig. B-23)	218, 255		221, 265		0.7	97, 193, 199
*3-Methyl- (Fig. B-24)	258	230	218, 283	245		68, 150, 193, 210, 236–238
3-(2-Aminoethyl)-	260		285			225
3-(Aminopropyl)-	260		284			225
O^4-Methyl-	267		276		0.8	210, 237
O^4-Ethyl-	269		220, 278		1.0	97, 193
5-Carboxymethyl-	263	232	290	246		45
5-Methylenesulfonate-			291	248		72
Disubstituted						
1,O^2-Dimethyl-	256 (215)	236	254	238	1.0	114, 210
1-Methyl-O^2-ethyl-	256 (215)		255	238		114

Table A (continued)

Compound	Acidic		Basic			
	λ_{max} (nm)	λ_{min} (nm)	λ_{max} (nm)	λ_{min} (nm)	pK_a	References[g]
URACIL *(continued)*						
1,3-Dimethyl-	266	234	266	234	−2.07	65, 97, 150, 193, 243
1,O^4-Dimethyl-	272				0.65	97, 210
1-Methyl,O^4-ethyl-	278	241	273	240		114
1,6-Dimethyl-	208, 268	234	266	241		236
O^2,3-Dimethyl-	213, 269				0.7	210, 237
3,O^4-Dimethyl-					2.0	210
3,6-Dimethyl-	205, 259	231	281	245		236
5,6-Dimethyl-	267	236	275	245		236
Trisubstituted						
1,5,6-Trimethyl-	206, 276	240	273	245		236
3-Methyl-5-(*N*-formylmethyl-amino)-6-amino-	261	235	263	240		35
***THYMINE** (Fig. B-25)						
Monosubstituted						
1-Methyl-			270	244		233, 237, 243
1-Ethyl-			273			241
1-(2-Diethylaminoethyl)-			265	248		162
O^2-Methyl-	217, 260					237
*O^2-Ethyl- (Fig. B-26)	216, 259	210, 235	269	243	0.8	199
3-Methyl-	266	237	290	248		49, 197, 233, 243
*3-Ethyl- (Fig. B-27)	265	237	289	247		197
3-(2-Diethylaminoethyl)-			288	244		162
O^4-Methyl-	274		283			237

1-Methyl-O^2-ethyl-	262 (215)	237	259	239		114
1,3-Dimethyl-	272					21, 52, 243
1,3-Di-(diethylamino-ethyl)-			269	245		162
1,O^4-Dimethyl-			280			237
1-Methyl-O^4-ethyl-	279	241	278	241		114
1,O^4-Di-(2-diethylamino-ethyl)-			274	245		162
O^2,3-Dimethyl-	217, 272					237, 238
*O^2,O^4-Diethyl- (Fig. B-28)	210, 263	207, 235				197
***HYPOXANTHINE** (Fig. B-29)						
Unmodified	248		263		1.98	40
Monosubstituted						
1-Methyl-	249		260			40, 140
3-Methyl-	253		226, 265		2.61	5, 40, 140
3-Ethyl-	254	266				164
3-Benzyl-	254		264 (277)			143
3-(2-Diethylaminoethyl)-	260	234	262	243		162
O^6-(3-Methyl,2-butenyl)-	247		262			123
7-Methyl-	250		262		2.12	40
9-Methyl-	250		254			40
Disubstituted						
1,7-Dimethyl-	252	232	267	237		137
1,7-Dibenzyl-	255		256			143
1,9-Dibenzyl-	263		259			143
3,7-Dimethyl-			267			16
3,7-Dibenzyl-	256		267			143
N^6,7-Dimethyl-	256		258			161
7,9-Dimethyl-	251			*c*		94
Trisubstituted						
7,9-Dimethyl,O^6-methoxy-	228 (250)					239
***ADENOSINE**[e] (Fig. B-30)						
Monosubstituted						
*1-Methyl- (Fig. B-31)	257	231	258 (265, 300)	233	8.2, 8.3, 8.8	43, 56, 67, 68, 116, 201

Table A (continued)

Compound	Acidic		Basic		pK_a	References[g]
	λ_{max} (nm)	λ_{min} (nm)	λ_{max} (nm)	λ_{min} (nm)		
ADENOSINE *(continued)*						
1-Methyldeoxy-	257	239	258	242		30, 201
2′-O,1-Dimethyl-	258	243	259 (266)	244		170
*1-Ethyl- (Fig. B-32)	259	235	261 (268, 300)	237		201
1-Ethyldeoxy-	259	231	260 (268)	236		201
1-(2-Chloroethyl)-	259	234	259	236		222
1-(2-Hydroxyethyl)-	258	233	259	232	8.3	222, 234
1-(2-Carboxyethyl)deoxy-	260	236	260 (268, 300)	232	≃9	26, 184
1-[2-(Ethylthio)ethyl]deoxy-	260		260 (268)			174
1-(2-Fluoroethyl)-	260	235	260	236		222
1-Benzyldeoxy-	259		259			10
1-*N*-Oxide	258		230, 272, 312		2.14	205, 206
2-Methyl-	258	230	264	227		56, 68, 176
2-Hydroxy- (Isoguanosine)	235, 283	252	251, 285	235, 265		34, 50
2-Methoxy-						32
2-Amino-	252, 291				4.3	79
3-Methyl-	270 (pH 7)					175
3-Methyldeoxy-	271 (pH 7)					53
*N^6-Methyl- (Fig. B-33)	262	231	266	223	4.0	12, 56, 68, 92, 95, 201, 230
2′-*O*,N^6-Dimethyl-	263	232	266	231		151, 170
N^6-Methyldeoxy-	262	231	266	226		30, 68, 95 201
*N^6-Ethyl- (Fig. B-34)	264	239	268	243		201
N^6-Ethyldeoxy-	263	237	268	241		201
N^6-Butyl-	263		267			46
N^6-Hydroxy-	266	236	unstable			19, 22, 23

N^6-Methoxy- (see Fig. B-64)	268	237	283	237		23
N^6-[Ethoxymethyl (2′,3′-*O*-isopropylidine)]-	263 (EtOH)					7
N^6-(2-Hydroxyethyl)-	263	233	267	232	3.1	234
N^6-(2-Carboxyethyl)deoxy-	266	236	269	236		184
N^6-Isopentenyl- (i^6A) (see Fig. B-63)	265	232	269	234	3.8	169, 171
N^6-[2-(Ethylthio)ethyl]deoxy-	263		268			174
N^6-Benzyl-	265	235	268	236		105, 143
N^6-Benzyldeoxy-	264		268			10
N^6-Carbamoyl-	276	241	296	245		185
N^6-(Methylcarbamoyl)-	280	245	284	296		185
N^6-(*N*-Threonylcarbonyl)- (t^6A)	271, 278		271, 279, 300			6
*7-Ethyl- [Fig. B-35(A)]	268	239		*c*		201
8-Methyl-	261		261			82
8-Methoxy-	261		259		3.85	56
8-(2-Hydroxy-2-propyl)-	258		260			204
2′-*O*-Methyl-[e]	257		259			15, 133, 244
3′-*O*-Methyl-[e]	260 (95% EtOH)	228				133
Disubstituted						
1-Methyl,2-hydroxy-	237, 283		253, 292			32
*1,N^6-Dimethyl- (Fig. B-36)	261	234	263 (300)	234		16, 201, 230
1-Methyl,N^6-(methylcarbamoyl)-	286	247	272	241		185
1,N^6-Ethano-	262	234	274 (310)	245		222
1,N^6-Etheno- (see Fig. B-65)	273	245	228, 265, 275 (258, 294)	250, 272	3.9, 4.0–4.2	90, 98, 183
3,5′-Cyclo-	272					29
N^6,N^6-Dimethyl-	268	233	276	237	3.6, 3.83, 4.5	43, 56, 68, 141, 151, 224
2′-*O*,N^6,N^6-Trimethyl-	268	235	276	239		170
3′-*O*,N^6,N^6-Trimethyl-	268	235	275	240		170

Table A (continued)

Compound	Acidic λ_{max} (nm)	Acidic λ_{min} (nm)	Basic λ_{max} (nm)	Basic λ_{min} (nm)	pK_a	References[g]
ADENOSINE *(continued)*						
N^6-Methyl,N^6-(*N*-threonyl)-carbonyl)- (mt^{6}A)	277, 284		277, 285			6
N^6-Ethyl,N^6-(*N*-threonyl-carbonyl)-	279, 285		280, 286			6
*N^6,7-Dimethyl- (Fig. B-37)	276	241	*c*			201
N^6-Methyl,7-ethyl-	276	242	*c*			201
***GUANOSINE**[e] (Fig. B-38)						
Monosubstituted						
*1-Methyl- (Fig. B-39)	258 (280)	230	255 (270)	228	1.9, 2.6	16, 43, 68, 203, 229
1-Methyldeoxy-	257 (278)	232	255 (270)	229		16, 44, 49
1-Ethyl-	261 (272)	232	258 (270)	239	2.8	194
1-Ethyldeoxy-	256		257			44
1-Butyldeoxy-	258 (282)		257 (280)			44
1-Benzyl-	258	231	256	229		158
N^2-Methyl-[b]	251–258 (280–290)	222–234	248–258 (270–275)	227–238		58, 68, 151, 203
N^2-Formyldeoxy-	250 (275)	226				187
N^2-[Ethoxymethyl-(2′,3′-*O*-isopropylidine)]-	255 (EtOH)					7
N^2-[2-(Ethylthio)ethyl]-deoxy-	257 (285)		258(274)			174
N^2-Carbamoyl-	256	230	255–270	234		185
N^2-Benzyl-	259 (282)		257 (270)			146
3-Methyl-	216, 243, 263 (H_2O)		250, 265			88, 148
O^6-Methyl-	284 (243)	259	243, 277	239, 261	2.4	194
O^6-Methyldeoxy-	284 (230)	252	243, 278	233, 261		49, 127

O^6-Ethyl-	244, 286	239, 260	247, 278	233, 261	2.5	194
*O^6-Ethyldeoxy- (Fig. B-40)	244, 288	234, 262	248, 280	237, 262		44, 197
O^6-(2-Hydroxyethyl)-	244, 288	232, 261	248	232		131, 220
O^6-Butyldeoxy-	246, 287	260	248, 280	261		44
7-Methyl-	257 (275)	230	*c*		6.7–7.3	68, 73, 95, 115, 194
7-Methyldeoxy-	256 (275)	229	*c*			44, 49, 66, 95
7-Ethyl-	258 (277)	238	*c*		7.2, 7.4	115, 194
*7-Ethyldeoxy- [Fig. B-41(A)]	257 (275)	229	*c*			44, 197, 199
7-Isopropyl-	256 (275)		*c*			123
7-Butyldeoxy-	257 (280)		*c*			44
7-(2-Oxoethyl)deoxy-	256 (280)	240	*c*			181
7-(2-Hydroxyethyl)-	258 (280)	232	*c*			130
7-(2-Fluoroethyl)-	258 (283)	232	*c*			130
7-(2-Chloroethyl)-	258	232	*c*			131
7-[2-(Ethylthio)ethyl]deoxy-	253 (280)		*c*			174
7-Benzyl-	258		*c*		7.2	10
7-(2-Aminoethyl)-	258	232	*c*			61
8-Methyl-	260 (273)		256		3.01	157
2′-*O*-Methyl[e]						172
3′-*O*-Methyl[e]						172
Disubstituted						
1,N^2-Dimethyldeoxy-	259		256			242
1,N^2-Etheno-	222, 272, 295		233, 280, 308			179
1,7-Dimethyl-	260 (270)	236	*c*			118, 194
*1,7-Diethyl- [Fig. B-42(A)]	263 (270)	237	*c*			194
1-Methyl,7-ethyl-	259 (277)	233	*c*			194
1-Ethyl,7-methyl-	261 (275)	235	*c*			194
N^2,N^2-Dimethyl-	265 (290)	237	262 (283)	240		58, 68, 203, 229
N^2,O^6-Dimethyldeoxy-	288		249, 284			44
*N^2,O^6-Diethyl- (Fig. B-43)	246, 292	239, 267	252, 281	237, 268		194, 197
3,5′-Cyclo-	247		266			76

Table A (continued)

Compound	Acidic		Basic			
	λ_{max} (nm)	λ_{min} (nm)	λ_{max} (nm)	λ_{min} (nm)	pK_a	References[g]
GUANOSINE *(continued)*						
Trisubstituted						
Δ^4-Imidazoline- (α)	225, 275 (300)		236, 283, 310			60
N^2,N^2,7-Trimethyl-	267, 300	239, 286	*c*			177
***CYTIDINE**[e] (Fig. B-44)						
Monosubstituted						
O^2-Methyl-	233, 262	221, 243	unstable		>9.5	197
*O^2-Ethyl- (Fig. B-45)	233, 262	221, 243	unstable		>9.5	195, 199
O^2-Ethyldeoxy-					>9.5	195, 196, 199
3-Methyl-	278	243	225, 267	212, 244	8.3–8.9	14, 67, 68, 93, 111, 151, 212, 214
3-Methyldeoxy-	280					242
*3-Ethyl- (Fig. B-46)	280	247	267	248	8.4	212
3-Ethyldeoxy-	280	245	268	247	8.6	212
3-(2-Hydroxyethyl)-	280	246	268	245		131
3-(2-Fluoroethyl)-	280	243	268	247		131, 221
3-(2-Diethylaminoethyl)-deoxy-	284	243	271			162
3-Benzyl-	281	242	267	244	7.7, 8.1	10, 103
N^4-Methyl-	217, 281	207, 243	237, 273	250	3.85, 3.92	214, 232
N^4-Methyldeoxy-	282	242	236, 270	229, 248	4.01	232, 242
2′-*O*,N^4-Dimethyl-	281	242	271	250		171
2′,3′,5′-Tri-*O*-methyl-N^4-methyl-	281	242	271	247		113
*N^4-Ethyl- (Fig. B-47)	281	244	272	253	4.2	212
N^4-Ethyldeoxy-	279	247	272	253	4.2	212

N^4-(2-Hydroxyethyl)-	283	245	238, 271	232, 250		131
N^4-Hydroxy- (see Fig. B-66)	216, 276	240			2.8	17
N^4-Methoxy-	216, 279	241				17
N^4-Benzyl-	285	249	238, 273	228, 247	3.5	103, 191
N^4-[Ethoxymethyl-(2′,3′-*O*-isopropylidine)]-	247, 272 (EtOH)					7
N^4-Acetyl-	245, 300		unstable			227
N^4-Phenyl-	294 (225)	249	292 (235)	248	3.2	189
N^4-Hydroxyphenyl-	290	247	234, 276 (325)	225, 256	3.4	189
N^4-Carbamoyl-	300	262	294	246		185
N^4-Methylcarbamoyl-	297 (240)		291			185
N^4-β-Naphthyl-	288	253				189
5-Methyl-	288	245	278	255	4.28	47
5-Methyldeoxy-	287	246	278	255	4.40	36, 47
5-Ethyldeoxy-			278	255		110
6-Methyl-	278	241	273	252	4.42	149, 235
6-Methyldeoxy-	278	241	273	252		235
2′-*O*-Methyl[e]	280	241	271	250		113, 133, 172
2′-*O*-Ethyl[e]	279	240	270	249		165
2′-*O*-Benzyl-[e]	280	241				103
3′-*O*-Methyl-[e]	280	240	272	250		55, 113, 133, 172
5′-*O*-Methyl-	280	241	271	249		113
5′-*O*-Ethyl-	280	241	271	250		113
Disubstituted						
1-Methyl,N^4-amino-	213, 280	241			4.85	17
O^2,2′-Cyclo-	232, 262	218, 243 (282)	unstable			96, 102, 231
O^2,5′-Cyclo-		*c*	*c*			29
3,N^4-Dimethyl-	285	249	276	249		212
*3,N^4-Diethyl- (Fig. B-48)	287	252	277	253		212
3,N^4-Di-(2-diethylamino-ethyl)deoxy-	284	245				162
3,N^4-Ethano	286	241	228, 282	255		25, 108, 147
3,N^4-Etheno- (see Fig. B-67)	288 (248, 302)		272 (281, 292)		3.7, 4.0	4, 90

Table A (continued)

Compound	Acidic		Basic		pK_a	References[g]
	λ_{max} (nm)	λ_{min} (nm)	λ_{max} (nm)	λ_{min} (nm)		
CYTIDINE *(continued)*						
N[4],*N*[4]-Dimethyl-	219, 285	245	279	238	3.7, 3.62	93, 214, 232
N[4],*N*[4]-Dimethyldeoxy-	287	245	278	238	3.79	232
**N*[4],*N*[4]-Diethyl- (Fig. B-49)	286	249	276	249		212
2′-*O*,*N*[4],*N*[4]-Trimethyl-	287	246	278	238		171
N[4],5-Dimethyl-			275 (234)	252		47
N[4],5-Dimethyldeoxy-	218, 287	246	275 (235)		4.04	47
Trisubstituted						
*3,*N*[4],*N*[4]-Triethyl- (Fig. B-50)	287	252	289	253		212
***URIDINE**[e] (Fig. B-51)						
Monosubstituted						
O[2]-Methyl-	223, 252	212, 236	252	237		20, 104, 114
**O*[2]-Ethyl- (Fig. B-52)	224, 253	212, 236	225, 253	238	0.5	104, 114
O[2]-Ethyldeoxy-					0.92	199
3-Methyl-	263	233	263	233		68, 115, 139, 151, 212
2′-*O*,3-Dimethyl-	261	234	261	234		55
3′-*O*,3-Dimethyl-	261	234	261	234		55
*3-Ethyl- (Fig. B-53)	262	235	264	237		114, 212
3-(2-Hydroxyethyl)-	261	235	262			78
3-Benzyl-	263	235	264	235		84, 158
**O*[4]-Methyl- (Fig. B-54)	275	235	275	239		114, 171
**O*[4]-Ethyl- (Fig. B-55)	274	237	274	239		114
O[4]-Ethyldeoxy-	215, 260	205, 240			0.66	199
5-Propyl-	268		270			107
5-Isopropyl-	268		270			107
5-Carboxymethyl	265	232	265	244		45
5-Methoxy-	280	243	278	250		1, 74

6-Methyl-	261	230	264	242		77, 215, 235, 236
6-Ethyl-	262					77, 215
6-Ethyldeoxy-	262					77
2′-*O*-Methyl-	261	231	262	244		55, 113, 133
3′-*O*-Methyl-	261	230	262	243		55, 113, 133
5′-*O*-Methyl-	262	230	262	242		113
5′-*O*-Ethyl-	262	230	262	242		113
Disubstituted						
O^2,2′-Cyclo-	221, 250	234	224, 253	237		125
O^2,3′-Cyclo-	230, 246	240	224, 253	240		125
O^2,5′-Cyclo-	252	224	253	226		20, 125
5,6-Dimethyl-	268	236	271	247		235
***THYMIDINE** (deoxy-)[e] (Fig. B-56)						
Monosubstituted						
O^2-Methyl-	228, 254	215, 234	255	237		114
*O^2-Ethyl- (Fig. B-57)	228, 255	217, 236	256	237	0.5	114, 199
O^2-Isopropyl-	229, 258	215, 238				112
3-Methyl-	266	238	267	239		49, 114
*3-Ethyl- (Fig. B-58)	269	239	270	240		114
3-(2-Diethylaminoethyl)-			270	242		162
O^4-Methyl-	279	239	280	241		114, 120
*O^4-Ethyl- (Fig. B-59)	279	241	279	241	−0.32	114, 199
O^4-Isopropyl-			281	242		112
2′-*O*-Methyl-ribo	266	237	266	248		64
Disubstituted						
O^2,3′-Cyclo-	254	220	254	219		138
O^2,5′-Cyclo-	251	219	252	219		125, 138
***INOSINE**[e] (Fig. B-60)						
Monosubstituted						
1-Methyl-	250	223	249		1.1	68, 140, 163
1-Methyldeoxy-	250		250 (265)			49
1-(2-Hydroxyethyl)-	250	226	250	226		78

Table A (continued)

Compound	Acidic		Basic			
	λ_{max} (nm)	λ_{min} (nm)	λ_{max} (nm)	λ_{min} (nm)	pK_a	References[g]
INOSINE *(continued)*						
1-Benzyl-	251		249			143
2-Methyl-	253		258			81
3-Methyl-	259 (H_2O)					86
O[6]-Methyl-	250		250			92, 140
*7-Methyl- (Fig. B-61)	252	221	*c*		6.4	137, 197
7-Ethyl-	252		*c*			164
8-Methyl-			250 (H_20)			82
2′-*O*-Methyl-[e]	248		251			15
Disubstituted						
1,7-Dimethyl-	265		*c*			137
***XANTHOSINE** (Fig. B-62)						
7-Methyl-	262	237	*c*			66, 197
3,5′-Cyclo-	240, 265		267			76
ADENYLIC ACID OR ADP[e,f]						
1-Methyl-	258	232	259 (268)	230		63, 192, 197
1-(2-Hydroxyethylthioethyl)-	261		261			192
1-*N*-Oxide						33
2-Hydroxy- (iso-GMP)	238, 280	253	250, 285	235, 267		59
2-Methyl-	259		263			176
N[6]-Methyl-	261	231	265	229		63, 192
N[6]-(2-Hydroxyethyl)-	263	233	266	230	3.5	136, 228
N[6]-(2-Hydroxyethylthio-ethyl)-			268			192
**N*[6]-Isopentenyl- (Fig. B-63)	264	233	268	235		218
**N*[6]-Methoxy- (Fig. B-64)	268	233	283	243		25, 200
8-Oxy-	265, 286		281			83

1,N^6-Etheno- (Fig. B-65)	220, 275	205, 245	228, 265, 275 (258)	230, 272		90, 197
2,N^6-Dimethyl-	263		269			70
8,5′-Cyclo-	263		267			208
GUANYLIC ACID OR GDP[e,f]						
1-Methyl-	258 (280)	230	256 (273)	230	<3	159, 229
N^2-Methyl-	263	237	263	240		160
N^2-(α-Ethoxypropionyl)-	261	226	unstable			187
O^6-Methyl-	245, 288	262	249, 281	263		57, 134
7-Methyl-	259, 279	230	*c*		6.9–7.2	73, 121
7-(2-Hydroxyethyl)-	259	232	*c*			129
8-Oxy-	263 (293)		273			83
8-Amino-	251, 290	230	260 (280)		4.7	69
N^2,N^2-Dimethyl-	265	232	263	237	≃3	62, 160, 229
N^2,N^2,7-Trimethyl-	262 (290)	237	*c*		7.4	160
CYTIDYLIC ACID OR CDP[e,f]						
3-Methyl-	276	242	223, 266	243	9.0, 9.2	8, 9, 132
3-Methyldeoxy-	279	245	267	246		71
3-(2-Hydroxyethyl)-	281	245	224, 267	245		108
N^4-Methyl-	217, 281	242	271	249	4.25	8, 9
*N^4-Hydroxy- (Fig. B-66)	222, 280	208, 243	240			24, 197
N^4-Methoxy-	220, 280	250	255			24, 197
N^4-Acetyl-	244, 302	225, 270	298	237		135
5-Methyl-	284		279			213
5-Methyldeoxy-	287	243	277	254	4.4, 4.5	31, 245
5-Ethyl			278	254		111
5-Ethyldeoxy-			278	255	4.65	110
5-Hydroxy-	216, 306	266	225, 320	275	3.8	39
6-Methyldeoxy-	278	241	272	252		235
O^2,2′-Cyclo-	232, 262	243	unstable		6.6	231
N^3-Methyl-N^4-methoxy-	225, 289		230, 282			24
N^4,N^4-Dimethyl-	219, 287	245	225, 278	238	4.0	8, 9
N^4,5-Dimethyldeoxy-			275 (240)	253		245
N^4-Ethyl-5-methyldeoxy-			278 (237)	255		245
*3,N^4-Etheno- (Fig. B-67)	288 (250)	228	273 (292)	234		90, 197

Table A (continued)

Compound	Acidic		Basic		pK_a	References[g]
	λ_{max} (nm)	λ_{min} (nm)	λ_{max} (nm)	λ_{min} (nm)		
CYTIDYLIC *(continued)*						
3,*N*[4]-Ethano-	286	241	228, 282	255		108
URIDYLIC ACID OR UDP[e,f]						
3-Methyl-	261	235	262	235		136, 197
3-Ethyl-	261		263			197
3-(2-Hydroxyethyl)-	263	234	263	234		108
3-(3-Amino-3-carboxypropyl)-	262	234	263	236		62
*5-Hydroxy- (Fig. B-68)	280	243	285 (305)	265		197
THYMIDYLIC ACID (deoxy-)[f]						
*3-Methyl- (Fig. B-69)	267	240	268	241		197
3-Ethyl-	267	241	268	242		197
INOSINIC ACID OR IDP[e,≈]						
2-Methyl-	253		258			81
2′-*O*-Methyl-[e]	248					216

* Spectra for starred compounds appear in Appendix B.

[a] The λ_{max} (nm), in those cases where more than one value has been reported, are either the most frequent value or an average of several values, the range being ±1–2 nm for the λ_{max}. Since the λ_{min} is more sensitive than the λ_{max} to impurities in the sample, the values of λ_{min} in the table are generally the lowest reported. Values in parentheses are shoulders or inflections. The cationic and anionic forms are either so stated by the authors or are arbitrarily taken at pH 1 and pH 13. The few exceptions are noted in the table. Individual values are given for pK_a except when there are multiple values that are similar. In that case, a range is given.

Complete spectra representing a range of derivatives are shown in Appendix B and reference to these is made in the table with an asterisk preceding the name of the compound. All spectra were obtained in the authors' laboratory from samples isolated from paper or thin-layer chromatograms or after HPLC separation. It is recognized that pH 1 or pH 13 are not ideal for obtaining the cationic or anionic forms when these pHs are close to a pK. Nevertheless, these conditions are useful for purposes of identification since the spectra are reproducible.

Additional data not here quoted are available in many of the references. These data include spectral characteristics in other solvents than H_2O and at other pH values, extinction coefficients, R_f values in various paper chromatographic systems, column chromatographic systems, methods of synthesis or preparation of derivatives, mass spectra, and NMR, optical rotatory dispersion, and infrared spectra.

The table includes primarily nucleic acid derivatives with aliphatic substituents and excludes those resulting from halogenation, thiolation, or reaction with bulky, aromatic mutagens or carcinogens.

[b] N^2-Alkyl guanines and guanosines do not exhibit sharp maxima or minima, particularly in basic solution, as shown in the spectra published by Hall (68), Smith and Dunn (203), and Singer and Fraenkel-Conrat (198). Therefore, some of the data are given as a range of values (see Figure B-10).

[c] All 7-alkyl purine nucleosides and nucleotides and 7,9-dialkyl purines are unstable in alkali and the imidazole ring opens at varying rates. For this reason spectral data obtained in alkaline solution do not represent the original compound and thus such data are omitted. The opening of the imidazole ring in alkali can be used as a means of identifying this class of alkyl compounds, and Figures 35, 37, 41, and 42 in Appendix B illustrate this point. Ring opening can lead to a number of derivatives (198) and the spectrum of the base from a major product of ring-opened 7-methyladenosine is shown in Figure B-6.

[d] Basic values are those of the dianion (pH 14).

[e] Alkylation of ribose does not cause any change in spectrum.

[f] Alkylated nucleoside diphosphates have the same spectral characteristics as alkylated nucleotides and the data are not separated. Alkylation of the phosphate group does not cause any change in spectrum. Figure 69 is in all respects identical to the spectra of 3-methylthymidylic acid (not shown).

[g] References to Appendix A:

1. Albani, Schmidt, Kersten, Geibel, and Luderwald, *FEBS Lett.* **70,** 37 (1976).
2. Al-Khalidi and Greenberg, *J. Biol. Chem.* **236,** 192 (1961).
3. Balsiger and Montgomery, *J. Am. Chem. Soc.* **25,** 1573 (1960).
4. Barrio, Secrist, and Leonard, *Biochem. Biophys. Res. Commun.* **46,** 597 (1972).
5. Bergmann, Levin, Kalmus, and Kwietny-Govrin, *J. Am. Chem. Soc.* **26,** 1504 (1961).
6. Boryski and Golankiewicz, *J. Carbohydr. Nucleosides Nucleotides* **6,** 497 (1979).
7. Bridson, Jiricny, Kemal, and Reese, *J. Chem. Soc. Chem. Comm.*, 208 (1980).
8. Brimacombe, *Biochim. Biophys. Acta* **142,** 24 (1967).
9. Brimacombe and Reese, *J. Chem. Soc. C*, 588 (1966).
10. Brookes, Dipple, and Lawley, *J. Chem. Soc. C*, 2026 (1968).
11. Brookes and Lawley, *Biochem. J.* **77,** 478 (1960).
12. Brookes and Lawley, *J. Chem. Soc.*, 539 (1960).
13. Brookes and Lawley, *J. Chem. Soc.*, 3923 (1961).
14. Brookes and Lawley, *J. Chem. Soc.*, 1348 (1962).
15. Broom and Robins, *J. Am. Chem. Soc.* **87,** 1145 (1965).
16. Broom, Townsend, Jones, and Robins, *Biochemistry* **3,** 494 (1964).
17. Brown, Hewlins, and Schell, *J. Chem. Soc. C*, 1925 (1968).
18. Brown, Hoerger, and Mason, *J. Chem. Soc. (London), 211 (1955).*
19. Brown and Osborne, *Biochim. Biophys. Acta* **247,** 514 (1971).
20. Brown, Todd, and Varadarajan, *J. Chem. Soc. (London)*, 868 (1957).
21. Bryant and Klein, *Anal. Biochem.* **65,** 73 (1975).
22. Budowsky, Sverdlov, and Monastyrskaya, *J. Mol. Biol.* **44,** 205 (1969).

Table A (continued)

23. Budowsky, Sverdlov, and Monastyrskaya, *Biochim. Biophys. Acta* **246,** 320 (1971).
24. Budowsky, Sverdlov, Shibaeva, Monastyrskaya, and Kochetkov, *Biochim. Biophys. Acta* **246,** 300 (1971).
25. Budowsky, Sverdlov, Spaskukotskaya, and Koudelka, *Biochim. Biophys. Acta* **390,** 1 (1975).
26. Chen, Mieyal, and Goldthwait, *Carcinogenesis* **2,** 73 (1981).
27. Chheda and Hall, *Biochemistry* **5,** 2082 (1966)
28. Chu, Brown, and Burdon, *Mutat. Res.* **23,** 267 (1974).
29. Clark, Todd, and Zussman, *J. Chem. Soc. (London)*, 2952 (1951).
30. Coddington, *Biochim. Biophys. Acta* **59,** 472 (1962).
31. Cohn, *J. Am. Chem. Soc.* **73,** 1539 (1951).
32. Cook, Bartlett, Green, and Quinn, *J. Org. Chem.* **45,** 4020 (1980).
33. Cramer, Randarath, and Schater, *Biochim. Biophys. Acta* **72,** 150 (1963).
34. Davoll, *J. Am. Chem. Soc.* **73,** 3174 (1951).
35. De, Mittelman, Jenkins, Crain, McCloskey, and Chheda, *J. Carbohydr. Nucleosides Nucleotides* **7,** 113 (1980).
36. Dekker and Elmore, *J. Chem. Soc.*, 2864 (1951).
37. Delia and Brown, *J. Org. Chem.* **31,** 178 (1966).
38. Dutta and Chheda, *J. Carbohydr. Nucleosides Nucleotides* **7,** 217 (1980).
39. Eaton and Hutchinson, *Biochim. Biophys. Acta* **319,** 281 (1973).
40. Elion, *J. Org. Chem.* **27,** 2478 (1962).
41. Elion, Burgi, and Hitchings, *J. Am. Chem. Soc.* **74,** 411 (1952).
42. Elion, Lange, and Hitchings, *J. Am. Chem. Soc.* **78,** 217 (1956).
43. Evans and Sarma, *J. Mol. Biol.* **89,** 249 (1974).
44. Farmer, Foster, Jarman, and Tisdale, *Biochem. J.* **135,** 203 (1973).
45. Fissekis and Sweet, *Biochemistry* **9,** 3136 (1970).
46. Fleysher, *J. Med. Chem.* **15,** 187 (1972).
47. Fox, Praag, Wempen, Doerr, Cheong, Knoll, Eidinoff, Bendich, and Brown, *J. Am. Chem. Soc.* **81,** 178 (1959).
48. Fox and Shugar, *Biochim. Biophys. Acta* **9,** 369 (1952).
49. Friedman, Mahapatra, Dash, and Stevenson, *Biochim. Biophys. Acta* **103,** 286 (1965).
50. Fuhrman, Fuhrman, Nachman, and Mosher, *Science* **212,** 557 (1981).
51. Fujii, Itaya, Wu, and Tanaka, *Tetrahedron* **27,** 2415 (1971).
52. Fujii, Saito, and Kawanishi, *Tetrahedron Lett.* **No. 50,** 5007 (1978).
53. Fujii, Saito, and Nakasaka, *J. Chem. Soc. Chem. Comm.*, 758 (1980).
54. Fujii, Sakamoto, Kawakatsu, and Itaya, *Chem. Pharm. Bull.* **24,** 655 (1976).
55. Furukawa, Kobayashi, Kanai, and Honjo, *Chem. Pharm. Bull.* **13,** 1273 (1965).
56. Garrett and Mehta, *J. Am. Chem. Soc.* **94,** 8532 (1972).
57. Gerchman, Dowbrowski, and Ludlum, *Biochim. Biophys. Acta* **272,** 672 (1972).

58. Gerster and Robins, *J. Am. Chem. Soc.* **87,** 3752 (1965).
59. Golas, Fikus, Kazimierczuk, and Shugar, *Eur. J. Biochem.* **65,** 183 (1976).
60. Goldschmidt, Blazej, and Van Duuren, *Tetrahedron Lett.* **No. 13,** 1583 (1968).
61. Gombar, Tong, and Ludlum, *Biochem. Pharmacol.* **29,** 2639 (1980).
62. Gray, *Can. J. Biochem.* **54,** 413 (1976).
63. Griffin and Reese, *Biochim. Biophys. Acta* **68,** 185 (1963).
64. Gross, Simsek, Raba, Limberg, Heckman, and RajBhandary, *Nucleic Acids Res.* **1,** 35 (1974).
65. Gukovskaya, Chervin, Sukhorukov, and Antonovskii, *Biofizika* 18, [illegible]57 (1973).
66. Haines, Reese, and Todd, *J. Chem. Soc.*, 5281 (1962).
67. Haines, Reese, and Todd, *J. Chem. Soc.*, 1406 (1964).
68. Hall, *The Modified Nucleosides in Nucleic Acids,* Columbia University Press, New York (1971).
69. Hattori, Frazier, and Miles, *Biochemistry* **14,** 5033 (1975).
70. Hattori, Ikehara, and Miles, *Biochemistry* **13,** 2754 (1974).
71. Hayashi, Yamauchi, and Kinoshita, *Bull. Chem. Soc. Jpn.* **53,** 277 (1980).
72. Hayatsu and Shiragami, *Biochemistry* **18,** 632 (1979).
73. Hendler, Furer, and Srinivasan, *Biochemistry* **9,** 4141 (1970).
74. Hillen and Gassen, *Biochim. Biophys. Acta* **562,** 207 (1979).
75. Holmes and Leonard, *J. Org. Chem.* **41,** 568 (1976).
76. Holmes and Robins, *J. Org. Chem.* **28,** 3483 (1963).
77. Holy, *Collect. Czech. Chem. Commun.* **39,** 3374 (1974).
78. Holy, Bald, and Hong, *Collect. Czech. Chem. Commun.* **36,** 2658 (1971).
79. Howard, Frazier, and Miles, *Biochemistry* **15,** 3783 (1976).
80. Ienaga and Pfleiderer, *Tetrahedron Lett.* **No. 16,** 1447 (1978).
81. Ikehara and Hattori, *Biochim. Biophys. Acta* **272,** 27 (1972).
82. Ikehara, Limn, and Fukui, *Chem. Pharm. Bull.* **25,** 2702 (1977).
83. Ikehara, Tazawa, and Fukui, *Biochemistry* **8,** 736 (1969).
84. Imura, Tsuruo, and Ukita, *Chem. Pharm. Bull.* **16,** 1105 (1968).
85. Institonis and Tamas, *Biochem. J.* **185,** 659 (1980).
86. Itaya and Matsumoto, *Tetrahedron Lett.* **No. 42,** 4047 (1978).
87. Itaya, Matsumoto, and Ogawa, *Chem. Pharm. Bull.* **28,** 1920 (1980).
88. Itaya and Ogawa, *Tetrahedron Lett.* **No. 32,** 2907 (1978).
89. Itaya, Tanaka, and Fujii, *Tetrahedron* **28,** 535 (1972).
90. Janik, Sommer, Kotick, Wilson, and Erickson, *Physiol. Chem. Phys.* **5,** 27 (1973).
91. Janion and Shugar, *Acta Biochim. Pol.* **12,** 337 (1965).
92. Johnson, Thomas, and Schaeffer, *J. Am. Chem. Soc.* **80,** 699 (1958).
93. Johnson, Vipond, and Girod, *Biopolymers* **10,** 923 (1971).
94. Jones and Robins, *J. Am. Chem. Soc.* **84,** 1914 (1962).

Table A (continued)

95. Jones and Robins, *J. Am. Chem. Soc.* **85,** 193 (1963).
96. Kanai, Kojima, Maruyama, and Ichino, *Chem. Pharm. Bull.* **18,** 2569 (1970).
97. Katritzky and Waring, *J. Chem. Soc. (London)*, 1540 (1962).
98. Kayasuga-Mikado, Hashimoto, Negishi, Negishi, and Hayatsu, *Chem. Pharm. Bull.* **28,** 932 (1980).
99. Kazimierczuk, Giziewicz, and Shugar, *Acta Biochim. Pol.* **20,** 169 (1973).
100. Kchromov, Sorotchkina, Nigmatullin, and Tikchonenko. *FEBS Lett.* **118,** 51 (1980).
101. Kenner, Reese, and Todd, *J. Chem. Soc.*, 855 (1955).
102. Kikugawa and Ichino, *J. Org. Chem.* **37,** 284 (1972).
103. Kikugawa, Sato, Tsuruo, Imura, and Ukita, *Chem. Pharm. Bull.* **16,** 1110 (1968).
104. Kimura, Fujisawa, Sawada, and Mitsunobu, *Chem. Lett.*, 691 (1974).
105. Kissman and Weiss, *J. Am. Chem. Soc.* **21,** 1053 (1956).
106. Kochetkov, Shibaev, and Kost, *Tetrahedron Lett.*, 1993 (1971).
107. Krajewska and Shugar, *Acta Biochim. Pol.* **19,** 207 (1972).
108. Kramer, Fenselau, and Ludlum, *Biochem. Biophys. Res. Commun.* **56,** 783 (1974).
109. Kriek and Emmelot, *Biochemistry* **2,** 733 (1963).
110. Kulikowski and Shugar, *J. Med. Chem.* **17,** 269 (1974).
111. Kulikowski and Shugar, *Biochim. Biophys. Acta* **374,** 164 (1974).
112. Kuśmierek, unpublished.
113. Kuśmierek, Giziewicz, and Shugar, *Biochemistry* **12,** 194 (1973).
114. Kuśmierek and Singer, *Nucleic Acids Res.* **3,** 989 (1976).
115. Lawley and Brookes, *Biochem. J.* **89,** 127 (1963).
116. Lawley and Brookes, *Biochem. J.* **92,** 19c (1964).
117. Lawley and Jarman, *Biochem. J.* **126,** 893 (1972).
118. Lawley, Orr, and Jarman, *Biochem. J.* **145,** 73 (1975).
119. Lawley, Orr, and Shah, *Chem. Biol. Interact.* **4,** 431 (1971/1972).
120. Lawley, Orr, Shah, Farmer, and Jarman, *Biochem. J.* **135,** 193 (1973).
121. Lawley and Shah, *Biochem. J.* **128,** 117 (1972).
122. Lawley and Thatcher, *Biochem. J.* **116,** 693 (1970).
123. Leonard and Frihart, *J. Am. Chem. Soc.* **96,** 5894 (1974).
124. Leonard and Fujii, *Proc. Natl. Acad. Sci. USA* **51,** 73 (1964).
125. Letters and Michelson, *J. Chem. Soc.*, 1410 (1961).
126. Lichtenberg, Bergmann, and Neiman, C, *J. Chem. Soc.*, 1676 (1971).
127. Loveless, *Nature (London)* **223,** 206 (1969).
128. Ludlum, *Biochim. Biophys. Acta* **174,** 773 (1969).

129. Ludlum, Kramer, Wang, and Fenselau, *Biochemistry* **14,** 5480 (1975).
130. Ludlum and Tong, *Biochem. Pharmacol.* **27,** 2391 (1978).
131. Ludlum and Tong, In *Nitrosoureas: Current Status and New Developments,* (Prestayko *et al.*, eds.), Academic Press, New York (1981), pp. 85–94.
132. Ludlum and Wilhelm, *J. Biol. Chem.* **243,** 2750 (1968).
133. Martin, Reese, and Stephenson, *Biochemistry,* **7,** 1406 (1968).
134. Mehta and Ludlum, *Biochemistry* **15,** 4329 (1976).
135. Michelson, *J. Chem. Soc.*, 3655 (1959).
136. Michelson and Grunberg-Manago, *Biochim. Biophys. Acta* **91,** 92 (1964).
137. Michelson and Pochon, *Biochim. Biophys. Acta* **114,** 469 (1966).
138. Michelson and Todd, *J. Chem. Soc.*, 816 (1955).
139. Miles, *Biochim. Biophys. Acta* **22,** 247 (1956).
140. Miles, *J. Org. Chem.* **26,** 4761 (1961).
141. Miyaki and Shimizu, *Chem. Pharm. Bull.* **18,** 1446 (1970).
142. Montgomery and Temple, Jr., *J. Am. Chem. Soc.* **79,** 5238 (1957).
143. Montgomery and Thomas, *J. Org. Chem.* **28,** 2304 (1963).
144. Montgomery and Thomas, *J. Am. Chem. Soc.* **85,** 2672 (1963).
145. Montgomery and Thomas, *J. Heterocycl. Chem.* **1,** 115 (1964).
146. Moschel, Hudgins, and Dipple, *J. Org. Chem.* **44,** 3324 (1979).
147. Murphy, Goldman, and Ludlum, *Biochim. Biophys. Acta* **475,** 446 (1977).
148. Nakatsuka, Ohgi, and Goto, *Tetrahedron Lett.* **No. 29,** 2579 (1978).
149. Notari, Witiak, DeYoung, and Lin, *J. Med. Chem.* **15,** 1207 (1972).
150. Nowak, Szczepaniak, Barski, and Shugar, *Z. Naturforsch.* **33c,** 876 (1978).
151. Ogilvie, Beaucage, Gillen, and Entwistle, *Nucleic Acids Res.* **6,** 2261 (1979).
152. Ortlieb and Kleihues, *Carcinogenesis* **1,** 849 (1980).
153. Pal, *Biochemistry* **1,** 558 (1962).
154. Pfleiderer, *Liebigs Ann. Chem.* **647,** 161 (1961).
155. Pfleiderer, *Liebigs Ann. Chem.* **647,** 167 (1961).
156. Pfleiderer and Nubel, *Liebigs Ann. Chem.* **647,** 155 (1961).
157. Pfleiderer, Shanshal, and Eistetter, *Chem. Ber.* **105,** 1497 (1972).
158. Philips and Horwitz, *J. Org. Chem.* **40,** 1856 (1975).
159. Pochon and Michelson, *Biochim. Biophys. Acta* **145,** 321 (1967).
160. Pochon and Michelson, *Biochim. Biophys. Acta* **182,** 17 (1969).
161. Prasad and Robins, *J. Am. Chem. Soc.* **79,** 6401 (1957).
162. Price, Gaucher, Koneru, Shibakawa, Sowa, and Yamaguchi, *Biochim. Biophys. Acta* **166,** 327 (1968).
163. Psoda and Shugar, *Biochim. Biophys. Acta* **247,** 507 (1971).

Table A (continued)

164. Rajabalee and Hanessian, *Can. J. Chem.* **49,** 1981 (1971).
165. Ransford, Glinski, and Sporn, *J. Carbohydr. Nucleosides Nucleotides* **1,** 275 (1974).
166. Reiner and Zamenhof, *J. Biol. Chem.* **228,** 475 (1957).
167. Roberts and Warwick, *Biochem. Pharmacol.* **12,** 1441 (1963).
168. Robins and Lin, *J. Am. Chem. Soc.* **79,** 490 (1957).
169. Robins, Hall, and Thedford, *Biochemistry* **6,** 1837 (1967).
170. Robins, MacCoss, and Lee, *Biochem. Biophys. Res. Commun.* **70,** 356 (1976).
171. Robins and Naik, *Biochemistry* **10,** 3591 (1971).
172. Robins, Naik, and Lee, *J. Org. Chem.* **39,** 1891 (1974).
173. Rogers and Ulbricht, *J. Chem. Soc. C,* 2364 (1961).
174. Sack, Fenselau, Kan, Kan, Wood, and Lau, *J. Org. Chem.* **40,** 3932 (1978).
175. Saito and Fujii, *J. Chem. Soc. Chem. Comm.,* 125 (1979).
176. Saneyoshi, Ohashi, Harada, and Nishimura, *Biochim. Biophys. Acta* **262,** 1 (1972.)
177. Saponara and Enger, *Nature (London)* **223,** 1365 (1969).
178. Sattsangi, Barrio, and Leonard, *J. Am. Chem. Soc.* **102,** 770 (1980).
179. Sattsangi, Leonard, and Frihart, *J. Org. Chem.* **42,** 3292 (1977).
180. Schalke and Hall, *J. Chem. Soc. Perkin Trans. I,* 2417 (1975).
181. Scherer, Van der Laken, Gwinner, Laib, and Emmelot, *Carcinogenesis* **2,** 671 (1981).
182. Schweizer, Chheda, Baczynkyj, and Hall, *Biochemistry* **8,** 3283 (1969).
183. Secrist, Barrio, Leonard, and Weber, *Biochemistry* **11,** 3499 (1972).
184. Segal, Maté, and Solomon, *Chem. Biol. Interact.* **28,** 333 (1979).
185. Serebryanyi and Mnatsakanyan, *FEBS Lett.* **28,** 191 (1972).
186. Shapiro, *Prog. Nucleic Acid Res. Mol. Biol.* **8,** 73 (1968).
187. Shapiro, Cohen, and Clagett, *J. Biol. Chem.* **245,** 2633 (1970).
188. Shapiro, Cohen, Shiuey, and Maurer, *Biochemistry* **8,**238 (1969).
189. Shapiro and Klein, *Biochemistry* **6,** 3576 (1967).
190. Shapiro and Shiuey, *Biochim. Biophys. Acta* **174,** 403 (1969).
191. Shapiro and Shiuey, *J. Org. Chem.* **41,** 1597 (1976).
192. Shooter, Edwards, and Lawley, *Biochem. J.* **125,** 829 (1971).
193. Shugar and Fox, *Biochim. Biophys. Acta* **9,** 199 (1952).
194. Singer, *Biochemistry* **11,** 3939 (1972).
195. Singer, *FEBS Lett.* **63,** 85 (1976).
196. Singer, *Nature (London)* **264,** 333 (1976).
197. Singer, unpublished.
198. Singer and Fraenkel-Conrat, *Biochemistry* **14,** 772 (1975).

199. Singer, Kröger, and Carrano, *Biochemistry* **17,** 1246 (1978).
200. Singer and Spengler, *FEBS Lett.* **139,** 69 (1982).
201. Singer, Sun, and Fraenkel-Conrat, *Biochemistry* **13,** 1913 (1974).
202. Smith, *Chem. Biol. Interact.* **16,** 275 (1977).
203. Smith and Dunn, *Biochem. J.* **72,** 294 (1959).
204. Steinmaus, Rosenthal, and Elad, *J. Am. Chem. Soc.* **91,** 4921 (1969).
205. Stevens and Brown, *J. Am. Chem. Soc.* **80,** 2759 (1958).
206. Stevens, Magrath, Smith, and Brown, *J. Am. Chem. Soc.* **80,** 2755 (1958).
207. Stewart and Harris, *Can. J. Chem.* **55,** 3807 (1977).
208. Stolarski, Dudycz, and Shugar, *Eur. J. Biochem.* **108,** 111 (1980).
209. Subbaraman, Subbaraman, and Behrman, *Biochemistry* **8,** 3059 (1969).
210. Sukhorukov, Gukovskaya, Nekrasova, and Antonovskii, *Biofizika* **19,** 790 (1974).
211. Sukhorukov, Gukovskaya, Sukhoruchkina, and Lavrrenova, *Biofizika* **17,** 5 (1972).
212. Sun and Singer, *Biochemistry* **13,** 1905 (1974).
213. Szer, *Biochem. Biophys. Res. Commun.* **20,** 182 (1965).
214. Szer and Shugar, *Acta Biochim. Pol.* **13,** 177 (1966).
215. Tanaka, Nasu, and Miyasaka, *Tetrahedron Lett.* **No. 49,** 4755 (1979).
216. Tazawa, Tazawa, Alderfer, and Ts'o, *Biochemistry* **11,** 4931 (1972).
217. Taylor and Loeffler, *J. Am. Chem. Soc.* **82,** 3147 (1960).
218. Thedford and Strauss, *Biochem. Biophys. Res. Commun.* **47,** 1237 (1972).
219. Tong and Gombar, *Cancer Res.* **41,** 380 (1981).
220. Tong, Kirk, and Ludlum, *Biochem. Biophys. Res. Commun.* **100,** 351 (1981).
221. Tong and Ludlum, *Biochem. Pharmacol.* **27,** 77 (1978).
222. Tong and Ludlum, *Biochem. Pharmacol.* **28,** 1175 (1979).
223. Townsend and Robins, *J. Chem. Soc.*, 3008 (1962).
224. Townsend, Robins, Loeppky, and Leonard, *J. Chem. Soc.*, 5320 (1964).
225. Ueda and Fox, *J. Am. Chem. Soc.* **85,** 4024 (1963).
226. Ueda and Fox, *J. Org. Chem.* **29,** 1770 (1964).
227. Uziel and Taylor, *J. Carbohydr. Nucleosides Nucleotides* **5,** 235 (1978).
228. Van Holde, Brahms, and Michelson, *J. Mol. Biol.* **12,** 726 (1965).
229. Venkstern and Baer, *Absorption Spectra of Minor Bases,* Plenum Press (1965).
230. Wacker and Ebert, *Z. Naturforsch.* **14b,** 709 (1959).
231. Walwick, Roberts, and Dekker, *Proc. Chem. Soc. (London),* 84 (1959).
232. Wempen, Duschinsky, Kaplan, and Fox, *J. Am. Chem. Soc.* **83,** 4755 (1961).
233. Wierzchowski, Litonska, and Shugar, *J. Am. Chem. Soc.* **87,** 4621 (1965).

Table A (continued)

234. Windmueller and Kaplan, *J. Biol. Chem.* **236,** 2716 (1961).
235. Winkley and Robins, *J. Org. Chem.* **33,** 2822 (1968).
236. Wittenburg, *Collect. Czech. Chem. Commun.* **36,** 246 (1971).
237. Wong and Fuchs, *J. Org. Chem.* **35,** 3786 (1970).
238. Wong and Fuchs, *J. Org. Chem.* **36,** 848 (1971).
239. Wong and Fuchs, *J. Chem. Soc. Perkin Trans. I,* 1284 (1974).
240. Yamauchi, Hayashi, and Kinoshita, *J. Org. Chem.* **40,** 385 (1975).
241. Yamauchi and Kinoshita, *J. Chem. Soc. Perkin Trans. I,* 391 (1973).
242. Yamauchi, Nakagima, and Kinoshita, *J. Chem. Soc. Perkin Trans. II,* 2787 (1980).
243. Yamauchi, Tanabe, and Kinoshita, *J. Org. Chem.* **41,** 3691 (1976).
244. Yano, Kan, and Ts'o, *Biochim. Biophys. Acta* **629,** 178 (1980).
245. Zmudzka, Bollum, and Shugar, *Biochemistry* **8,** 3049 (1969).
246. Oesch and Doerjer, *Carcinogenesis* **3,** 663 (1982).

APPENDIX B

Acid, Neutral, and Basic Spectra of Bases, Nucleosides, Nucleotides, and 55 Modified Derivatives

INTRODUCTION TO FIGURES

For all figures except those noted below, the spectrum in water is the solid line (——); in 0.1 N HCl, the dashed line (----); and in 0.1 N KOH, the dotted line (· · · · ·). The absorption maxima at each pH are given in the figures. These are not necessarily the same as in Appendix A, which gives average data.

In several instances the spectrum in water is omitted due to the low solubility of the compound. In some cases, the spectrum in water and in 0.1 N HCl are identical and only one of these spectra is shown.

As noted in footnote *c* to Appendix A, all 7-alkyl purine nucleotides and 7,9-dialkyl purines undergo imidazole ring opening in alkali. Thus, in Figures 35, 37, 41, and 42, the spectra in alkali are of the partially or totally ring-opened derivative. In Figure 37, the dotted line is that of imidazole ring-opened N^6,7-dimethyladenosine and the broken line (–·–·) is the spectrum of this derivative in 0.1 N HCl. Figures 35(B), 41(B), and 42(B) are the spectra of other imidazole ring-opened derivatives. Figure 6(B) shows the spectrum of the base derived from imidazole ring-opened 7-methyladenosine and can be compared with Figure 6(A) (7-methyladenine), the base derived from intact 7-methyladenosine.

Figure 31 shows the absorption spectra of 1-methyladenosine in 0.01 M pH 5.7 Tris buffer (– – –), 0.01 M pH 7.8 Tris buffer (–·–·), 0.01 M pH 8.1 Tris buffer (——), and 0.1 N KOH (·····). This figure illustrates in a rough way the use of spectrophotometer data to determine dissociation constants. (The pK_a of 1-methyladenosine is about 8.5.) O^2-alkyl-, 3-alkyl-, and O^4-alkyl uridines have pK_as below 1 and a shift in spectra to the acidic form can only be demonstrated in strong acid (10 N H_2SO_4). The dashed line in Figures 52, 53, and 55 are the spectra in 10 N H_2SO_4. In Figure 26, the spectrum at pH 0 is shown by the broken line (–·–·). It should be noted that *O*-alkyl derivatives are dialkylated at varying rates in acid or alkali (199). For this reason, Figure 45 shows the spectrum of *O*-ethylcytidine only at pH 1–7 since dealkylation is almost instantaneous in alkali.

The spectra of 3-methylthymidylic acid are identical to those of the spectra for the methyl ester shown in Figure 69.

Alkylation of the ribose and/or phosphate does not cause a spectral change and data for compounds additionally alkylated at one or more ribose positions are included in Appendix A, but spectra are not shown.

Methyl and ethyl derivatives have the same spectra, except that ethyl derivatives occasionally have a higher λ_{max}. Examples are Figures 2 and 3; 31 and 32; 33 and 34; 54 and 55.

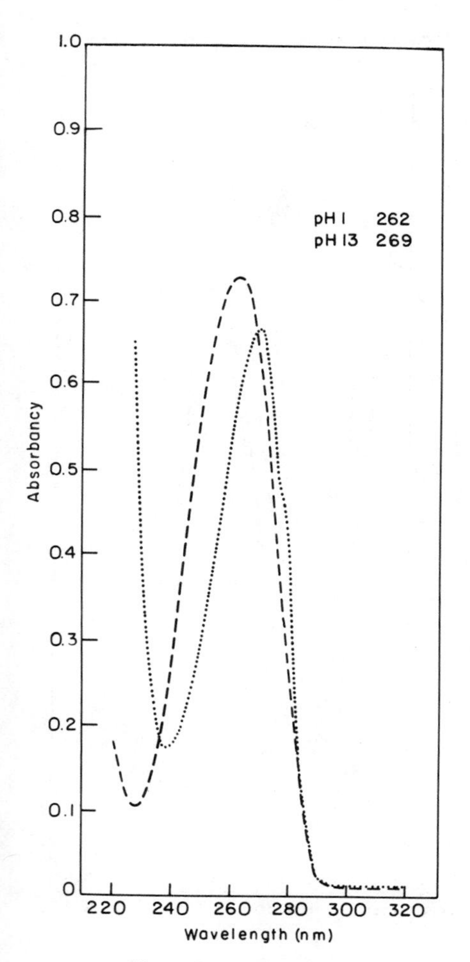

Figure B-1. Adenine.

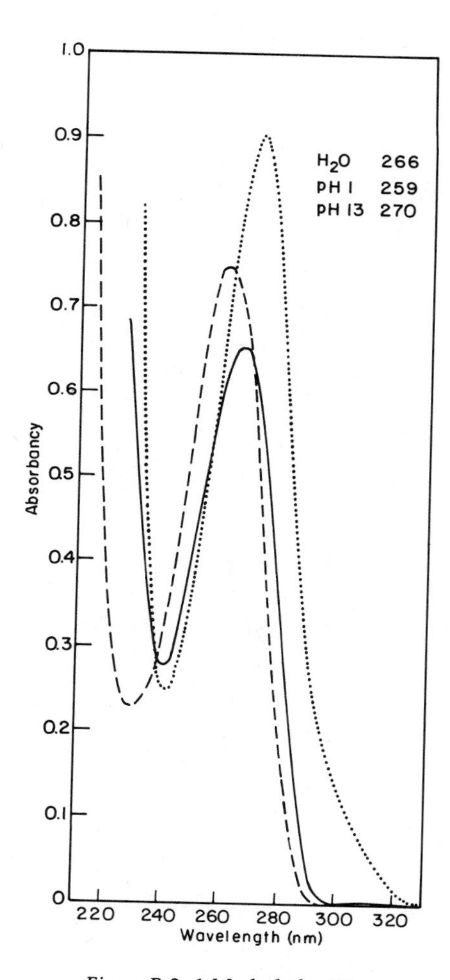

Figure B-2. 1-Methyladenine.

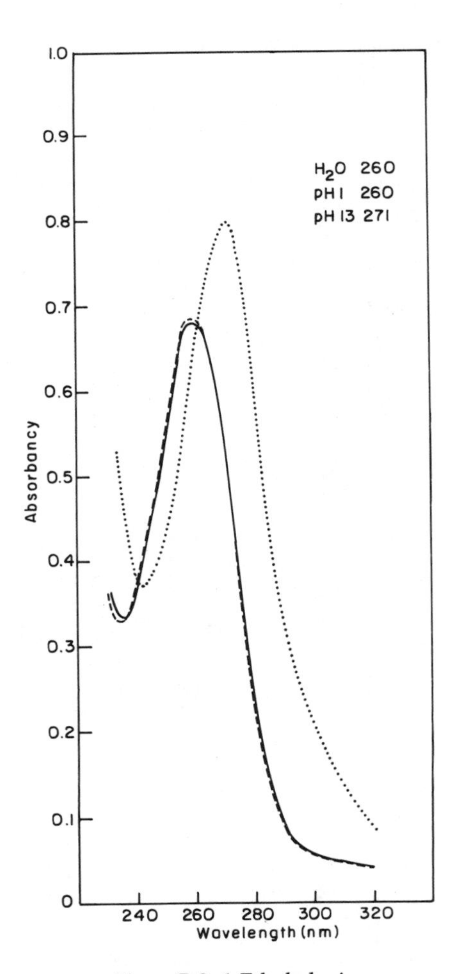

Figure B-3. 1-Ethyladenine.

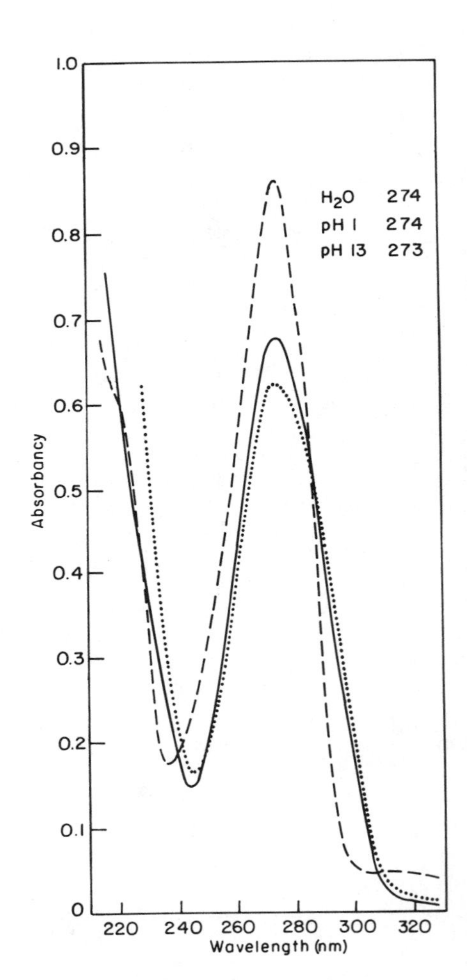

Figure B-4. 3-Methyladenine.

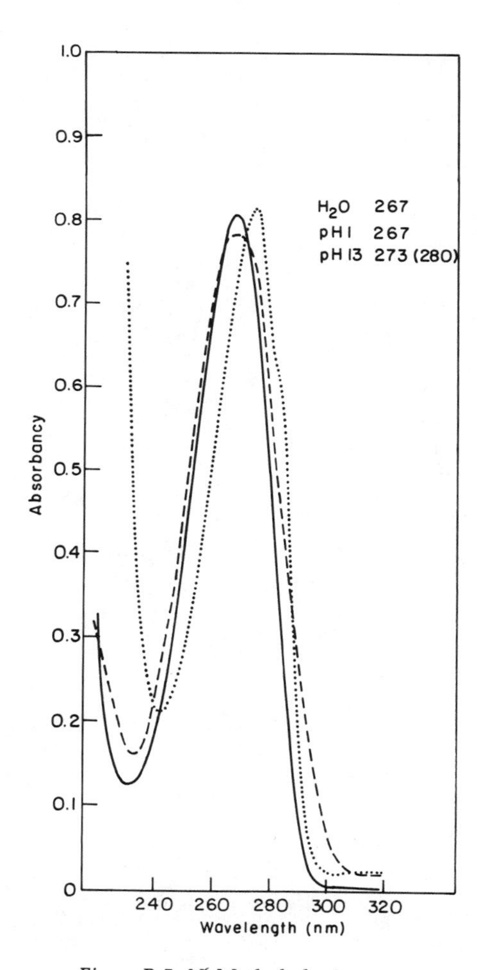

Figure B-5. N^6-Methyladenine.

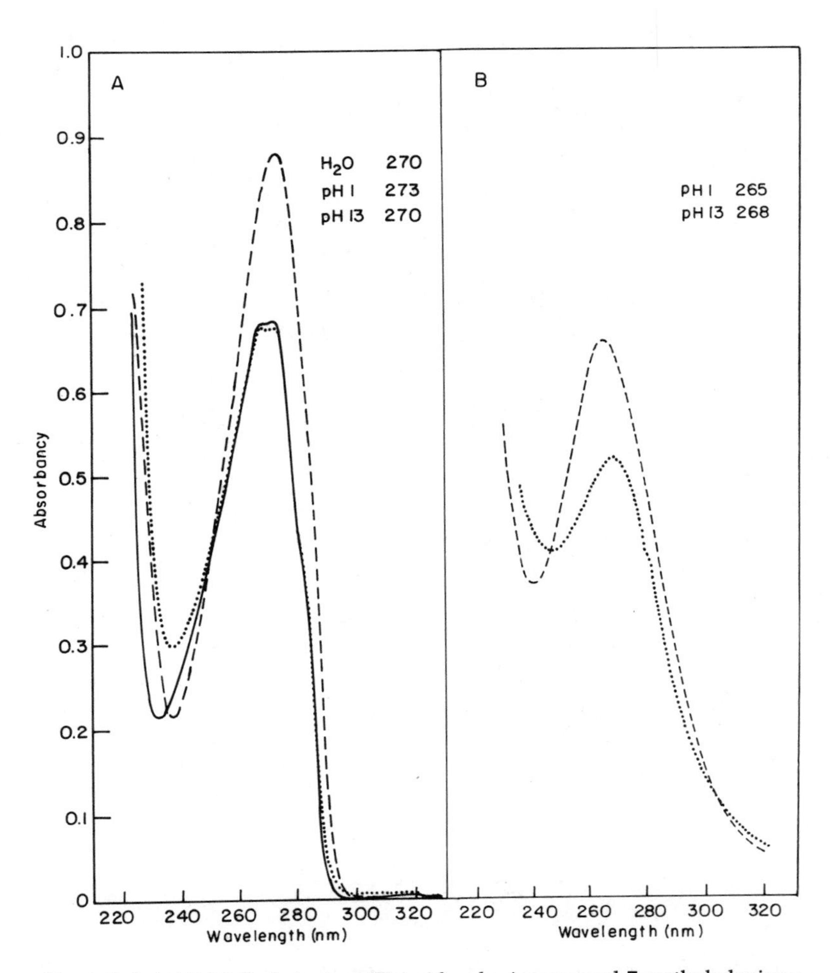

Figure B-6. (A) 7-Methyladenine; (B) imidazole ring-opened 7-methyladenine.

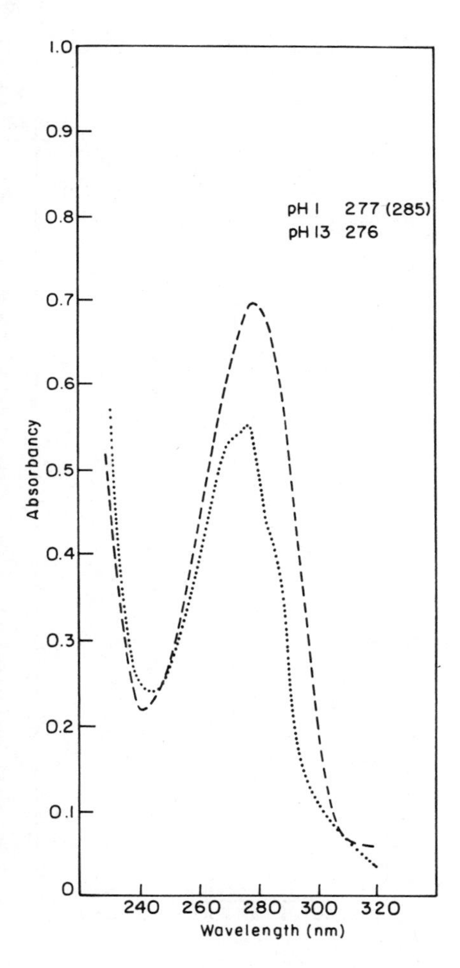

Figure B-7. N^6-Methyl, 7-ethyladenine.

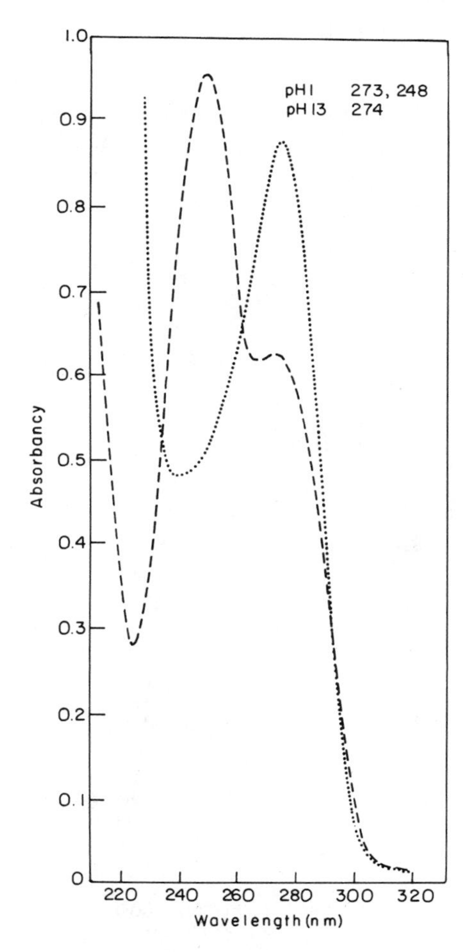

Figure B-8. Guanine.

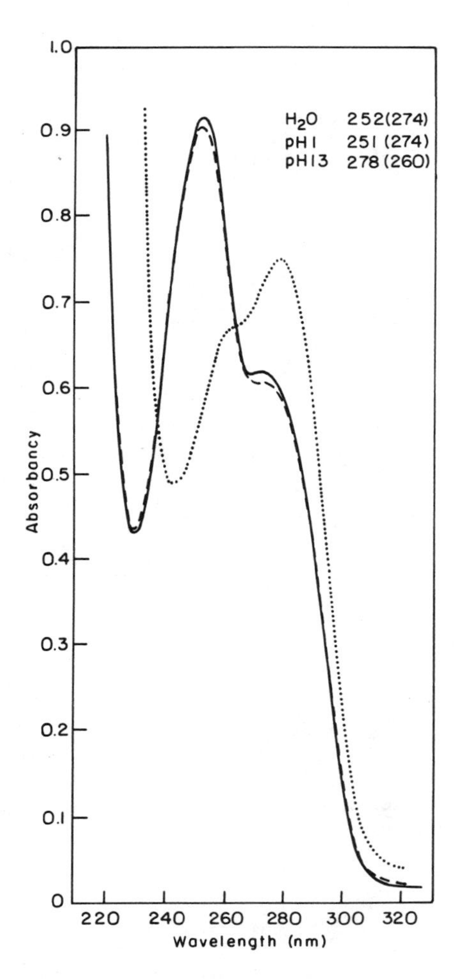

Figure B-9. 1-Ethylguanine.

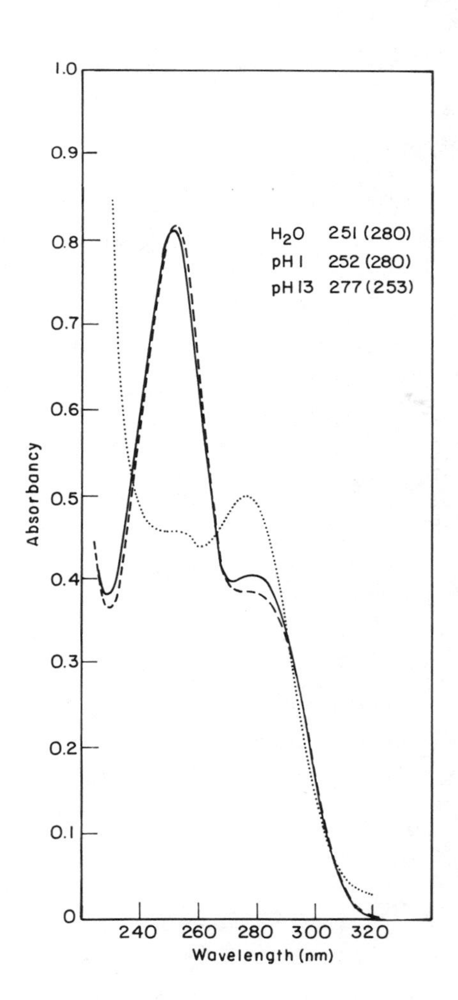

Figure B-10. N^2-Ethylguanine.

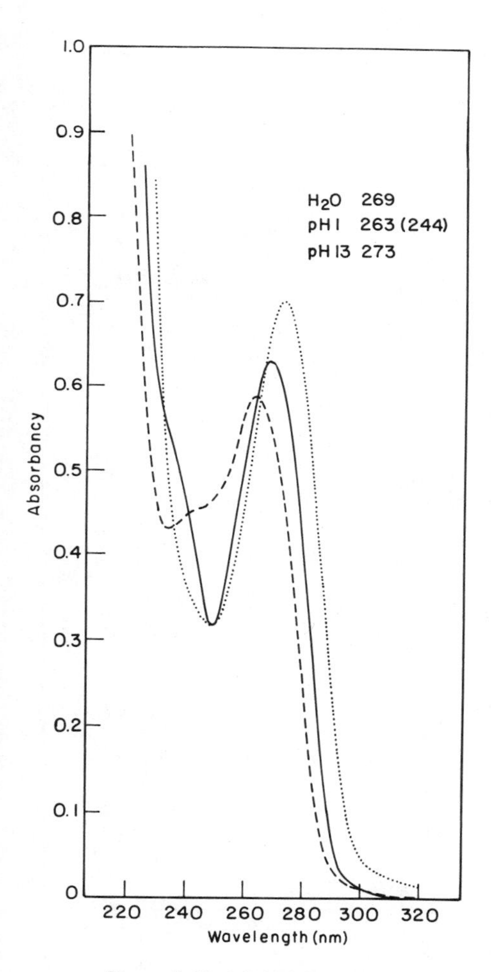

Figure B-11. 3-Ethylguanine.

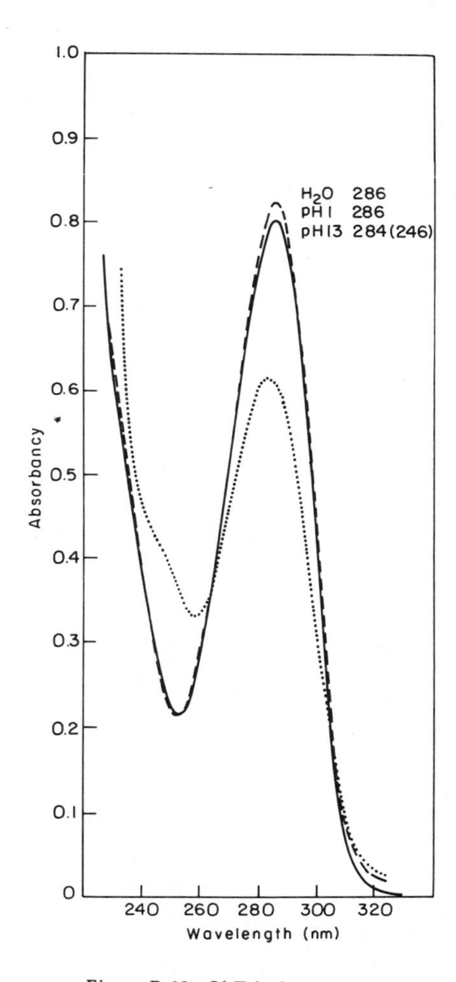

Figure B-12. O^6-Ethylguanine.

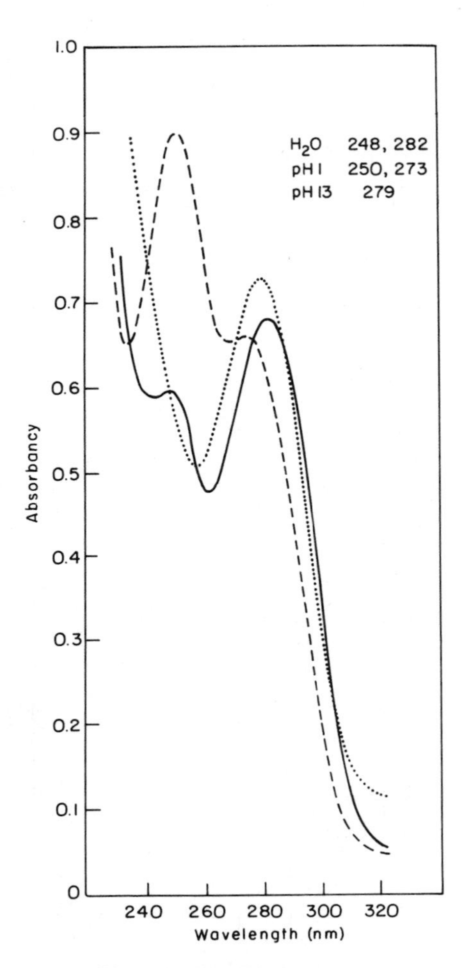

Figure B-13. 7-Ethylguanine.

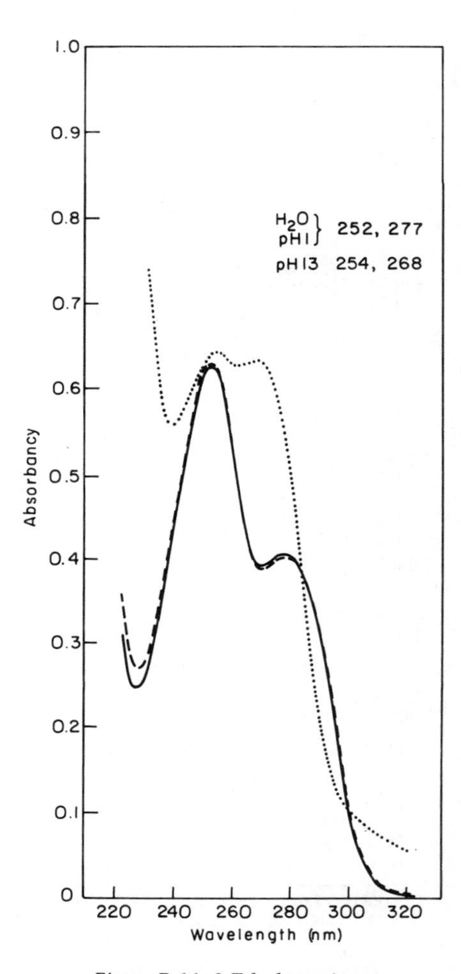

Figure B-14. 9-Ethylguanine.

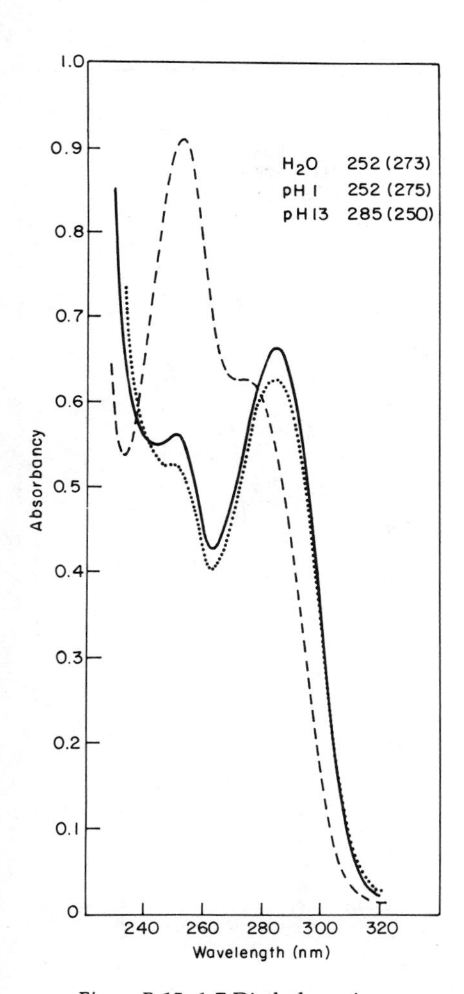

Figure B-15. 1,7-Diethylguanine.

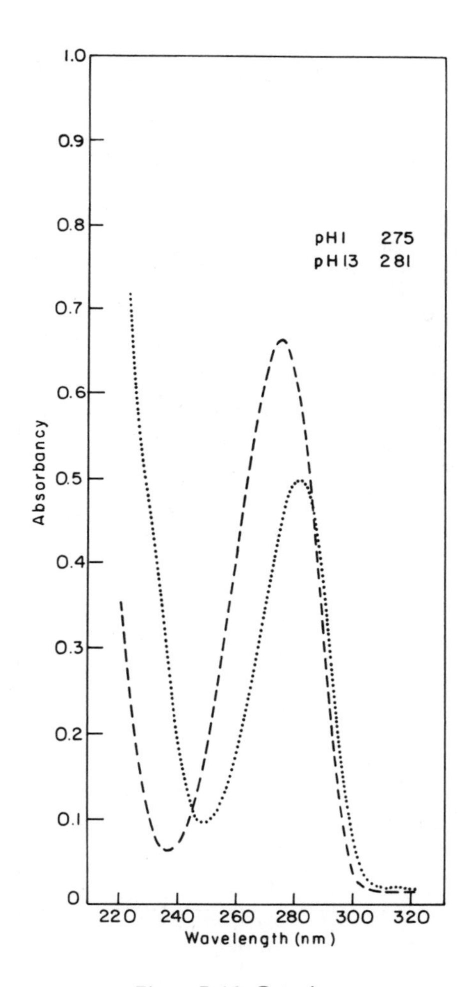

Figure B-16. Cytosine.

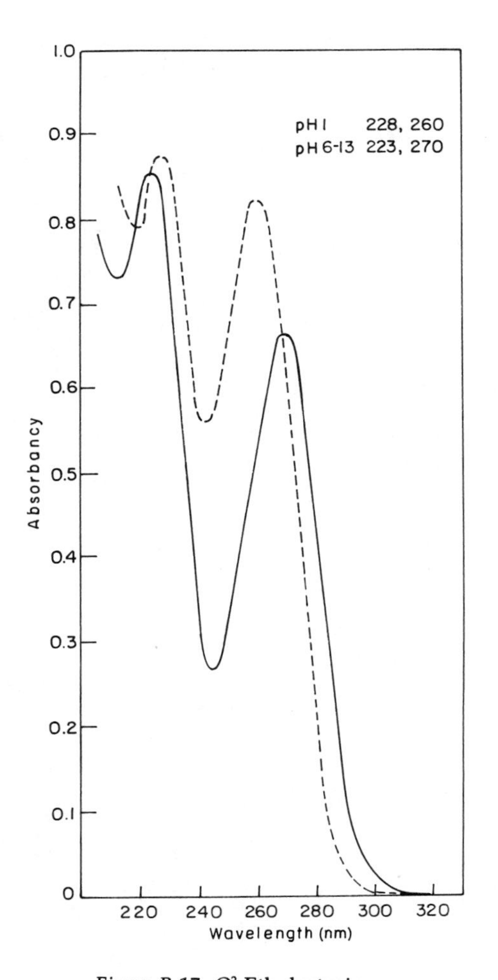

Figure B-17. O^2-Ethylcytosine.

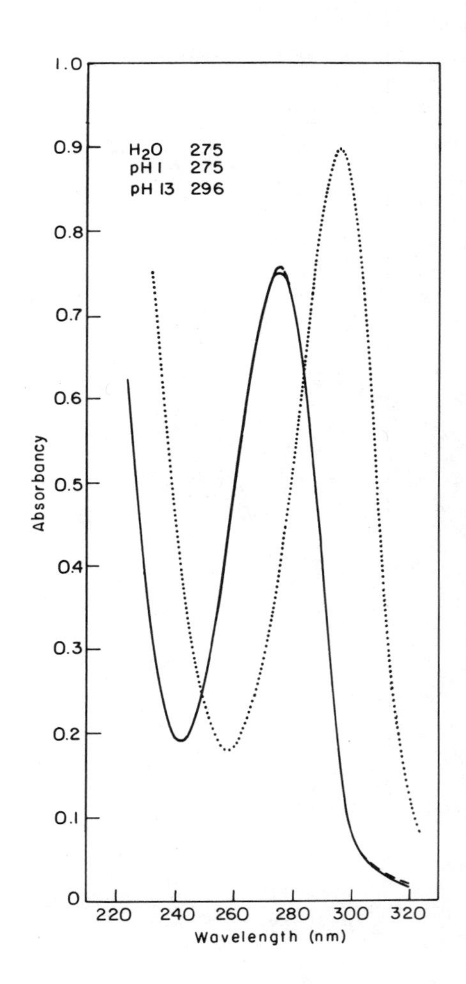

Figure B-18. 3-Ethylcytosine.

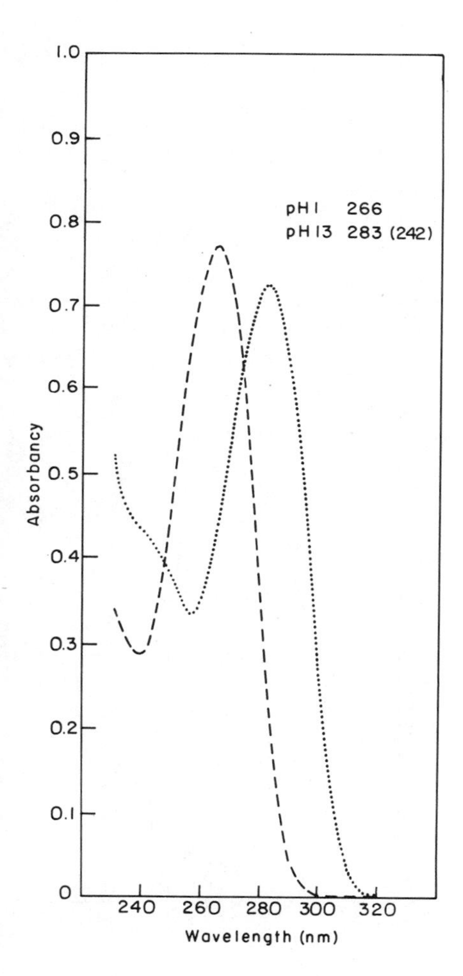

Figure B-19. Xanthine.

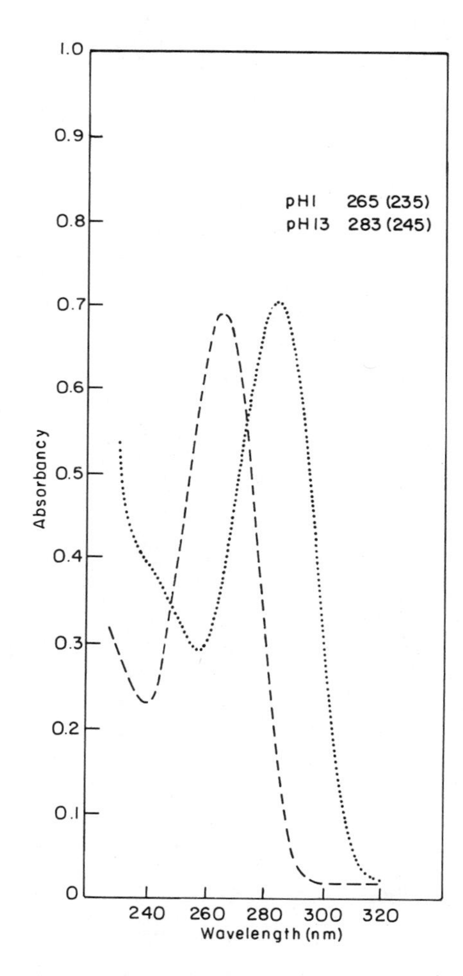

Figure B-20. 1-Methylxanthine.

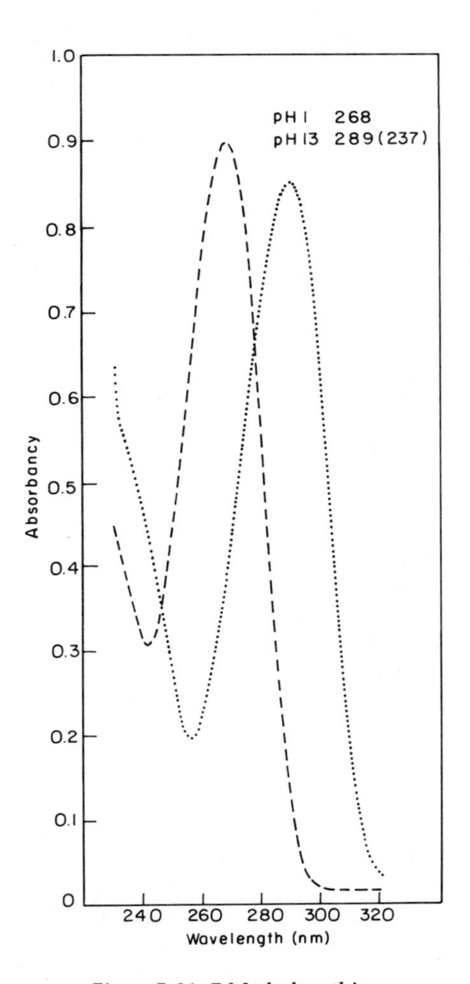

Figure B-21. 7-Methylxanthine.

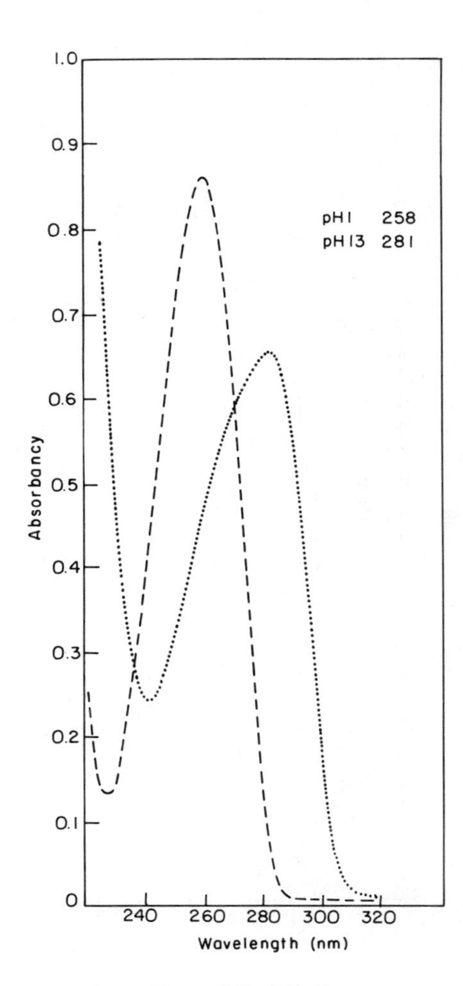

Figure B-22. Uracil.

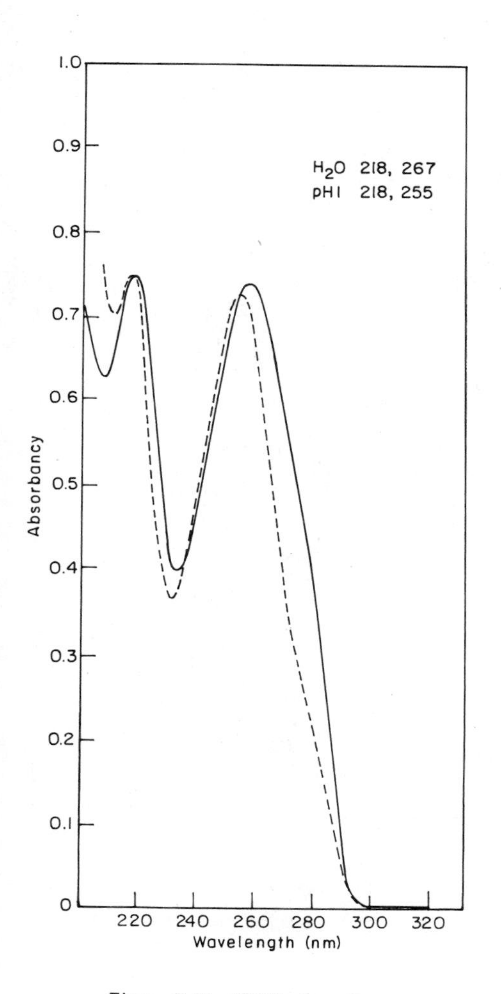

Figure B-23. O^2-Ethyluracil.

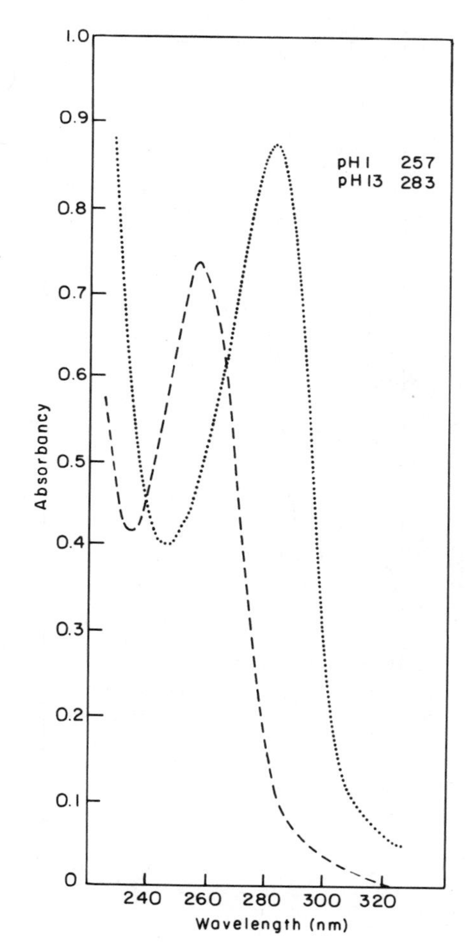

Figure B-24. 3-Methyluracil.

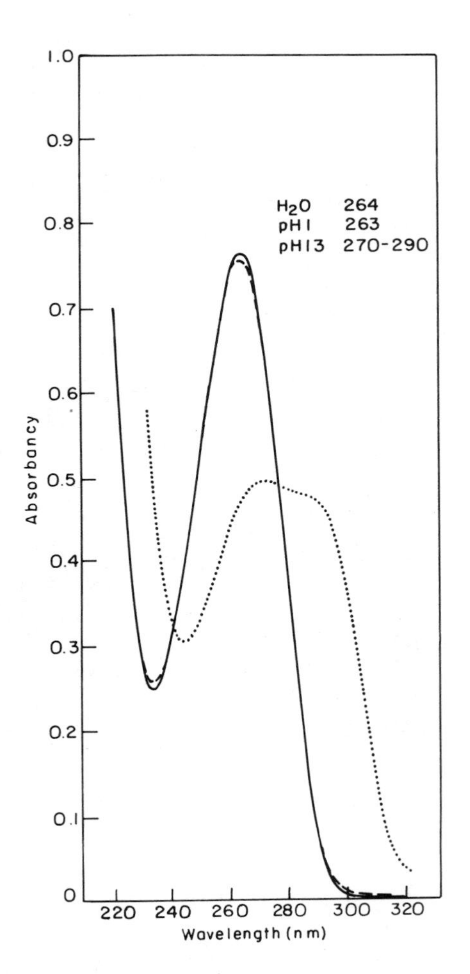

Figure B-25. Thymine.

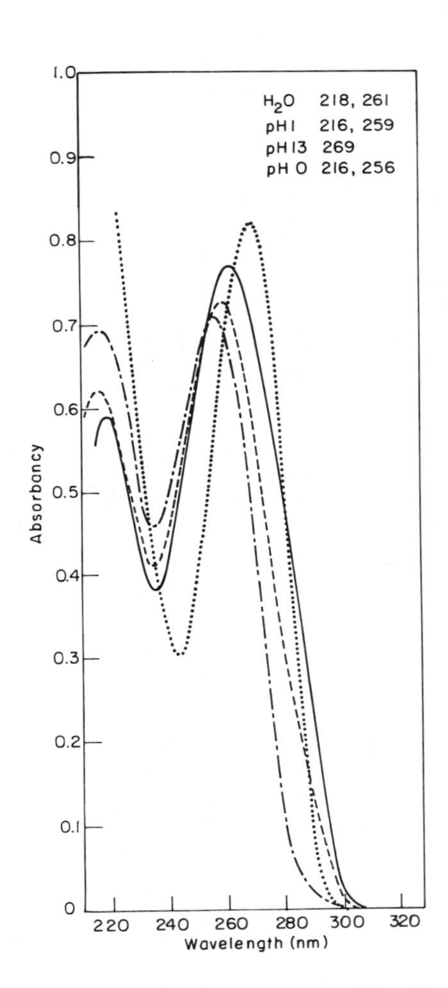

Figure B-26. O^2-Ethylthymine. [pH 0 (–·–·)]

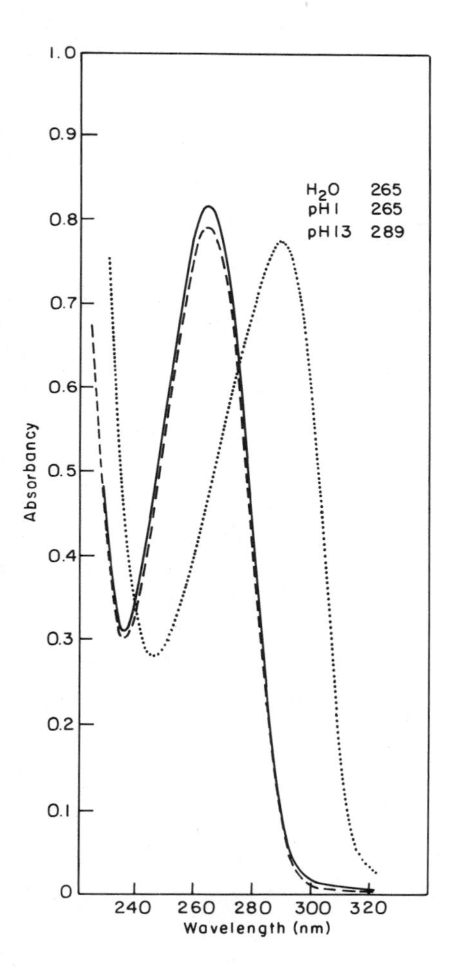

Figure B-27. 3-Ethylthymine.

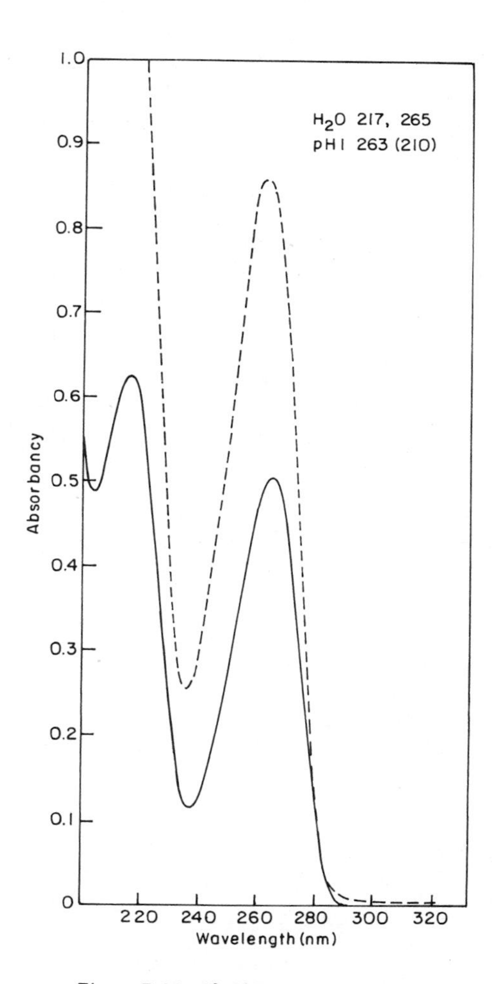

Figure B-28. O^2,O^4-Diethylthymine.

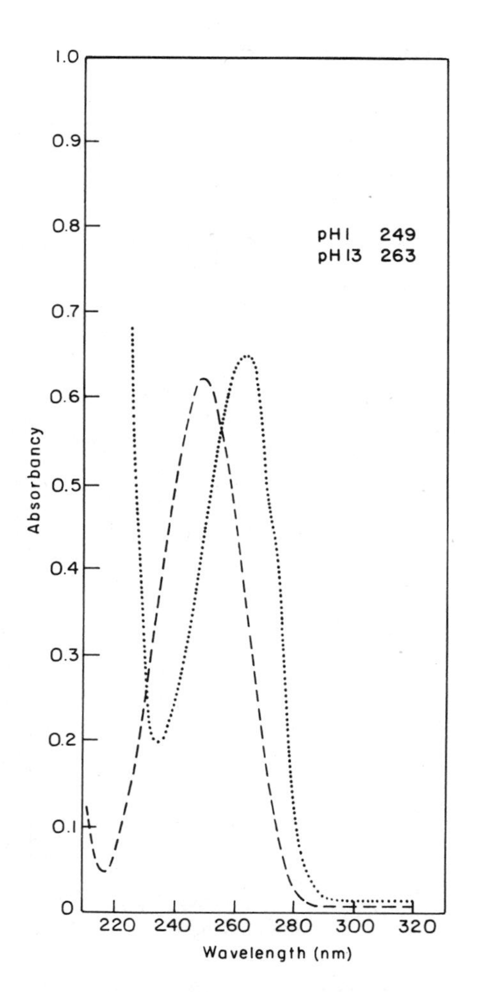

Figure B-29. Hypoxanthine.

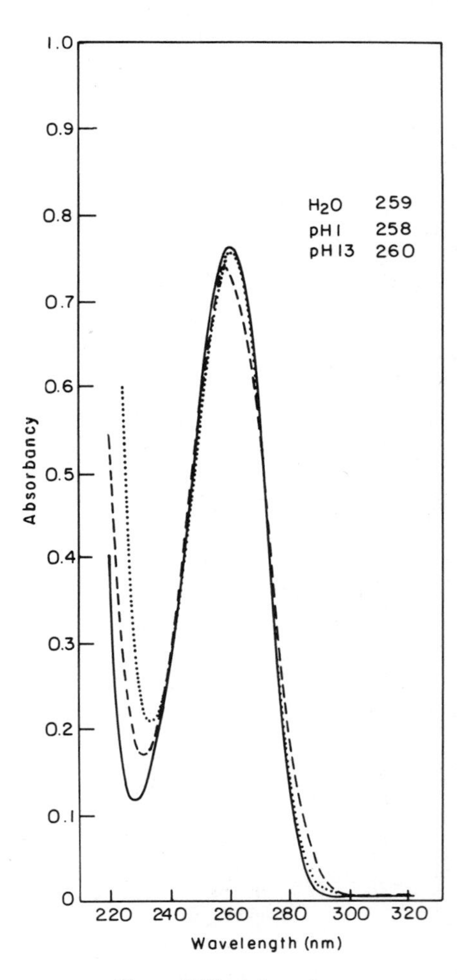

Figure B-30. Adenosine.

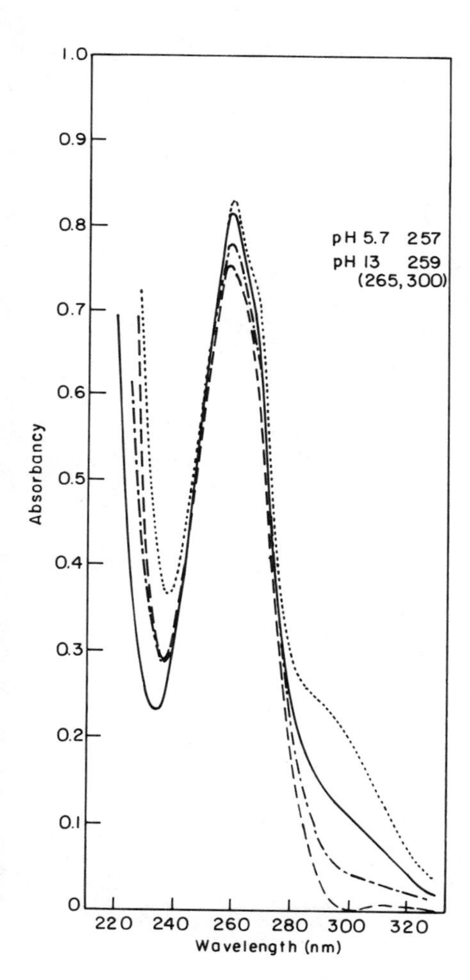

Figure B-31. 1-Methyladenosine. [pH 5.7 (– –), pH 7.8 (–·–·), pH 8.1 (——), pH 13 (· · ·)]

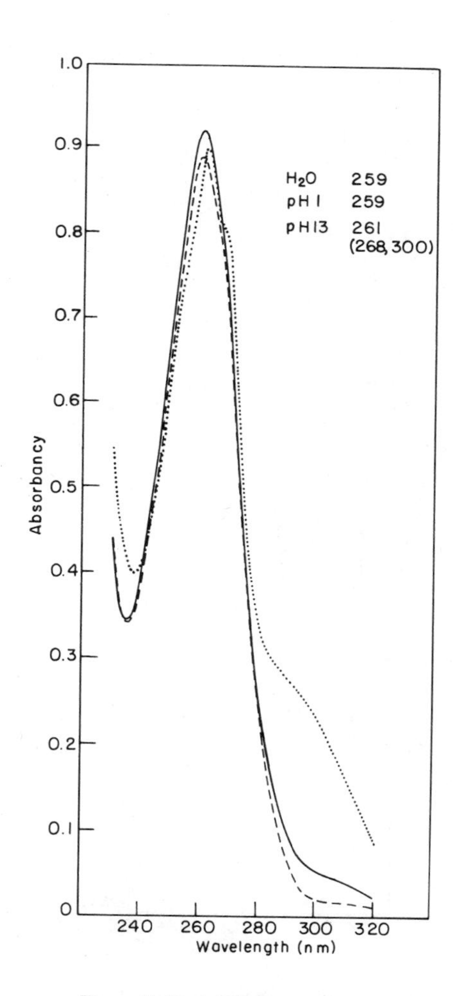

Figure B-32. 1-Ethyladenosine.

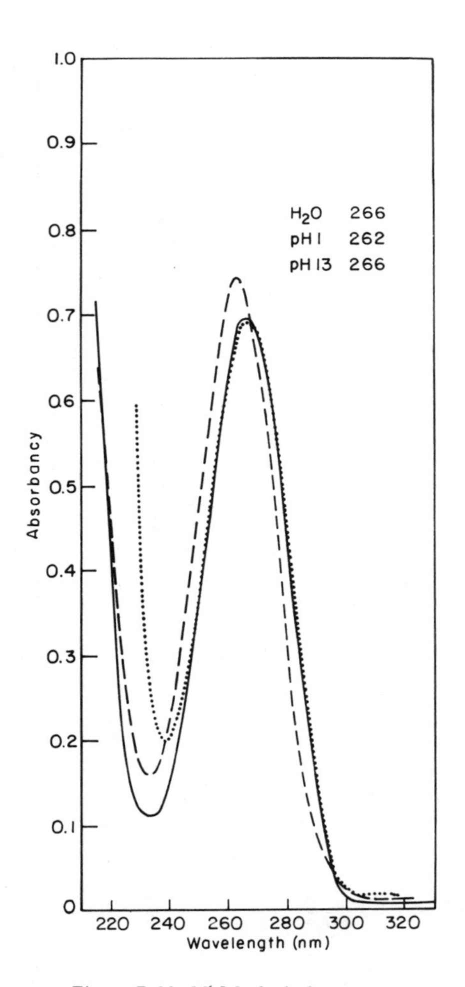

Figure B-33. N^6-Methyladenosine.

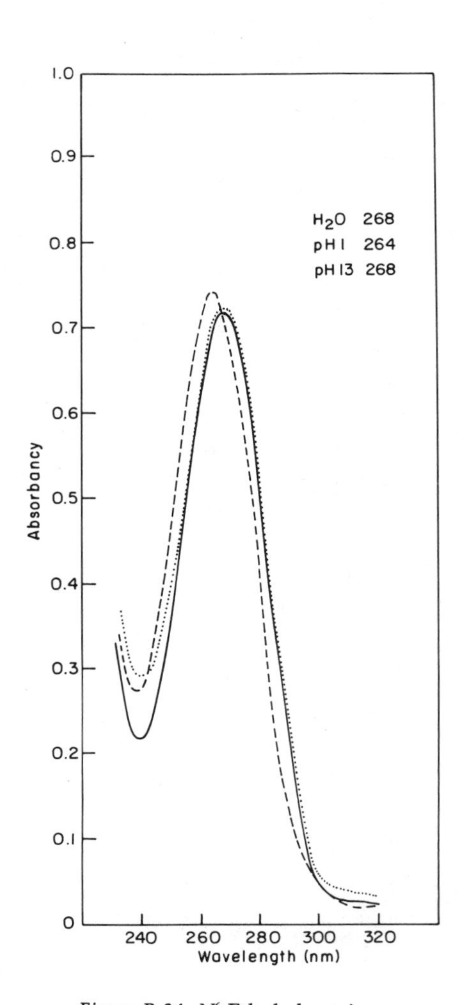

Figure B-34. N^6-Ethyladenosine.

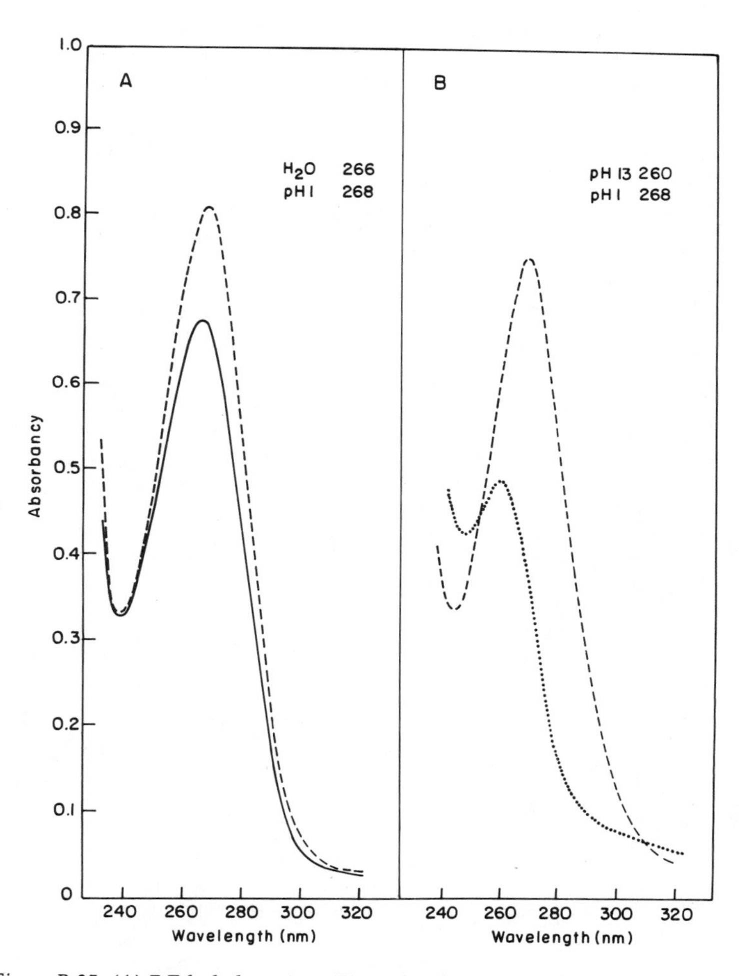

Figure B-35. (A) 7-Ethyladenosine; (B) imidazole ring-opened 7-ethyladenosine.

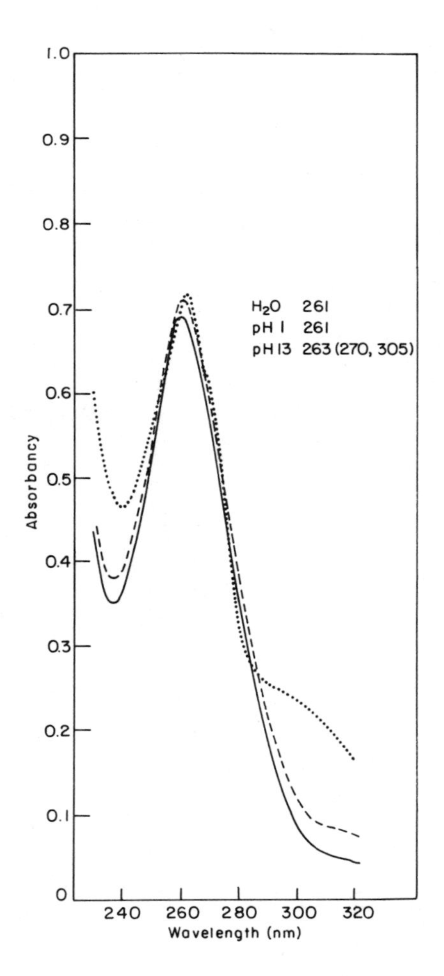

Figure B-36. 1,N^6-Dimethyladenosine.

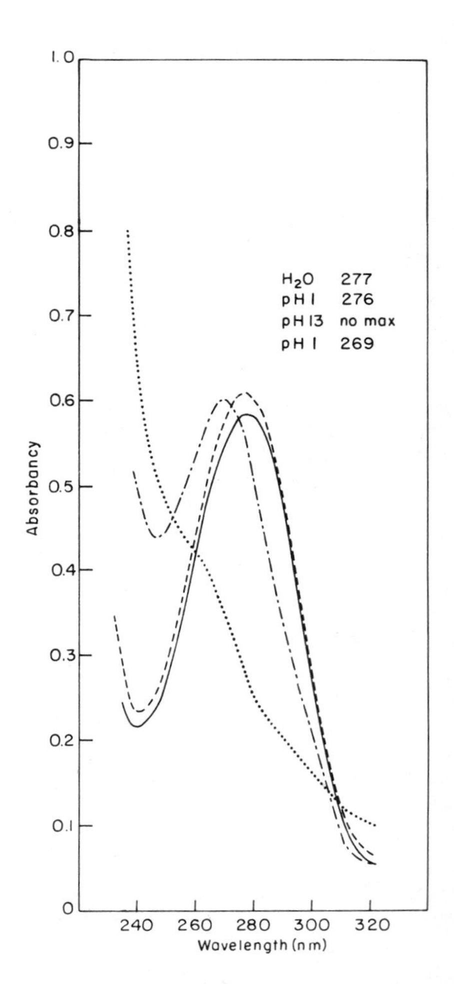

Figure B-37. N^6,7-Dimethyladenosine and its imidazole ring-opened derivative at pH 1 (–·–·) and pH 13 (····).

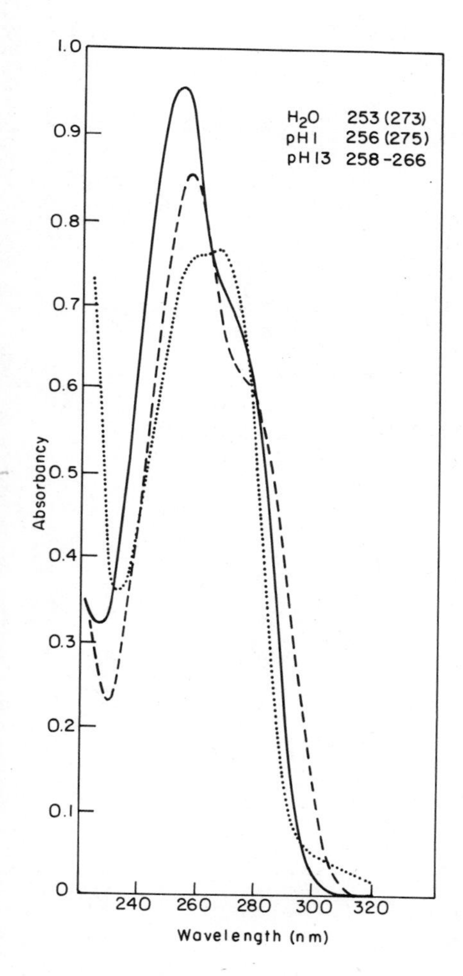

Figure B-38. Guanosine.

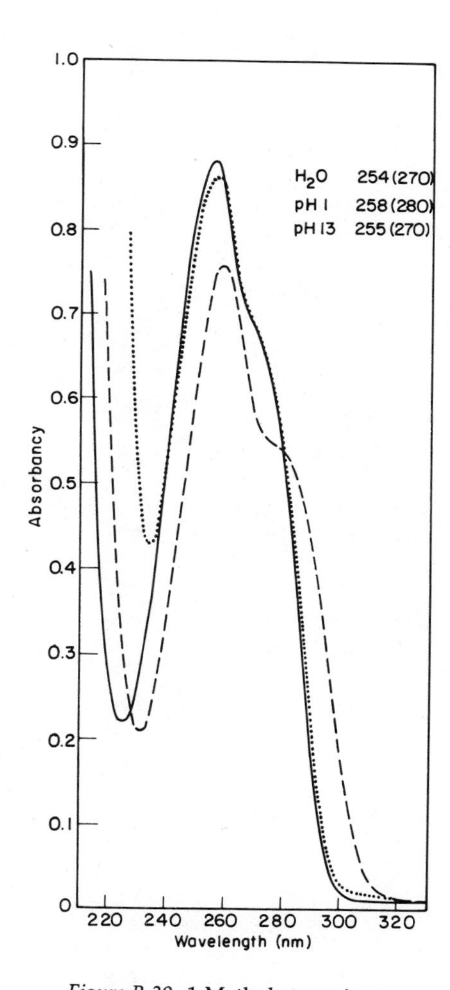

Figure B-39. 1-Methylguanosine.

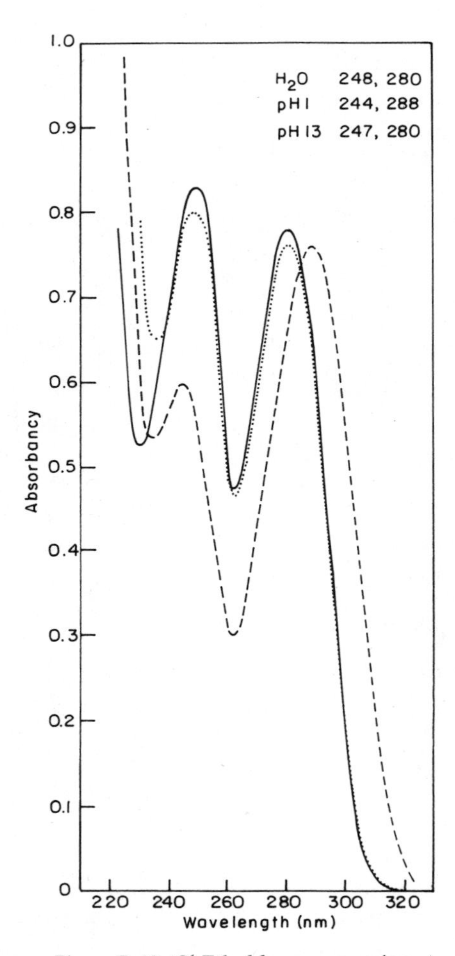

Figure B-40. O^6-Ethyldeoxyguanosine.

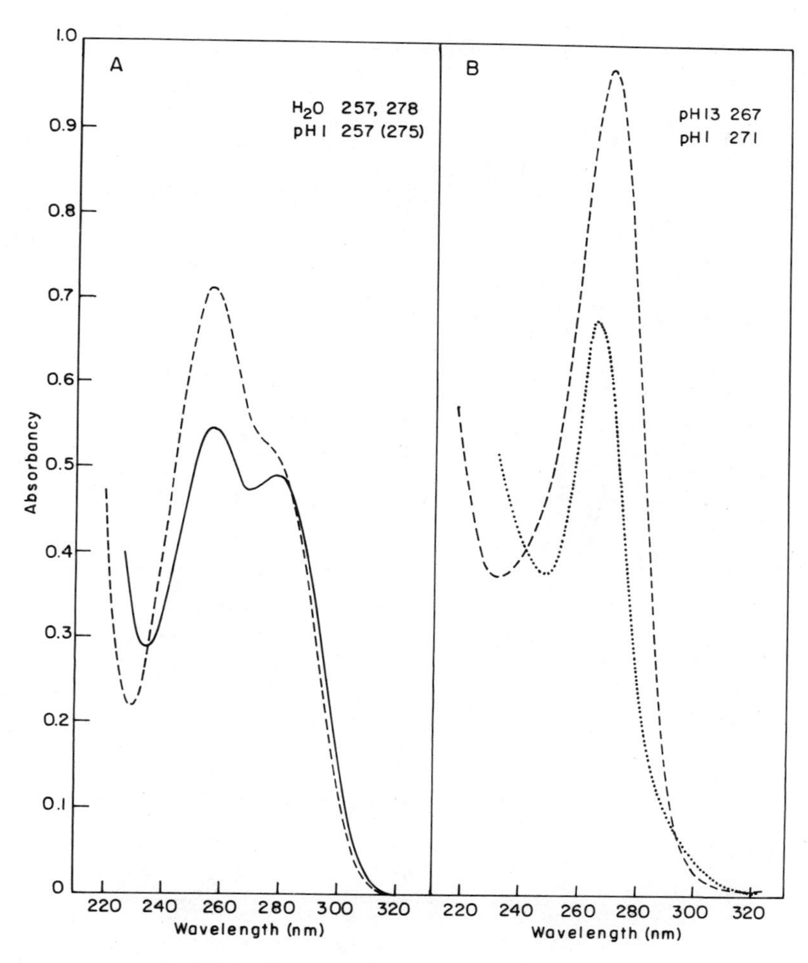

Figure B-41. (A) 7-Ethyldeoxyguanosine; (B) imidazole ring-opened 7-ethyldeoxyguansine.

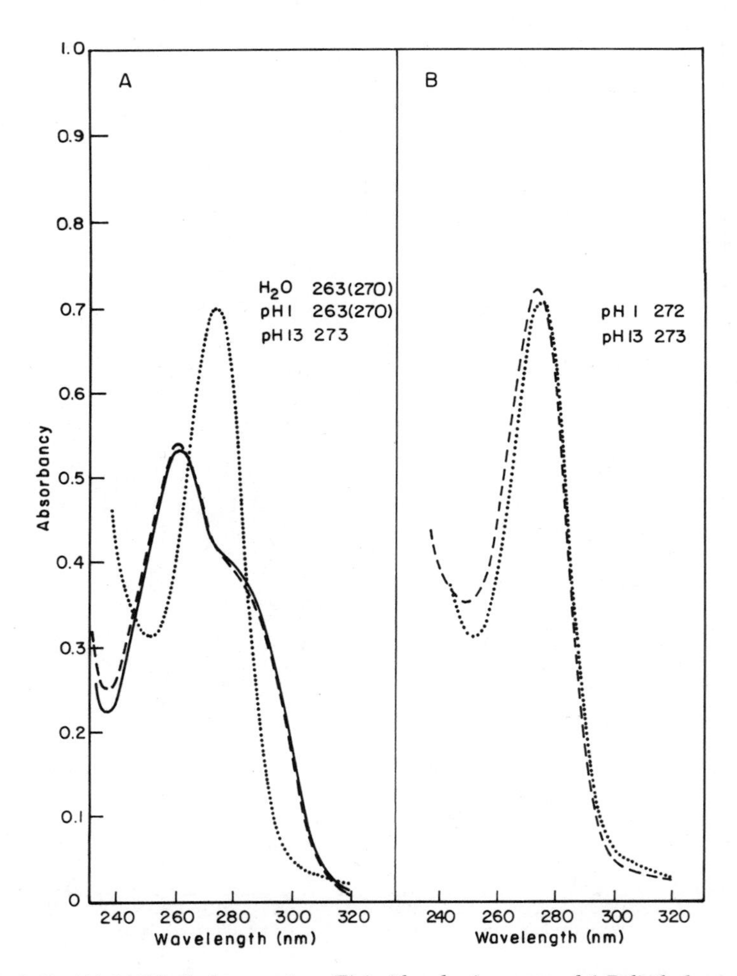

Figure B-42. (A) 1,7-Diethylguanosine; (B) imidazole ring-opened 1,7-diethylguanosine.

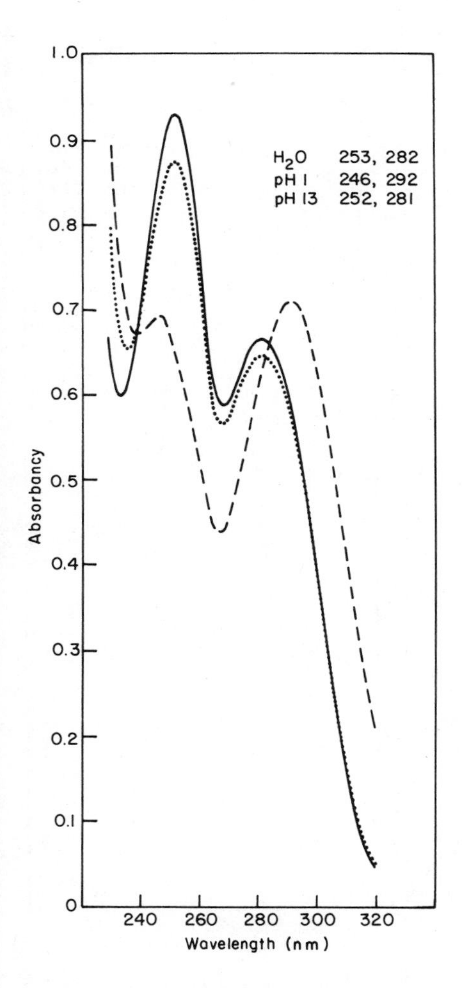

Figure B-43. N^2,O^6-Diethylguanosine.

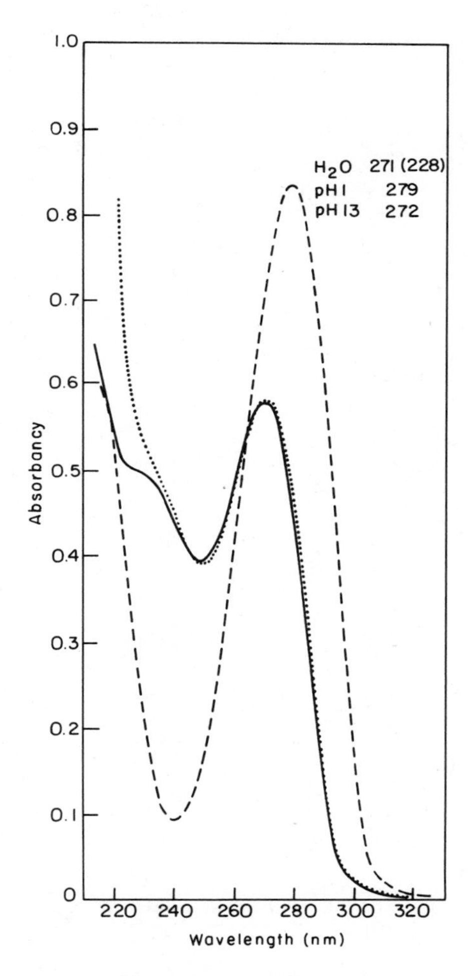

Figure B-44. Cytidine.

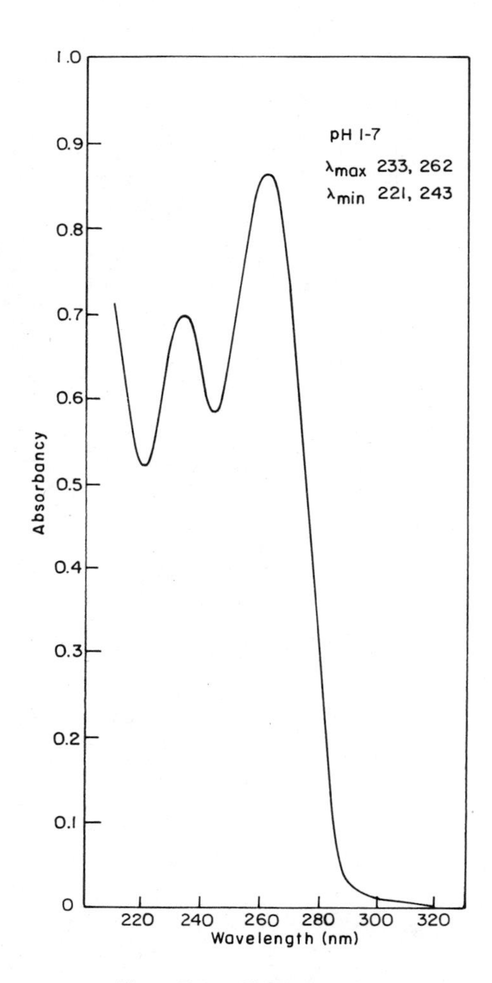

Figure B-45. O^2-Ethylcytidine.

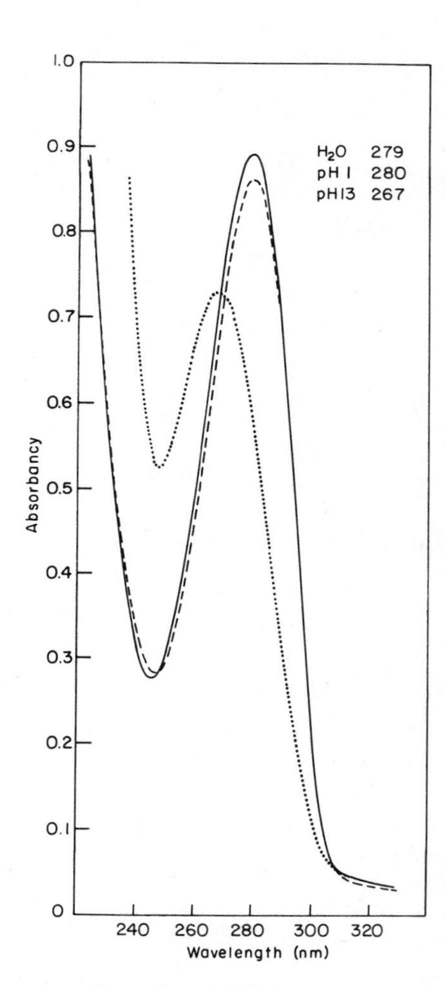

Figure B-46. 3-Ethylcytidine.

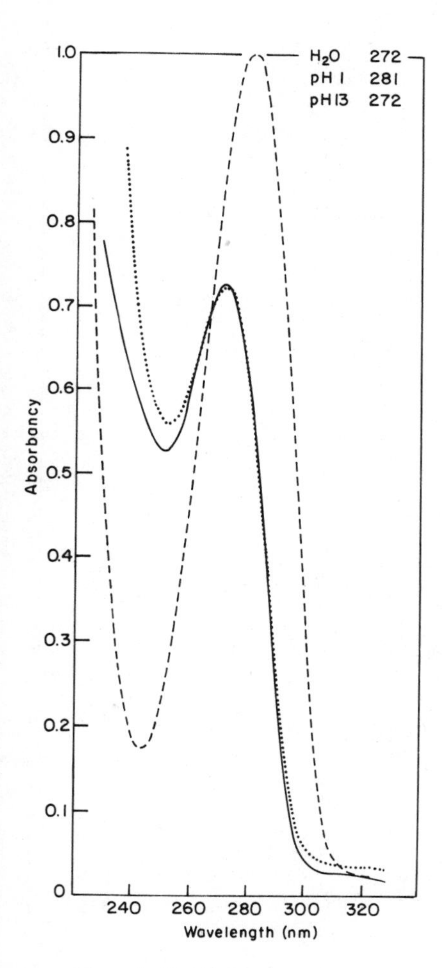

Figure B-47. N^4-Ethylcytidine.

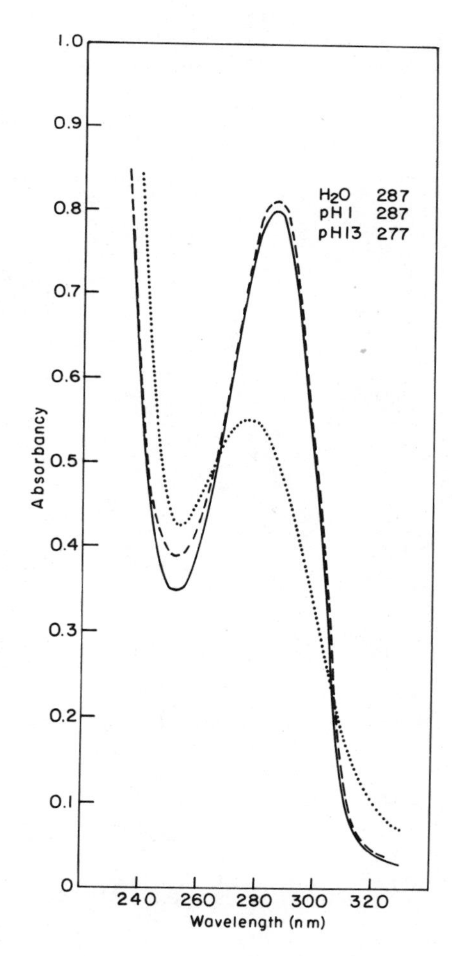

Figure B-48. 3,N^4-Diethylcytidine.

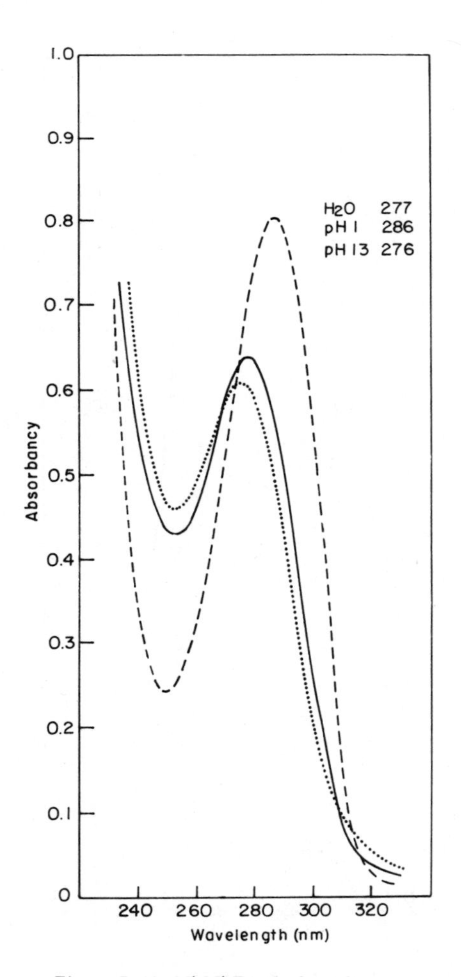

Figure B-49. N^4,N^4-Diethylcytidine.

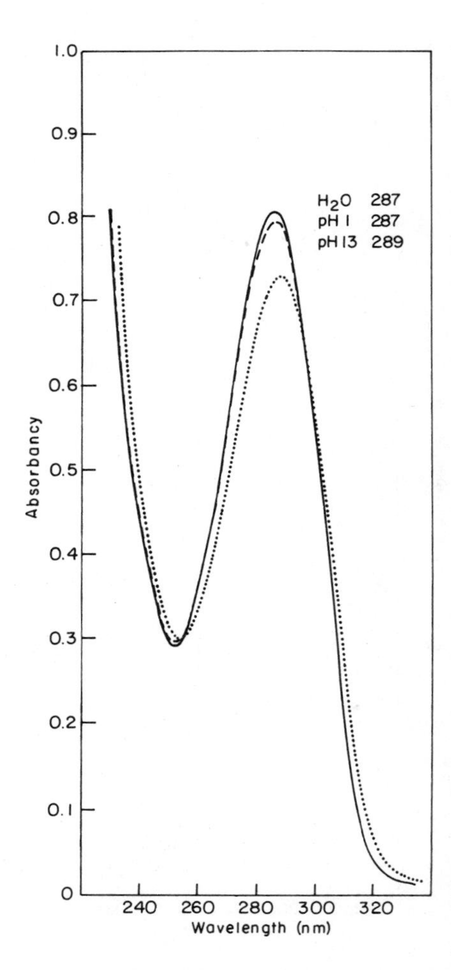

Figure B-50. 3,N^4,N^4-Triethylcytidine.

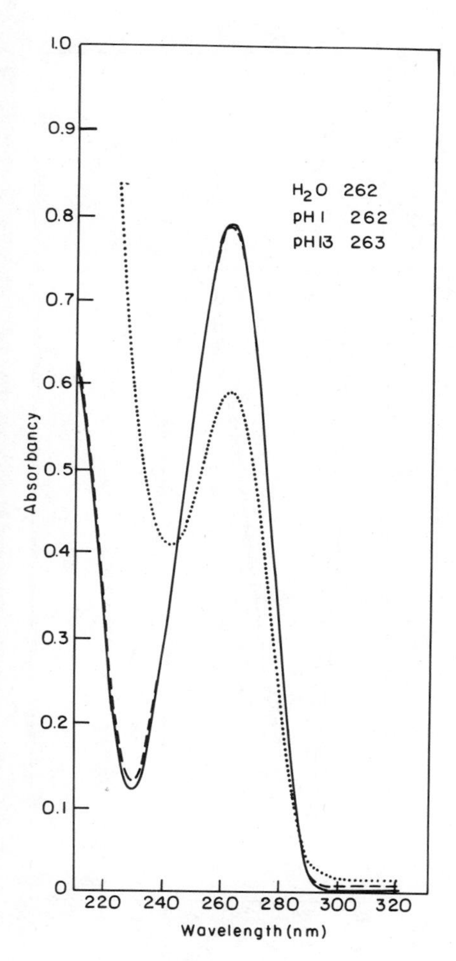

Figure B-51. Uridine.

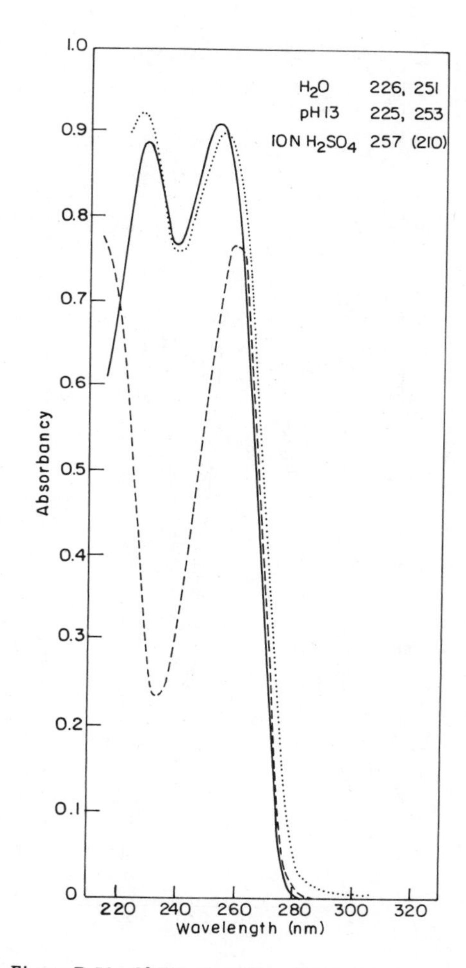

Figure B-52. O^2-Ethyluridine. [10 N H_2SO_4 (– –)]

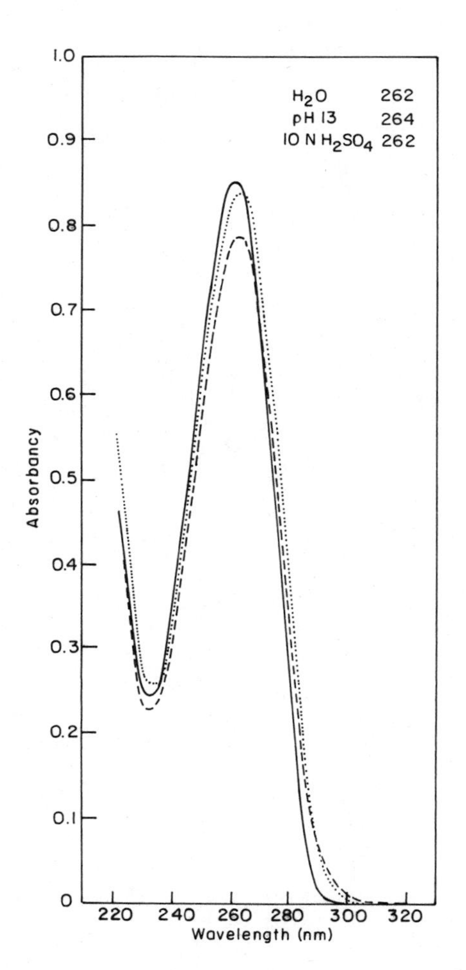

Figure B-53. 3-Ethyluridine. [10 N H_2SO_4 (– –)]

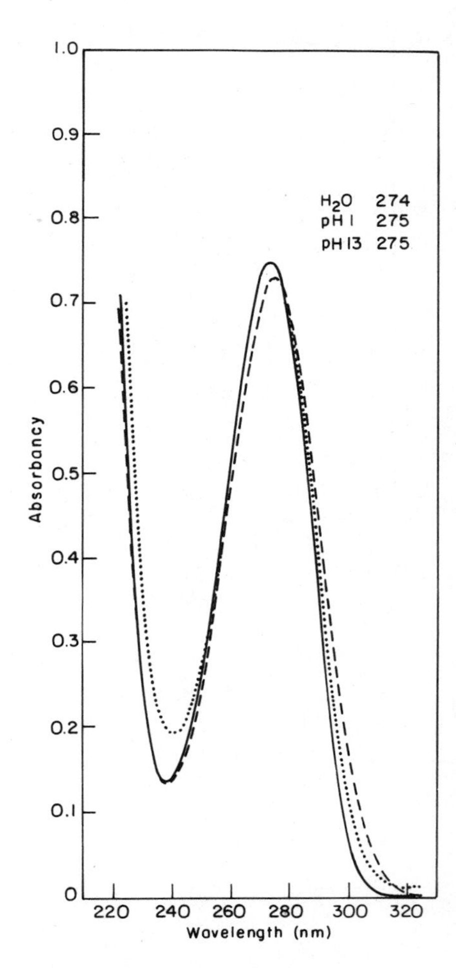

Figure B-54. O^4-Methyluridine.

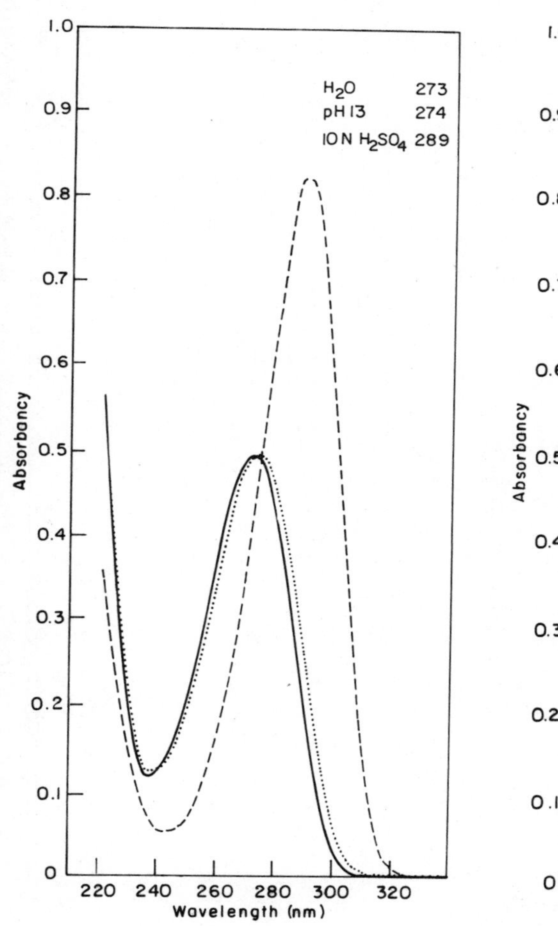

Figure B-55. O^4-Ethyluridine. [10 N H_2SO_4 (– –)]

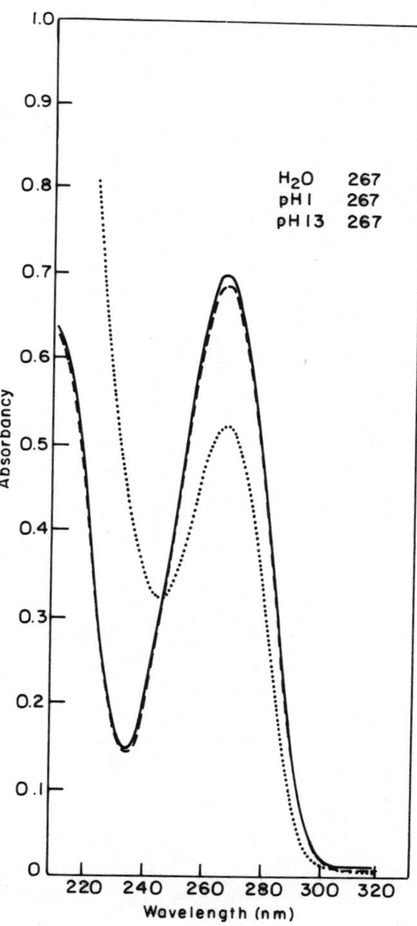

Figure B-56. Thymidine.

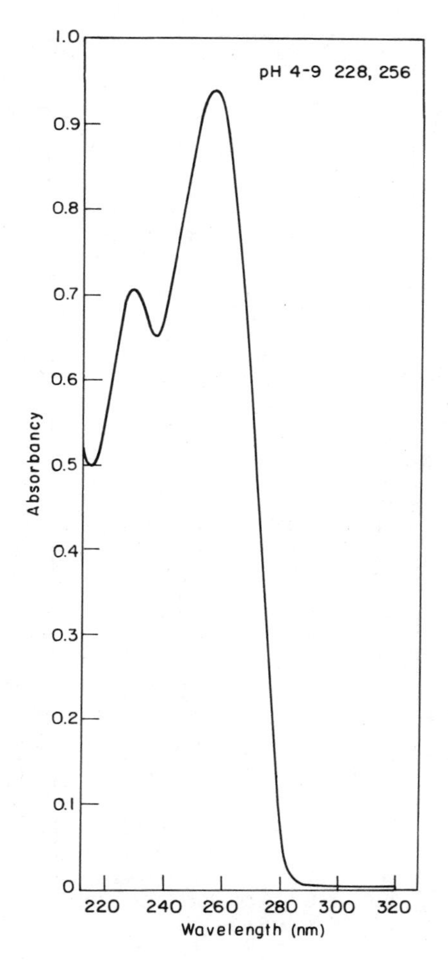

Figure B-57. O^2-Ethylthymidine.

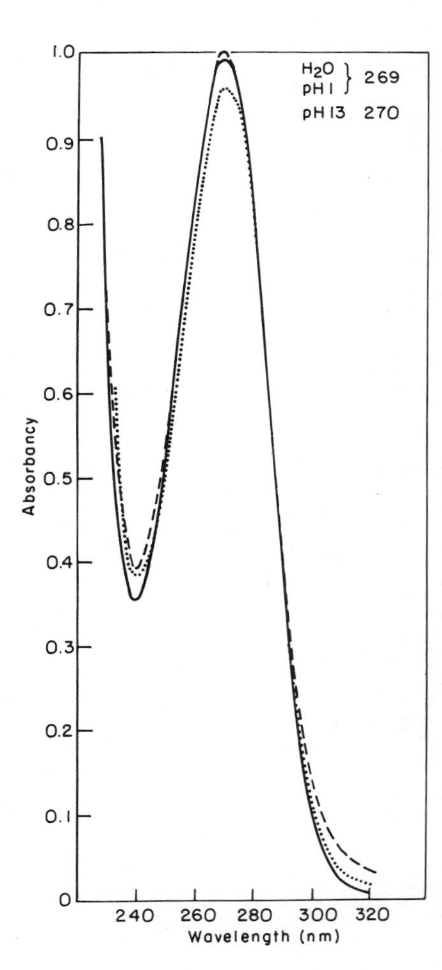

Figure B-58. 3-Ethylthymidine.

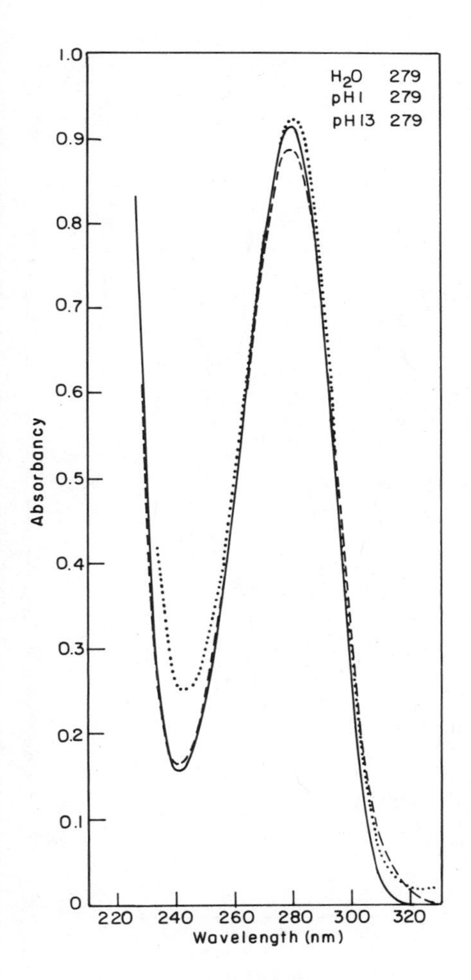

Figure B-59. O^4-Ethylthymidine.

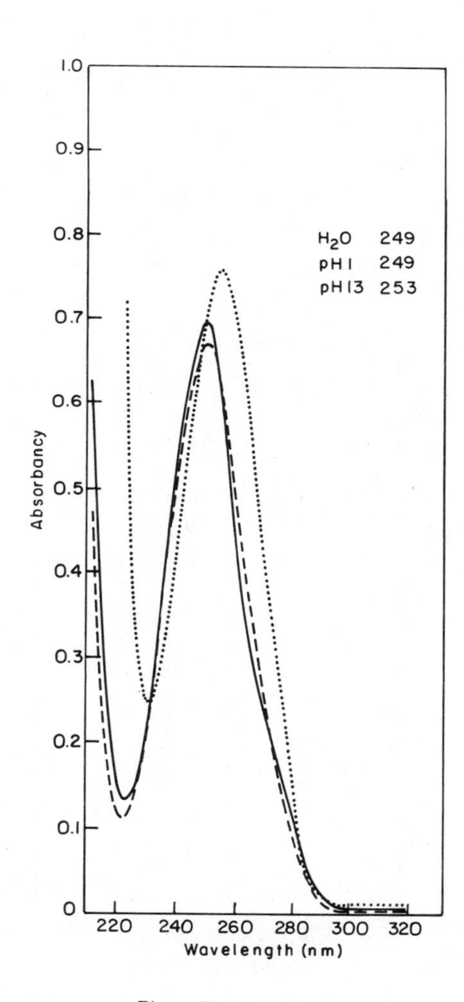

Figure B-60. Inosine.

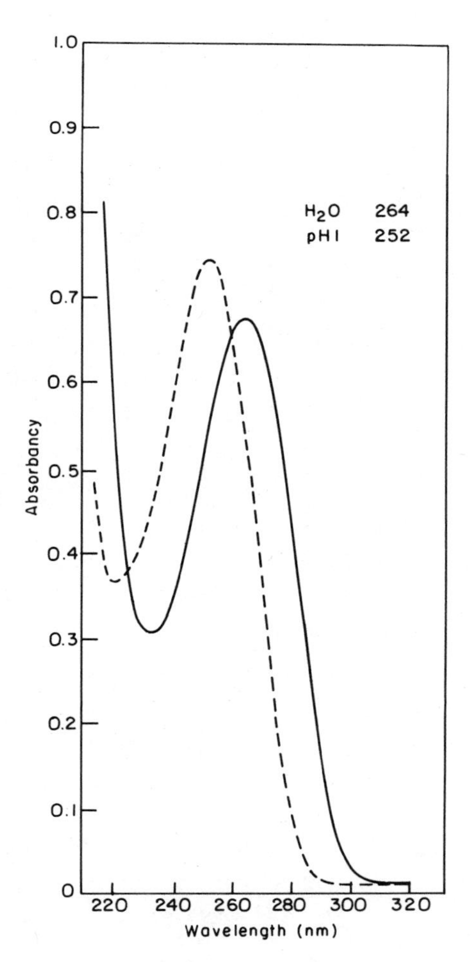

Figure B-61. 7-Methylinosine.

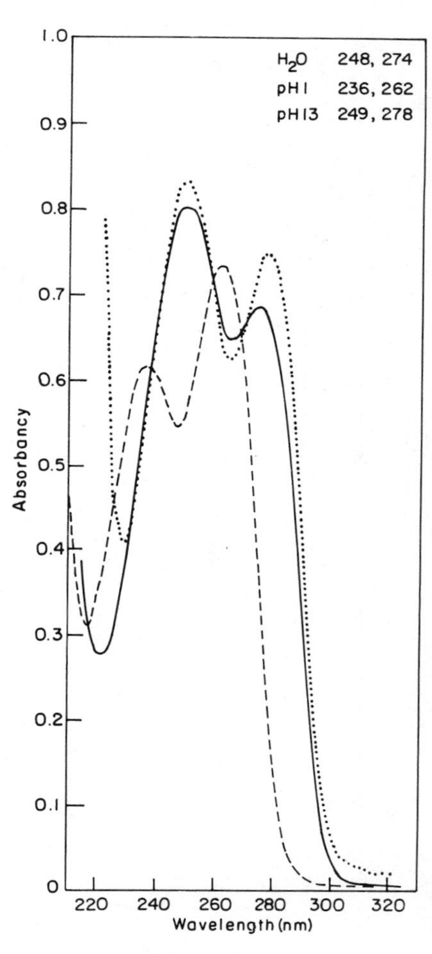

Figure B-62. Xanthosine.

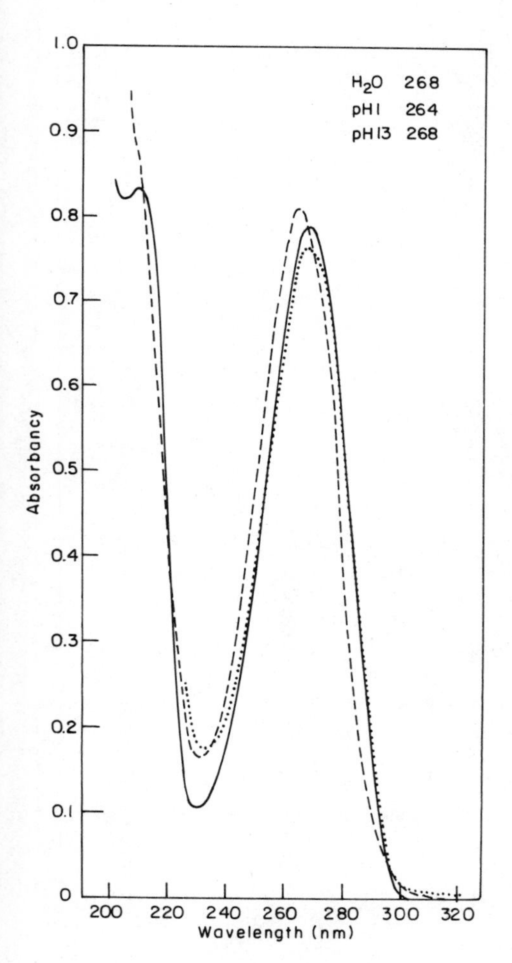

Figure B-63. N^6-Isopentenyl ADP.

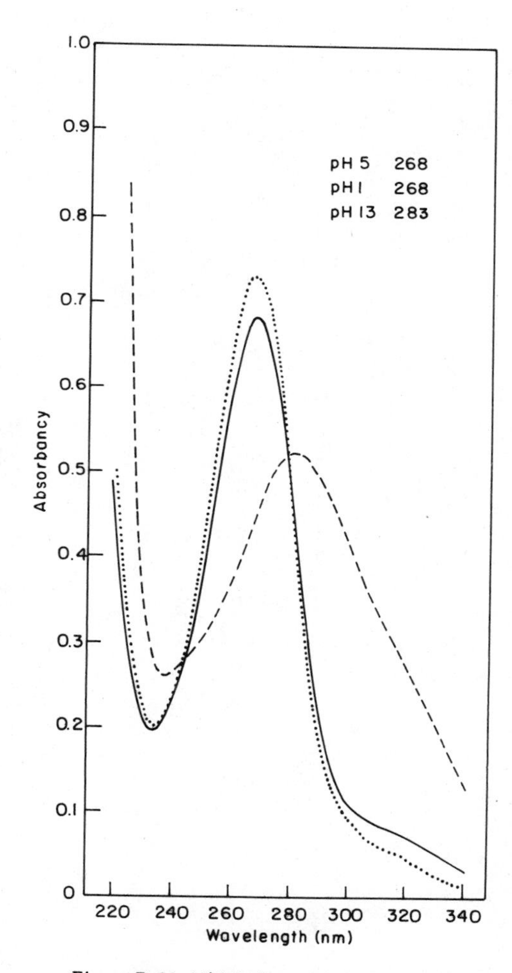

Figure B-64. N^6-Methoxyadenylic acid.

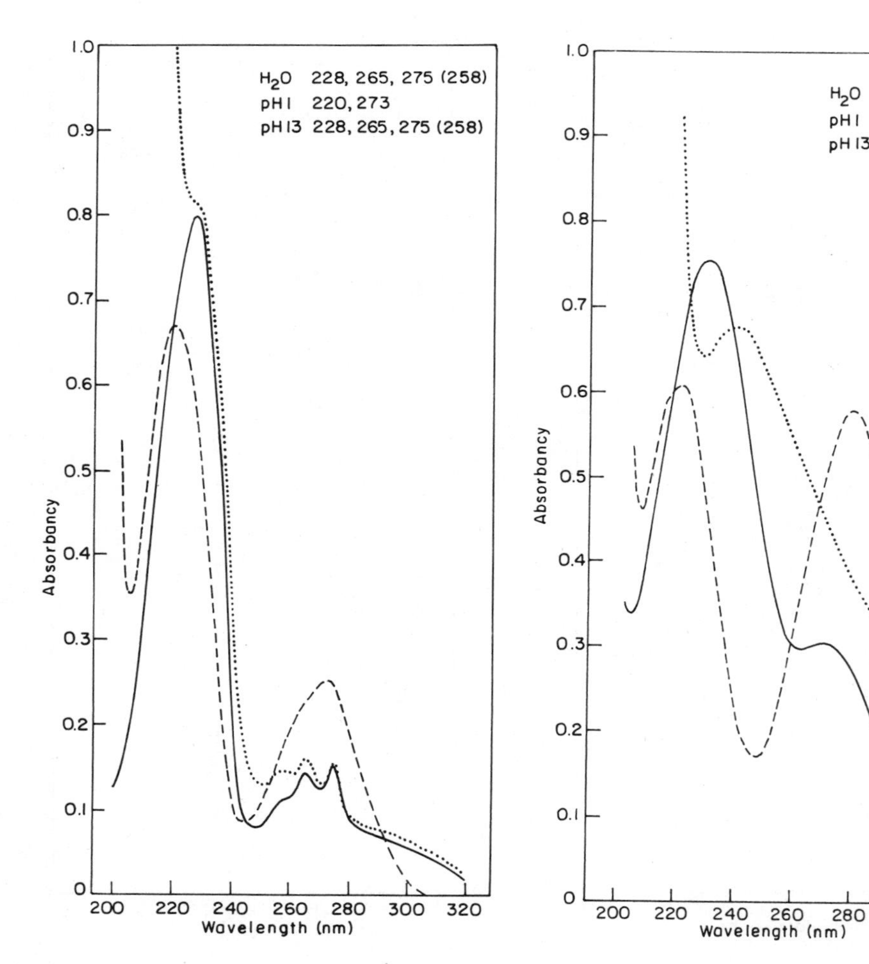

Figure B-65. 1,N^6-Etheno ADP.

Figure B-66. N^4-Hydroxy CDP.

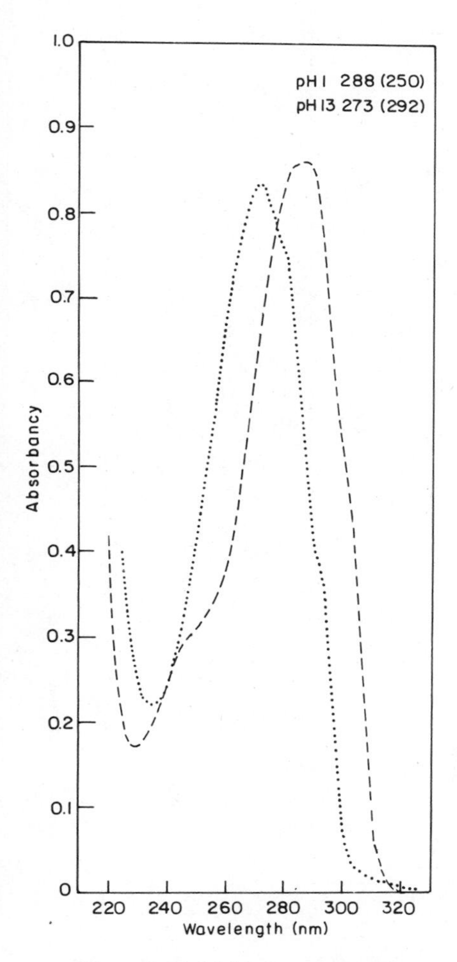

Figure B-67. 3,N^4-Etheno CDP.

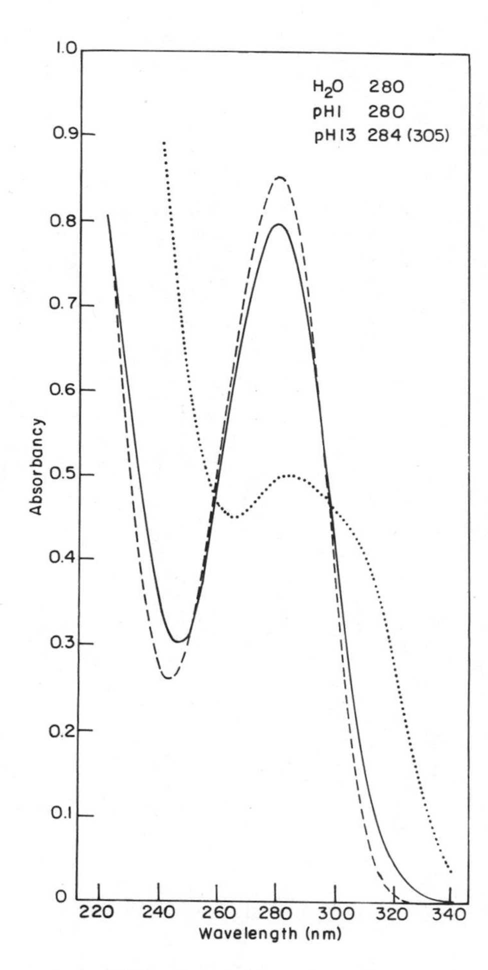

Figure B-68. 5-Hydroxy UDP.

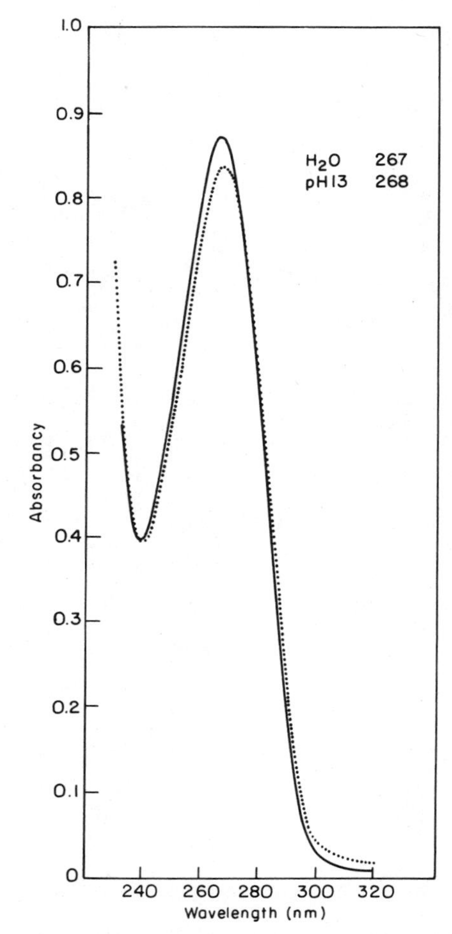

Figure B-69. Methyl ester of 3-methyldeoxythymidylic acid.

Index

This index does not include complete chemical terms for all metabolic intermediates, nor terms only used in the Appendices.

The letters A, C, G, U, and T are generally used to denote the common base components of nucleic acids, regardless of whether they are nucleosides, nucleotides, di- or triphosphates, or components of nucleic acids. Abbreviations for methyl and ethyl derivatives are given in different forms in the text: e.g., 1-methyl A(dA), m^1A, m^1dA or 1-MeA, 1-MedA. Only the first is used in the index.

3 1162 00236 4926